Instrumentation and Metrology in Oceanography

Instrumentation and Metrology in Oceanography

Marc Le Menn

ISTE

WILEY

First published 2012 in Great Britain and the United States by ISTE Ltd and John Wiley & Sons, Inc.

ISTE Ltd
27-37 St George's Road
London SW19 4EU
UK

www.iste.co.uk

John Wiley & Sons, Inc.
111 River Street
Hoboken, NJ 07030
USA

www.wiley.com

Library of Congress Cataloging-in-Publication Data

Le Menn, Marc.
Instrumentation and metrology in oceanography / Marc Le Menn.
p. cm.
Includes bibliographical references and index.
ISBN 978-1-84821-379-1
1. Oceanographic instruments. 2. Oceanography--Measurement. I. Title.
GC41.L44 2012
551.46028'4--dc23

2012025490

British Library Cataloguing-in-Publication Data
A CIP record for this book is available from the British Library
ISBN: 978-1-84821-379-1

FSC
www.fsc.org
MIX
Paper from responsible sources
FSC® C013604

Table of Contents

Preface

The ocean is generally defined as "a vast stretch of salt water that covers the greatest part of the globe". However, the simplicity of this definition must not hide the complexity of this environment, which oceanography attempts to decrypt. More than just a stretch of salty water, it is a place with living rules, where study over takes the biology. Within the science of oceanography, there are fields of study focusing on geology and chemistry, but it is the problems posed by thermal, optical and dynamic properties that physical oceanographers seek to answer.

Numerous models and theories attempt to describe and predict these properties. To initiate and reinforce these, some measurements are necessary. The complexity of this environment and its hostility affect another field: instrumentation. In order to examine the complexity of this environment, specific technological developments are necessary. These developments fall under the field of oceanographic instrumentation.

This work mainly aims to describe the oceanographic instrumentation used to determine the physical properties of the ocean through *in situ* measurements. With the development of space satellite technology, new systems have appeared that have enabled us to discover these properties, creating a new field (that we will not discuss in this book): space oceanography. Space is a favored place from which we can observe the ocean, as it permits large-scale study.

The instruments loaded on satellites use the properties of electromagnetic waves to measure the thermal, optical and dynamic characteristics of the ocean. Unfortunately, these waves are very rapidly absorbed or reflected by the oceanic environment, so if ocean-predicting models use "spatial" data, the collection of *in situ* observations remains an essential source of diagnostics to validate the models, and it is always necessary to develop complementary equipment that permits the study of deep water layers.

There are multiple fields of application for these studies. They concern the evolution of fundamental knowledge on the movement of water masses, the creation of thermal or density anomalies, and the coupling between the ocean and the atmosphere that lead to a better understanding of climate change. These also include the development of acoustic technologies that have multiple civil and military applications. The ocean is a favored environment for these technologies; however the propagation of acoustic waves is dependent on its physical characteristics, so the use of tools that permit emission and reception (sonar, echo sounders, Doppler current meters, etc.) must be optimized by measuring the properties of the environment if we want to achieve ultimate accuracies.

These accuracies cannot be achieved without the use of metrology. It is unusual to associate oceanography with metrology, but this is the other field that this book attempts to describe. Metrology is officially "the science of measurement and its applications", but it is, above all, the science that allows the referencing of measured data, and referencing is an essential part of oceanography when we want to ensure the accuracy and replicability of measurements. Oceanography is probably the area of physics where the requirements of the subject matter are greatest because the acquisition and interpretations of variables are often at the limits of our know-how.

This book, first and foremost therefore, is for instrumentalists and metrologists who want to know more about measurements in oceanography. It is also for scientists who want to gather information on oceanographic instrumentation, and for all those for which this is an area of interest.

All of the chapters of this book are enriched with practical and theoretical details. Sections include recent relationships for the calculation of the physical properties of sea water, the measurement of certain physicochemical properties (carbonates, fluorimetry, etc.), suspended particulate matter, ways of positioning and probing in water, instrumented buoys, underwater platforms, etc. We also discuss the processes of calibration of these instruments, without which these measurements would not have all the legitimacy we attach to them.

Marc Le Menn
September 2012

Chapter 1

What We Measure and What We Process

1.1. The quantities we want to know

Measurements made in physical oceanography have two main aims: to improve our fundamental knowledge of the ocean and the functioning of our planet; and to optimize the use of acoustic tools, which also sometimes help us to gain knowledge of the ocean.

The improvement of fundamental knowledge is integrated with the more general topic of climate change. A global research program was launched in 1979 to try and model this evolution: the *World Climate Research Program* (*WCRP*). The WCRP is funded by the *World Meteorological Organization* (*WMO*) and *UNESCO*. To maximize the efforts of different countries with regard to oceanographic measurements, in 1982 the American National Science Foundation launched another program called the *World Ocean Circulation Experiment* (WOCE). In 1989 it created the WOCE Hydro-graphic Program Office, whose aim was to coordinate, supervise and ensure the quality of measurements taken. To ensure the quality of data collected, it was requested that "*the standards to approach in terms of accuracy and reproducibility for the one-time survey, are to be the highest possible under current measuring techniques*". With "repeat surveys, these standards must be approached sufficiently closely to achieve the appropriate regional goals".

There are several applications of this. The data collected will serve to determine the long-term evolution of ocean circulation. To be able to detect small changes over time, it is necessary to make measurements with low resolution and high reproducibility. The data must be comparable from one country to another and in a common format. Therefore, the measurements should be performed with great accuracy and be comparable to common references. Finally, it should be noted that at great depths the thermal stability of water masses is huge, so to detect small changes over time and space and in order that centers or institutions other than the one that carried out the measurements can verify such changes, the measurements must be made with high resolution, high reproducibility and high accuracy.

In 2003, the WOCE program was relayed by an intergovernmental initiative called the *Global Earth Observation System of Systems*, which aimed to provide the means to overcome the lack of observations concerning key factors that influence the Earth's climate. In Europe, another initiative pursuing the same objectives also emerged called the *Global Monitoring for Environment and Security* (or GMES).

When using acoustic instruments, the specifications are less stringent. Oceanographic measurements performed using these specifications aim to determine the propagation speed or velocity of sound in water either locally or along an acoustic path. It is the temporal and spatial variability of physical characteristics of the environment and the lack of accuracy of velocity references that limit the accuracy of measurements, and that reduced the requirements on the measurement of influence quantities in comparison to those displayed by the WOCE program.

1.1.1. *Velocity and density*

Whether the measurements are performed to gain environmental knowledge or aim to use acoustic instruments, they have a common point: the measured quantities are the same. Indeed, oceanic circulation largely depends on the *density* ρ of water masses (the ratio between the mass of a sample and the volume it occupies). Similarly, velocity c is dependent on density. In adiabatic conditions, it is defined by,

$$c = \sqrt{\frac{\partial p}{\partial \rho}} \qquad [1.1]$$

where p corresponds to the pressure exerted by the acoustic wave, which results in rapid oscillations in density throughout the medium.

It is well-known that the density of a fluid depends on its temperature, pressure and composition. With variation in density, the velocity will also depend on these parameters. Temperature and pressure are therefore the primary variables to be measured.

The composition of sea water varies locally depending on the quantities of dissolved salts. These amounts change the *salinity*. Fochhammer introduced this concept for the first time in 1865. Salinity is a chemical parameter of sea water, and a characteristic of water masses, but its definition has continued to evolve since the creation of the concept, as it is a difficult parameter to measure in absolute terms.

The first accurate definition of the salinity of sea water dates from 1902 and was produced by Forch, Knudsen and Sörensen. It follows a protocol of measurement that they had developed showing that:

> "salinity is the quantity of solid material in grams contained in a kilogram of sea water, having converted the carbonates to oxides, bromide and iodide ions having been replaced by their equivalents in chloride, the organic materials having been oxidized."

This is the definition of absolute salinity, S_a, which is expressed in $g.kg^{-1}$. From this definition, we see that it is not only dissolved salts that are measured but all dissolved matter, so the definition of absolute salinity, carried on to present day, is as follows:

> "*absolute salinity is the fraction of mass of dissolved materials contained in a sample of sea water, under reference conditions (t = 25°C and normal atmospheric pressure).*"

Due to the difficulty of implementation, to date this definition has not been applied in routine measurements. To circumvent these difficulties, a *Practical Salinity Scale* was defined in 1978 (see section 1.2.3). This scale, which is called *PSS-78*, calculates a practical salinity S from the simultaneous measurement of temperature, pressure and electrical conductivity in a sample of sea water. The third common variable to be measured is therefore *electrical conductivity*. It is used to calculate the value of salinity that occurs in density and velocity calculations.

The PSS-78 is based on conductivity measurements – and thus on the speed at which ions move – and does not take into account all dissolved materials, hence biases exist between practical salinity S_p, which we can measure, and absolute salinity. These biases affect the calculation of density. To illustrate this point, let us take the simple example (cited in the TEOS-10 guide), where a small amount of pure water is exchanged with an equivalent mass of silicate compound that is mainly non-ionic. If this exchange takes place at constant temperature and pressure, the value of conductivity will also remain constant, while the absolute salinity and density will have increased. Conversely, if we replace a mass of silicates with the same mass of salt (NaCl), the absolute salinity and density will in principle remain unchanged, while the conductivity and therefore practical salinity will increase.

Until recently, the physical properties of sea water were calculated from the *Equations of State* 1980 or EOS-80, which were based on PSS-78. In 2009, these properties were redefined by a UNESCO working group called *SCOR/IAPSO WG 127* (Scientific Committee on Oceanic Research/International Association for the Physical Sciences of the Ocean/Working Group 127). The group created the *International Thermodynamic Equation of Sea water*, known as TEOS-10. TEOS-10 is based on the Gibbs potential, which is a scalar function that represents the thermodynamic state of a system. For sea water, it is dependent on absolute salinity, temperature and pressure. This approach is similar to that defined for pure water in the 1970s by the *International Association for the Properties of Water and Steams* (IAPWS-95).

The Gibbs potential of sea water is denoted as $g(S_a, t, p)$. Here, g is in relation to the specific enthalpy h, which is a measure of the calorific content of a system, and entropy η that translates the amount of disorder or inaccessible non-mechanical energy:

$$g = h - (273.15 + t)\eta \qquad [1.2]$$

$g(S_a, t, p)$ is defined by:

$$g(S_a, t, p) = g^w(t, p) + g^S(S_a, t, p) \qquad [1.3]$$

where $g^w(t, p)$ is the pure water part, which is determined by equations from IAPWS-95 (although it is advisable to program the codes given in the IAPWS-09 as they are faster) and $g^S(S_a, t, p)$ is the saline part. The equations for calculating these quantities (and the coefficients of these equations) are provided in Annexes G and H of the TEOS-10 guide. For more information on the TEOS-10, refer to [IOC 10], which describes all of the basic thermodynamic properties that can be calculated from the Gibbs function and the derived variables of these properties.

Among the properties commonly calculated by oceanographers, we have the *specific volume* V, which is the inverse of density:

$$V = \frac{1}{\rho} = V(S_A, t, p) = g_p = \left.\frac{\partial g}{\partial p}\right|_{S_a, T} \quad \text{in m}^3.\text{kg}^{-1} \qquad [1.4]$$

It is now defined as a derivative of the Gibbs potential with respect to pressure for a constant absolute salinity and *in situ* temperature T. Density then becomes:

$$\rho = \rho(S_A, t, p) = (g_p)^{-1} = \left(\frac{\partial g}{\partial p}\Big|_{S_a, T}\right)^{-1} \text{ in kg.m}^{-3} \qquad [1.5]$$

It is important to note that the maximum variation in sea water density, or in specific volume, is only 7% over the whole ocean surface. To improve the accuracy of numerical calculations of specific volumes made in engineering units, oceanographers use a derived quantity called the *specific volume anomaly* δ, which is defined by the relationship:

$$\delta = V(S_a, t, p) - V(S_{S0}, 0, p) \qquad [1.6]$$

where $V(S_{S0}, 0, p)$ is the specific volume of a "standard" ocean, that is to say it would have a salinity of S_{S0} = 35.16504 g.kg^{-1} and a temperature of 0°C. δ is expressed in units of 10^{-8} m^{3}kg^{-1}. The density anomaly or *density excess* $\gamma(S_a, t, p)$ is, in turn, defined by relationship [1.7], and is expressed in kgm^{-3}.

$$\gamma(S_A, t, p) = \frac{1}{V(S_A, t, p)} - 1000 = g_p^{-1}(S_A, t, p) - 1000 \qquad [1.7]$$

where γ is the letter recommended by UNESCO to designate excess density. Many oceanographers, however, continued to use the old notation, σ. To compare the values of density, they plot so-called "sigma-t" curves taking $p = 0$ in the calculation of $V(S_A, t, p)$ in order to normalize relatively to atmospheric pressure. Comparisons based on sigma-t are biased when calculated for deep layers, so they are performed by calculating the excess of potential density called σ_θ. θ designates the temperature of a sea water sample relative to pressure $p = 0$, or *potential temperature.*

$$\sigma_\theta = \gamma(S_A, \theta, 0) = \frac{1}{V(S_A, \theta, 0)} - 1000 \text{ in kgm}^{-3} \qquad [1.8]$$

A slight increase in water temperature with increasing depth can be observed, which is due to compression. Water is a compressible fluid, so the volume it occupies decreases when subjected to increased pressure. This compression is accompanied by an increase in temperature. It is useful to calculate a variable that represents what the increase may be when compression is performed without heat exchange with the medium, i.e. adiabatically. This variable is called the *rate of adiabatic variation of temperature* or *adiabatic lapse rate*, $\Gamma(S, t, p)$. It is calculated from relationship [1.9] where T represents absolute temperature, $T = t + 273.15$ in

Kelvin (K), and C_p^0 is the specific heat of sea water at constant pressure for $S_a = 35.16504$ gkg^{-1} and $\theta = 25°C$.

$$\Gamma(S_A,t,p) = \left.\frac{\partial t}{\partial p}\right|_{S_a,\eta} = \left.\frac{\partial V}{\partial \eta}\right|_{S_a,p} = \left.\frac{(T+\theta)}{C_p^0}\frac{\partial V}{\partial \Theta}\right|_{S_a,p} \quad [1.9]$$

with: $C_p^0 = 3991.86795711963$ J kg^{-1}K^{-1}.

In this expression, Θ designates the *conservative temperature*, which corresponds to potential enthalpy $h^0(S_a, \theta, 0)$ divided by C_p^0. Potential enthalpy is the enthalpy that a plot of fluid would have if brought back to a reference pressure of 0 dbar in an isentropic manner (with constant and isohaline entropy or without the dissipation of mechanical energy). It allows evaluation of the "real" thermal content of the ocean. Therefore, TEOS-10 enables us to calculate and to use the conservative temperature rather than the potential temperature, as it better represents the thermal content of sea water (by a factor of two).

$\Gamma(S, t, p)$ is expressed in °C Pa^{-1}. Knowing its value makes it possible to define the notion of *potential temperature* $\theta(S_a, t, p, p_r)$ as the temperature that a water sample would have if taken from a depth of pressure p, brought up to the surface in an adiabatic manner without a change in salinity, which would then be placed under a reference pressure, p_r. It is calculated using the variable $\Gamma(S_a, t, p)$ and the *in situ* measured temperature t_{in}:

$$\theta(S_A,t,p,p_r) = t_{in} + \int_p^{p_r} \Gamma(S_A,\theta(S_a,t,p,p_r),p')dp' \quad [1.10]$$

The evaluation of $\theta(S_A, t, p, p_r)$ is carried out by a Newton-Raphson type iterative technique.

1.1.2. *Pressure and depth*

Pressure intervenes in the calculation of several oceanographic parameters, but it is not a representative quantity of the environment in the eyes of oceanographers: it is interesting to know depth so that measurements of other parameters can be positioned vertically. This quantity is not directly measurable; it is calculated through pressure measurements. Pressure is related to depth though the intermediary of the hydrostatic relationship:

$$p = p_0 + \rho g z \qquad [1.11]$$

where p_0 designates surface pressure, z the position on the vertical axis and g designates the acceleration due to gravity. In relationship [1.11], g depends on depth (the closer we get to the center of the Earth, the higher g becomes). This phenomenon is expressed by equation [1.12], where g_0 is the acceleration of gravity at the surface of the ocean (which depends on latitude, φ) and γ_g is the average gradient of gravity.

$$g(z) = g_0(\varphi) + \gamma_g z \qquad [1.12]$$

As discussed in section 1.1.1, ρ also depends on pressure and we prefer to talk in terms of specific volume V. If the depth varies by dz, the pressure will vary by dp such that:

$$dp = \frac{1}{V(p)} g(z)dz \qquad [1.13]$$

Assuming hydrostatic equilibrium, relationship [1.12] linked to relationship [1.4] gives:

$$V(p)\, dp = (g_0(\varphi) + \gamma_g z)dz \qquad [1.14]$$

If we integrate equation [1.14], we get the depth P that we are looking for:

$$P = \frac{\varphi_0 - \int_0^p V(p)dp}{\left(g_0(\varphi) - \frac{1}{2}\gamma_g P\right)} \qquad [1.15]$$

In practice, depth variable P, which is still present in the expression, is replaced with pressure variable p. Error is in the order of 5 cm for P = 10,000 m. ϕ_0 is the gravitational potential energy or *geopotential* at sea level. Its value is obtained by multiplying $g_0(\varphi)$ by the height of the free surface above the geoid. Integration of $V(p)$ is performed by considering the successive variations in specific enthalpy h (defined by relationship [1.2]) between the layers of pressure p^i and p^{i+1}:

$$\int_0^p V(p')dp' = \sum_{i=1}^{n-1}\left[\hat{h}\left(S_A^i, \Theta^i, p^{i+1}\right) - \hat{h}\left(S_A^i, \Theta^i, p^i\right)\right] \qquad [1.16]$$

where Θ designates the conservative temperature defined in [1.9] and $\hat{h} = V$, knowing that it is possible to demonstrate that $\hat{h} = \partial h/\partial p = V$ for constant values of S_A and Θ.

The precise knowledge of depth requires not only pressure measurements, but also temperature and absolute salinity measurements. $g_0(\varphi)$ can be obtained if we know latitude φ, which comes from relationship [1.17] and includes a corrective term involving pressure p.

$$g_0(\varphi, p) = 9.7803185(1 + 5.278895 \times 10^{-3} sin^2(\varphi) + 2.3462 \times 10^{-5} sin^4(\varphi)) + \frac{1}{2} \times 2.184 \times 10^{-6} p \quad [1.17]$$

1.1.3. *Speed and movement*

Improving our basic knowledge of the ocean also requires us to measure the speed of oceanic current at different depths. This variable can be measured directly using purely mechanical means or purely acoustic means (see section 2.5). Instruments used can be fixed vertically so the horizontal movement of the surrounding water can be measured. When acoustic means of measurement are used, corrections in velocity are sometimes necessary. These are carried out using temperature and conductivity sensors, but the bulk of the measurement is based on knowledge of frequency shift caused by the Doppler effect. This technique allows the measurement of speeds, either at a single point or over several levels of water.

We need to know the movement of water masses at depth or over long distances as well as the speed of marine currents. Such variations can be estimated by calculations based on the characterization of temperature and salinity of water masses. Movement is another variable that requires the measurement of temperature and salinity for an evaluation to be made. This technique is particularly useful when the ocean region being studied has a water intrusion with particular characteristics. Temperature–salinity diagrams can then be established. With these diagrams, it is possible to know the spatial displacement of a point of minimum or maximum salinity.

The variation in the position of water masses can also be assessed by measuring the movement of *trackers*. In fact, we can distinguish between what we call a *typical water*, which is a large volume of water of uniform temperature and salinity, and a *mass of water*, which corresponds to a large volume of water in which the properties vary within defined limits. The analysis of concentrations of certain chemical, biological or radioactive elements allows us to characterize and follow the spatial and sometimes temporal evolution of certain typical water and water masses. To do this, water samples are taken during oceanographic studies, and the samples are analyzed using laboratory chemical analysis techniques. These physicochemical techniques are not further described or addressed in this book as they lie within another field, but certain trackers have been the object of the development of specific instruments for *in situ* use. This is the case, for example, for dissolved oxygen and several other constituents. An overview is given in section 2.11.

Finally, we can use instruments to directly study the movement of water. These are floats or drifting buoys (see sections 3.2 and 3.3). These measurement processes comply with the fluid mechanics principles followed by Lagrange, which place the observer in the moving frame rather than the phenomenon being observed from a fixed point. To do this, Lagrangian buoys are "anchored" in the water mass that is being studied. For buoys on the water surface, the greatest difficulty lies in how they are anchored. The location is provided by ARGOS type satellite systems (see section 2.7). For buoys that are located underwater, the anchoring is done by varying their volume. Once dropped, these buoys descend to their immersion equilibrium within a few hours and move with the water mass in which they find themselves. Their location is provided by acoustic means that function over long distances (500 to 1,000 km) or by satellite. After a pre-programmed time, they rise to the surface, the data is transmitted and their location is found. Nowadays, the majority of these buoys are also equipped with temperature, pressure, conductivity or physicochemical sensors (see section 3.3).

1.1.4. *Time and space*

In oceanography, the knowledge of variables that are measured or calculated is only partial if they are not correctly positioned in time and space. Bearing in mind the periods of evolution of natural phenomenon, temporal positioning is not a constraint for the instrumental developments except in cases where synchronization of several elements is required (see section 2.6). Due to satellite systems, positioning also is no longer a constraint. Excluding the ARGOS system or Iridium or Orbcom type communication systems used to follow drifting buoys, tools such as a *global positioning system* (*GPS*) can provide instant coordinates of a surface carrier anywhere in the world, as long as it is equipped with a receiver whose output can be queried by embedded data acquisition systems. In the case of GPS, clock frequencies are transmitted to the mobile devices that require positioning. In the case of the ARGOS system, it is the mobile devices that emit a frequency of which the shift (Doppler effect), which is dependent on the distance separating it from the satellite, is measured to decipher the position (see section 2.7).

However, not all instruments are equipped with ARGOS emitters or GPS receivers. Many operate autonomously, especially those used in moorings. They are equipped with frequency-stabilized oscillators from which the data acquired can be dated. As these instruments are usually submerged for long periods of time, a stable clock is needed so that the dating of data is not subjected to temporal drift (see section 2.6.2).

1.2. Linking of essential quantities in oceanography

As explained in section 1.1, the need for the comparability of measurement results between different laboratories or even different countries, and over long periods of time, requires uniting quantities that we can measure to common references, and these must be made with great accuracy. Temperature is a basic variable of the International System of Units (SI). Its connection to the international metrological network is therefore direct. For this variable, we can obtain high-accuracy measurements because the field of ocean temperatures is restrained. It ranges from –2°C to +35°C. The specifications from the WOCE program are therefore high, as an accuracy of 2 mK needs to be approached with a repeatability value of 0.5 mK.

Pressure is an SI-derived unit, but the importance of this variable means that there are numerous types of calibration means and its connection is direct. Its unit is the Pascal (Pa). As 1 Pa is a small value, in oceanography we prefer to use one of its multiples, the bar (1 bar = 10^5 Pa), and sometimes even the dbar (1 dbar corresponds to approximately a 1 m water column). In the ocean, the range of this variable is large, going from 0 to 600 bar, but can be divided into sub-domains depending on the seabeds and the objectives of studies. The specifications of the WOCE program are less stringent for pressure than those for temperature; however the measurements must approach an accuracy of 3 dbar and a repeatability of 0.5 dbar.

Electrical conductivity is more problematic. On one hand, it is a derived unit that is not usually measured, but on the other hand it is merely a means by which to access knowledge of the salinity of water in oceanography. Its attachment to physical standards is indirectly achieved by prior knowledge of reference salinities. Its unit is the Siemens/meter (S m^{-1}), but we prefer to express it as mS cm^{-1}. It ranges from 2 mS/cm for brackish waters to 75 mS/cm for warmer and saltier waters. Here again, the specifications of the WOCE program are extreme, as the accuracy of conductivity measurements must approach 2 µS/cm and their repeatability 0.5 µS/cm.

Velocity is a quantity that is related to speed measurements, according to its unit. The connection of the propagation speed of sound to reference values, however, can only be made by an intermediary of complex polynomial formulas. The multiplicity of these relationships and their divergence for certain temperatures and pressures affect the accuracy of measurements that can be achieved, especially since none of them currently seem to be absolutely accurate.

Time is the variable for which joining is probably the easiest and the most immediate. This is due to the advent of *global navigation satellite systems*. If, for common applications, we consider a call to the speaking clock proven and sufficient to calibrate time-measuring instruments, clocks on board satellites allow the referencing and synchronization of receivers that can be separated by considerable distances with far greater accuracy. These points are discussed in sections 1.2.5 and 2.6.

Techniques for joining instruments measuring speed are outlined throughout section 2.5, first for rotor current meters and second for Doppler effect current meters considering aspects of the measurements of speeds, direction and correction for the inclination of instruments.

1.2.1. *Temperature*

Temperature is an intensive macroscopic quantity, meaning it is a variable that has the same value for all or part of a system at equilibrium. If we place a thermometer in an adiabatic setting, it will show its own temperature. This temperature is that of the surroundings, assuming there is thermal equilibrium between the thermometer and its surroundings.

We cannot add or subtract two temperatures. We can only classify the bodies from hottest to coldest depending on their temperatures. It is therefore not strictly a measurable variable and so it is treated as a stake-out quantity: we can only grade a physical phenomenon by temperature. This is how we constructed a temperature scale based on variations of other quantities selected to be references.

1.2.1.1. *The International Temperature Scale*

The first definition of a measurable temperature scale was proposed in 1852 by Sir William Thomson, alias Lord Kelvin, based on experimental work carried out by Carnot. This scale is therefore based on Carnot's principle applied to thermal motors:

– di-thermal, which means functioning with hot and cold sources;

– reversible motors, which means where the transformations consist of series of equilibrium states.

This principle says that:

> "if a perfect fluid evolves reversibly between a hot source at temperature T_1 and a cold source at temperature T_2, it receives a quantity of heat $Q_2 > 0$ from the hot source (counted positively) and yields a quantity of heat $Q_1 < 0$ to the cold source."

We show that this heat exchange does not depend on the nature of the thermal motor or on the agent of transformation (hot air, water vapor, water, alcohol, etc.). This principle allows the establishment of a relationship of proportionality between the amounts of heat exchanged and two functions dependent on T_1 and T_2:

$$\frac{Q_1}{Q_2} = -\frac{f(T_1)}{f(T_2)} \qquad [1.18]$$

Due to signing conventions, this ratio is always negative. To define $f(T)$, we simply need to arbitrarily choose a continuous and monotonous function. If we choose $f(T) = \alpha T$, it will be defined to within a constant. To determine α, it is sufficient for us to identify the value of the temperature of a remarkable and reproducible phenomenon.

To find this phenomenon, we lean towards the physical properties of the bodies that surround us, knowing that the state of a pure substance depends on three parameters: pressure P, temperature T and volume V. At constant pressure, if we add heat to a solid body, it changes state and begins to melt. The amount of heat Q that we must apply to a solid mass m in order for it to pass to the liquid state is called the latent melting heat L. It can be calculated by the relationship:

$$Q = m \times L \qquad [1.19]$$

where L is a characteristic of the considered body. The temperature at which the change occurs is called the melting temperature. If constant pressure is maintained and a near-constant heat is applied, a pure substance will stay in equilibrium at its melting temperature, so long as there is solid to be melted (see Figure 1.1). This is a remarkable and reproducible phenomenon that can be used to define $f(T)$.

There are temperature and pressure conditions that have been determined at which the three states – liquid, solid and gas – can be present. This is the case, for example, for distilled water at 0.01°C. This particular point on the phase diagram is called the triple point (see Figure 1.2). At this point, the three variables P, V and T, can only take one possible value that characterizes the substance. Triple points are therefore even more reliable than melting points when identifying temperatures.

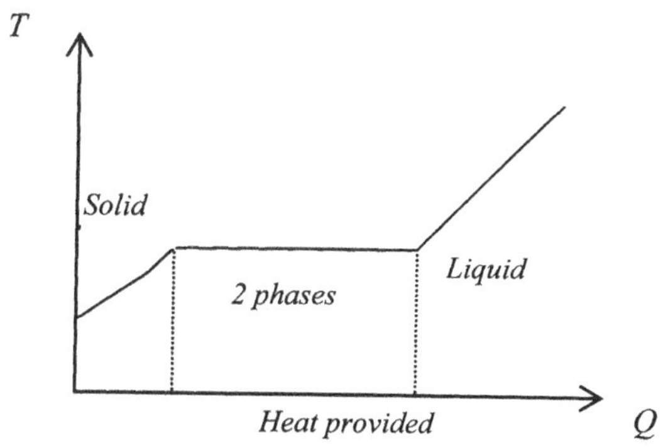

Figure 1.1. *The principle of function for melting fixed-point cells of pure substance, used to define the International Temperature Scale. At constant pressure, the amount of melting obtained by introducing heat is remarkable and reproducible, providing an ideal measuring condition*

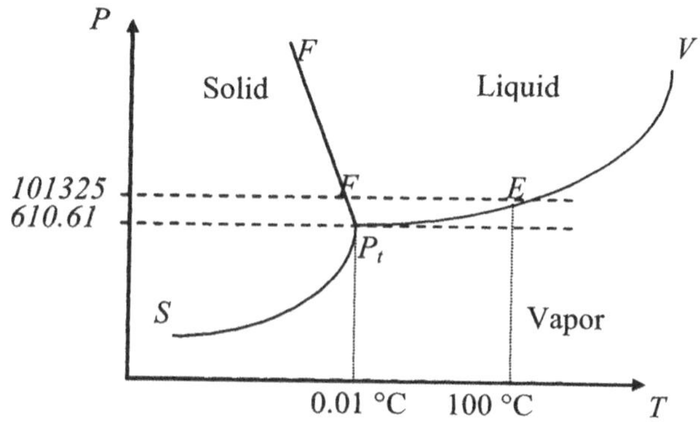

Figure 1.2. *Phase diagram for distilled water. Curves F, V and S correspond to the equilibrium curves for melting, vaporization and sublimation, respectively. At the triple point, P_t, the three phases coexist*

The temperature of the triple point of water was proposed as a reference value by Giauque. It was conventionally set at 273.16 K, knowing that the melting temperature of ice under normal atmospheric pressure is 273.15 K and the temperature at which water boils is 100°C above this value. From this, a unit could be associated to temperature. This unit is the Kelvin:

Kelvin is a 1/273.16 fraction of the thermodynamic temperature of the triple point of water.

Due to habit, the use of degrees Celsius remains authorized, but temperatures expressed as °C must include the symbol 't'. Here, we have:

$$t\,(°\mathrm{C}) = T\,(\mathrm{K}) - 273.15 \qquad [1.20]$$

From this definition and relationship [1.18], it is possible to define the temperature T of any body by measuring the ratio of amounts of heat:

$$T = -\,273.16 \times Q\,/\,Q_{ref} \qquad [1.21]$$

Here Q_{ref} is the amount of heat exchanged to reach the triple point of water. From this definition, we can express it in terms of thermodynamic temperature instead of thermodynamic temperature scale, since it has an origin and relationship based on a ratio (instead of a reference cue). Thermodynamic temperature is therefore a measurable variable.

This principle has helped with the creation of the International Practical Temperature Scale, or IPTS, which is based on the measurement of temperature intervals and on a stake out from an origin. The values obtained using this method must be the closest possible to thermodynamic temperature values. The IPTS definition has therefore evolved over time to reflect technological and scientific developments. These evolutions are managed by the International Weights and Measures Committee. The first scale dates from 1927. It was revised in 1948, then in 1968, in 1976 (for the range 0.5 K to 30 K) and in 1990. With the revisions, the areas of use have been expanded and the accuracy of the scale has been improved.

In 1990, the term "Practical" disappeared and we now speak of the ITS-90. This scale is based on:

a) *Temperature values assigned to a certain number of equilibrium states of high reproducibility*. These are fixed points of definition. These fixed points are distributed between 13.81 K and 1,084.62°C (there are 18 fixed points plus the choice of two points between 3 and 5 K). These are the phase transition points of metals or pure liquids. Two of these points are in the ranges of oceanographic temperatures: the triple point of water (Pt_{H2O}) for which the temperature is +0.01°C; and the melting point of gallium (Pf_{Ga}) at +29.7646°C. The reference probes having to measure negative temperatures must, strictly speaking, also be calibrated at the triple point of mercury (Pt_{Hg}), which is –38.8344°C. There are also two secondary fixed points that enter into the field of oceanographic temperatures: the triple point of phenoxybenzene at 26.8625°C and that of ethylene carbonate at 36.3135°C. These can also be used to check the calibration of temperature sensors, however they are less well-known than the ITS fixed points of definition and their use is not a requirement. (Note that metrologists have a habit of expressing themselves in Kelvin when they speak of low temperatures or temperatures below 0.01°C, and otherwise use degrees Celsius.)

b) *Interpolation instruments*. ITS-90 consists of four temperature ranges that partially overlap. Each range definition includes a reference instrument. The range that includes oceanographic temperatures goes from 13.8033 K to 961.78°C. In this interval, the instrument of reference is the platinum resistive thermometer. To be recognized as a reference, this instrument must meet a number of criteria. The main one concerns the purity of platinum. The purity is monitored by Pf_{Ga} and Pt_{Hg}. To evaluate this, we measure the ratios of resistance or reduced resistance $W(t)$:

$$W(29.7646) = R_{29.7646} / R_{0.01} \text{ and } W(-38.8344) = R_{-38.8344} / R_{0.01} \qquad [1.22]$$

We must find:

$$W(29.7646) \geq 1.11807 \text{ and } W(-38.8344) \geq 0.844235 \qquad [1.23]$$

for the instrument to be considered a Standard Platinum Resistive Thermometer (SPRT).

c) *To use these instruments*, ITS-90 has additionally defined the interpolation relationships in each range. These relationships are used to calculate the difference between the measured ratio $W(T_{90})$ and the reference ratio $W_r(T_{90})$. In fact, given its size, the ITS range of interest to oceanography is divided into sub-ranges that may overlap. One of these ranges from –38.8344°C to + 29.7646°C. Its interpolation relationship is:

$$W(T_{90}) - W_r(T_{90}) = a[W(T_{90}) - 1] + b[W(T_{90}) - 1]^2 \qquad [1.24]$$

where a and b are coefficients to be determined during the calibration of platinum resistive probes in the fixed points of this sub-range. The value of $W(T_{90}) - W_r(T_{90})$ is directly obtained during this calibration. When using the probe to measure an unknown temperature, the value of $W_r(T_{90})$ is extracted from equation [1.24] and is introduced into a polynomial relationship defined for a specified sub-range. In fact, ITS-90 defines one polynomial from 13.8033 K to 273.16 K,

$$T_{90} / 273.16 = B_0 + \sum_{i=1}^{15} B_i \left[\frac{W_r(T_{90})^{1/6} - 0.65}{0.35} \right]^i \qquad [1.25]$$

and a second polynomial [1.24] from 273.16 K to 964,78°C, where coefficients B_0 and B_i (i = 1 to 15), D_0 and D_i (i = 1 to 9) are provided in the ITS-90 reference publications.

$$T_{90} - 276.15 = D_0 + \sum_{i=1}^{9} D_i \left[\frac{W_r(T_{90}) - 2.64}{1.64} \right]^i \qquad [1.26]$$

As these polynomials are difficult to handle, the WOCE Hydrographic Program (WHP) defined a simplified relationship [1.27], which allows us to calculate a temperature t_{90} with accuracy greater than 0.1 mK in the area of oceanographic interest, which ranges from –2°C to +35°C.

$$t_{90} = 0.010015 + 250.7140(W_r - 1) + 9.71421(W_r - 1)^2 \qquad [1.27]$$

Having improved the accuracy of the thermodynamic scale approach over the years, there are significant differences between the ITS and the IPTS. These differences and their correction relationships are given in the official International Bureau of Weights and Measures (*Bureau International des Poids et Mesures, BIPM.* See www.bipm.org) publications. Also, when processing data measured by a temperature sensor, it is important to know which temperature scale the sensor was related to, or to which temperature scale the data should be related. These corrections are particularly important in oceanography, as we seek high accuracy and may need to use data recorded decades ago. This is accentuated by the fact that temperature is used to calculate other quantities, such as salinity or velocity. In fact, the calculation relationships may have been established at a time before the ITS-90 was used. The instruments used to make temperature measurements, having been linked to the ITS-90, must be corrected before being introduced into these calculation relationships. To simplify this process, the WHP suggests a linking relationship between the IPTS-68 and ITS-90 as follows:

$$t_{90} = 0.99976 \times t_{68} \text{ or } t_{68} = 1.00024 \times t_{90} \qquad [1.28]$$

It ensures that the error introduced by use of these relationships is less than 0.5 mK in the area of –2°C to +35°C.

1.2.1.2. *Uniting instruments to the ITS*

SPRTs are composed of a platinum filament free of any constraints in order to avoid introducing biases from the support and the sensor temperature (see Figure 1.3). This means that instruments must be handled with care and their use is limited to the laboratory. Other technologies therefore had to be developed to help achieve *in situ* measurements with high resolution and high accuracy.

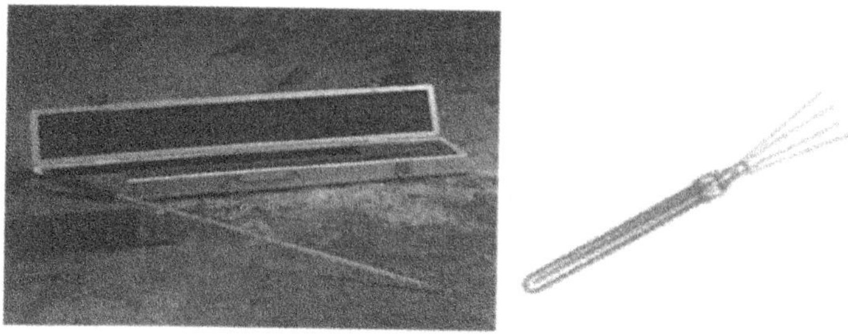

Figure 1.3. *A SPRT in its transport case and a reference platinum probe in a sealed glass tube (Courtesy of HartScientific, A Fluke Company)*

Similarly, fixed reference points (see Figure 1.4) are in the form of sealed cells containing a central shaft in which the SPRT can be placed. If the probe to be calibrated meets the SPRT dimensional criteria, it can be directly linked to the ITS-90 using fixed point cells. This calibration technique is called *ITS fixed point calibration*. It is not applicable to all types of thermometric sensors. In the case of oceanographic instruments, temperature sensors are often secured to the instrument and have small dimensions in order to limit the response time. They do not therefore meet the SPRT dimensional criteria and their linkage to the ITS has to be made indirectly. These instruments are calibrated by comparisons to sensors (whether SPRT or other) that are referenced to the ITS. This calibration technique is called *calibration by comparison*.

In the case of calibration by comparison, the sensor to be calibrated is positioned close to a reference probe in an enclosure for which temperature is stabilized. This enclosure is generally a tank filled with fresh water, sea water or a more viscous liquid such as glycol. Sea water has the advantage of allowing the simultaneous calibration of temperature and conductivity sensors, and even velocity sensors, that are fitted to the instrument. The tank is large in order to be able to contain entire instrument (see Figure 1.6). During calibration, the device is placed in the same thermal conditions as when used *in situ* and eventual biases introduced by thermal drift of electronics are considered.

The WHP imposes significant constraints on the linkage of instruments. For temperature, the accuracy of measurements must be of 0.002°C with a repeatability of 0.0005°C. To satisfy these constraints, the thermal stability of calibration baths must be less than or in the order of 0.001°C during measurement. Sophisticated regulation systems must be used to achieve such high performances. Calibration is done at several points (4 to 8), even when the sensor response is linear, in order to increase the accuracy of comparisons. These points are distributed over the instrument's range of use.

Calibration by comparison provides measurement uncertainties that are greater than fixed-point calibration, as it is necessary to take into account:

– the uncertainty of the reference probe;

– the uncertainty of the instrument to be calibrated; and

– the uncertainty caused by the thermal instability of the tank.

The sensors and electronics can drift over time and deviate from their calibration values; to accommodate the accuracy constraints imposed by the WHP, it is necessary to establish calibration periodicities. The instruments from which we expect the greatest accuracy are generally calibrated before and after use at sea. The calibration is either to provide coefficients that allow linearization of sensor response, in which case the data acquired prior to its calibration cannot be corrected, or to provide slope and offset values that are obtained by measuring differences compared to the reference probe, in which case it is possible to establish a drift map of the sensor or control card and correct data that have previously been acquired and processed.

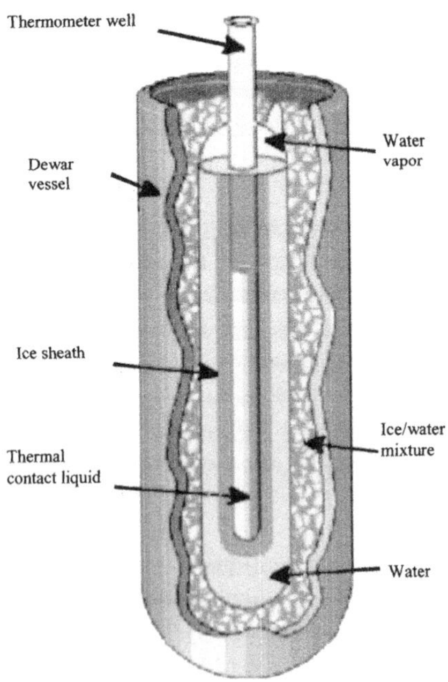

Figure 1.4. *Representation of a triple point cell of water stored in a Dewar flask filled with crushed ice for conservation. (© Crown Copyright 2006. Reproduced with the permission of the Controller of HMSO and the Queen's Printer for Scotland)*

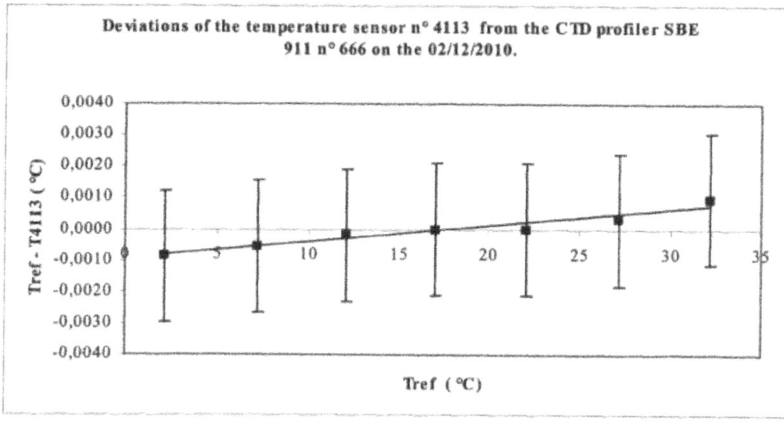

Figure 1.5. *Example of a correction curve obtained after calibration by comparison of a temperature sensor of a CTD profiler. The Y-axis shows the values of measured deviations with respect to a reference probe. The X-axis shows reference temperatures*

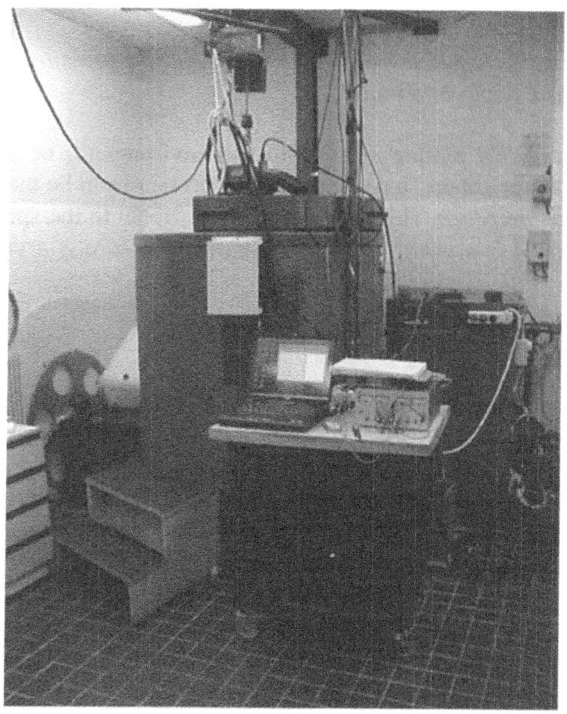

Figure 1.6. *Overview of a calibration facility dedicated to oceanographic instruments (© SHOM)*

1.2.2. *Pressure*

A liquid or gaseous substance contained in a chamber exerts a force on the walls of the chamber, called the pressure force. If we place a sensor at a certain depth in the chamber, we can see that the pressure force imposed on it is independent of the direction in which the sensor is oriented. Static pressure p is therefore a scalar quantity. It can be defined on the force dF that it exerts on a surface element dS:

$$p = dF/dS \qquad [1.29]$$

This quotient is independent of orientation, but is dependent on position within the chamber. If we lower the sensor, the measured value of p increases. This increase is due to the effect of gravitational force on the fluid. If the upper level of the chamber is placed at atmospheric pressure p_0, the sensor will measure the absolute static pressure given by equation [1.30] where ρ is the density of the fluid, g is the gravitational acceleration, and z is the vertical distance of the sensor with respect to level 0 where we determine p_0.

$$p = p_0 + \rho gz \qquad [1.30]$$

If we measure the value $p - p_0$, we have the *relative pressure*. This measurement is used to estimate depth or altitude.

If the sensor is in motion relative to the environment, or if the environment moves relative to the sensor, a weak additional pressure can be measured. This is the force of *dynamic pressure*. This pressure is proportional to the speed of movement v and to ρ. As speed is a vector quantity, dynamic pressure is a vector quantity. We therefore have:

$$p = p_0 + \rho gz + \rho v^2/2 \qquad [1.31]$$

The dynamic pressure term is an additional error term when we only want to measure static pressure.

Measured pressure values are used directly in the calculation relationships of certain quantities (salinity, density, velocity, etc.) but they are primarily used to determine depth (see section 1.12) using relationship [1.15] or variations in depth, i.e. the tide. The measurement ranges of sensors used to measure depth range from 0 to 6,000 dbar (the explorable limit of the ocean floor). The measurement ranges of tide gauges are limited at most to 250 dbar in order to maintain a measurable resolution in the order of or less than 1 mm of water.

Linkage of these pressure measurements to international standards is done directly by using an instrument called a *dead weight tester*. It is based on the same definition as that of the quantity pressure (see equation [1.29]). A force is applied by the intermediary of calibrated masses M that we place on a piston with a section S that generates pressure on a fluid (gas or oil):

$$p = gM/S \qquad [1.32]$$

The pressure exerted on this fluid is directly transmitted to the sensor to be calibrated by a flexible pipe or capillary tube that is connected to it.

The masses, M, placed on the scale must be corrected for the difference in buoyancy. To do this, an additional term involving the density of air ρ_a and density of steel that makes up the masses ρ_m, is introduced into relationship [1.30]. Moreover, it is necessary to take into account the thermal expansion of the piston and the cylinder that is represented by coefficient α. Relationship [1.30] then becomes:

$$p = \frac{gM}{S}\left(1 - \frac{\rho_a}{\rho_m}\right)[1 - \alpha T] \qquad [1.33]$$

In fact, section S results in the calibration coefficient of the pressure balance called $K_n(20)$, as it is measured at 20°C, and at normal gravity or reference gravity g_n. This measurement is carried out in a primary laboratory and is one of the elements for the uniting of the dead weight tester to the metrological chain of pressure. The $K_n(20)$ also includes a corrective term for buoyancy. The expression [1.33] becomes:

$$p = (g/g_n)K_n(20)[1 - \alpha T]M \qquad [1.34]$$

For high pressures, the deformation of the piston with applied pressure is taken into account by an intermediary coefficient λ given by the manufacturer. Finally we get:

$$p = (g/g_n)K_n(20)[1 - \alpha T][1 - \lambda P]M \qquad [1.35]$$

Pressure thus generated is a reference pressure. Its value is not measured but is calculated. Other corrections are added to the calculation, such as the difference in height between the sensor to be calibrated, the reference level of the scale and, optimally, the difference between atmospheric and normal pressure. If all of the precautions are taken, we can calibrate sensors with a relative uncertainty of 4×10^{-5}, which corresponds to 0.25 dbar in a range of 6,000 dbar.

Figure 1.7. *A dead weight tester with its flexible pipe allowing linkage to the sensor to be calibrated, its temperature gauge to correct for piston-cylinder expansion, and its calibrated mass box (© SHOM)*

1.2.3. *Conductivity and salinity*

1.2.3.1. *History and definition of salinity*

The definition of salinity given in section 1.1.1 is that of absolute salinity, which is not directly measurable. It has, however, allowed Knudsen to simultaneously connect salinity to chlorinity, *Cl* (expressed in ‰), and to establish that:

$$S_k\text{‰} = 0.030 + 1.8050 \times Cl\text{‰} \qquad [1.36]$$

Chlorinity is defined as the mass (in g) of halogens contained in 1 kg of sea water, bromide and iodide ions having been replaced by their equivalents in chloride.

The halogens are chlorine, fluorine, bromine and iodine. The validity of this formula rests on a concept developed by Dittmar in 1884 [DIT 84], which claimed that the proportion of major constituents of sea water, the ions in this case, is the same for all oceans. Relationship [1.36], which gave salinity a relative unit (‰), was used until the 1960s, so that it was not possible to determine an equivalent mass per kilogram. In 1962, it was modified following work carried out on chlorinity, conductivity ratio, salinity and density. From 1962 to 1969, *S* was:

$$S = 1.806\,55 \times Cl\ (‰) \qquad [1.37]$$

Knowledge of the atomic masses of chlorine and silver molecules evolved, making it necessary to find a way to eliminate variations in chlorinity. Thus, salinity was redefined in 1969. From 1969 to 1978, we used the following relationship where Ag is the mass of silver (in grams) necessary to precipitate the halogens:

$$S = 1.806\,55 \times 0.328\,523 \times Ag\ (‰) \qquad [1.38]$$

The formulas that link salinity and chlorinity were not compatible if the constituents of sea water varied from one sample to another. The consequence of these defects was sensitive at the level of density calculations. At the time, however, experimental work had begun to show that density could be deducted from conductivity measurements, with an uncertainty that was 10 times smaller than that produced using chlorinity measurements.

In 1978, salinity was once again redefined. This new definition took into account the older one in order to retain some consistency in measured values and acknowledge the conductivity measurements. The definition of the Practical Salinity Scale adopted in 1978 (or PSS-78) is still in effect, and is as follows:

The practical salinity, S_P, of a sample of sea water is the ratio, called K_{15}, of the electrical conductivity of this sample at a temperature of 15°C and a pressure of 101,325 Pa, over that of a solution of potassium chloride (Kcl) containing 32.4356 g of KCl/kg of solution, under the same temperature and pressure conditions.

This scale is (theoretically) valid for $2 < S_P < 42$. Practical salinity is a quantity defined from a ratio. It does not have a unit, contrary to the old definition where salinity was based on the percentage of chlorinity. It does not always give an equivalent to mass, but it allows the linkage of oceanographic measurements to ratios of practical, measurable quantities. The value 32.4356 g of KCl/kg of solution was chosen because it gave conductivity that was identical to that of normal water of salinity 35‰, determined by relationship [1.38]. The value $K_{15} = 1$ corresponds to a practical salinity of 35. The general relationship between practical salinity and K_{15} is the following:

$$S_P = 0.0080 - 0.1692K_{15}^{1/2} + 25.3851K_{15} + 14.0941K_{15}^{3/2} - 7.0261\ K_{15}^{2} + 2.7081K_{15}^{5/2} \qquad [1.39]$$

The absolute conductivity of a solution corresponding to $K_{15} = 1$, measured at $t = 15°C$ and $p = 101{,}325$ Pa, is 42.933 mS/cm according to Poisson [POI 78] and 42.914 mS/cm according to Culkin and Smith [CUL 80]. It was 42.914 mS/cm that was formally adopted (and is programmed into a number of instruments), although this value is questionable.

There are huge difficulties in taking absolute conductivity measurements as a large number of parameters must be mastered: environment temperature, air humidity, purity of the KCl solutions used and the elements necessary to their preparation, such as reagent weighing and calibration of the cell constant of the conductivity meter used. In 2005, Kawano *et al.* [KAW 05] showed that the quality of reagents used had a significant effect (in the order of 0.0012 for salinity) on the accuracy of measurements. In 2009, a working group (the P111 of the *Consultative Committee for the Amount of Substance*) was responsible for evaluating the uncertainty that could arise from the absolute conductivity of sea water collected from the Atlantic and corresponding salinity values, in order to ensure the traceability of practical salinity measurements to the SI. They concluded that traceability could only be guaranteed with a relative uncertainty of 0.05%, which corresponds to ±0.02 for $S_P = 35$.

It is important to note that practical salinity does not accurately reflect the content of dissolved components of sea water. Notably, as conductivity cells do not allow the measurement of non-ionic components, differences remain between S_P and S_a. Depending on the areas of measurement, errors can arise in the order of 0.195 *parts per thousand* (ppt) or ‰, which is 50 to 100 times higher than the uncertainty that can be achieved making *in situ* S_P measurements.

For this reason, the TEOS-10 defines several salinities. Following work done by Millero in 2008 on standard sea water (see section 1.2.3.3), the notion of a reference salinity, S_R, was introduced. It can be calculated in g/kg by the following relationship:

$$S_R = \left(\frac{35.16504}{35}\right) S_P \qquad [1.40]$$

for standard water, $S_R = S_A$ and for $S_P = 35$, $S_R = 35.16504 \pm 0.007$ g/kg (according to Millero *et al.* [MIL 08a]). However, waters for which the composition (in mass fraction of constituents) is different to that of standard water, S_A is no longer proportional to S_P and it is necessary to introduce the correction δS_A. To do this, in 2010 McDougall *et al.* [MAC 10] built an algorithm that estimated δS_A depending on the geographic area of measurement, but the TEOS-10 has authorized the use of other means. We then have:

$$S_A = S_R + \delta S_A(p, \lambda, \varphi) \qquad [1.41]$$

where p designates pressure, λ longitude and φ latitude (in ° North). A substantially different relationship must be applied in the case of the Baltic Sea (see [FEI 10]). According to McDougall *et al.*, the largest correction δS_A that we can provide for deep-sea measurements is 0.027 g/kg (which corresponds to an error of 0.020 kg/m^3 over density). The standard uncertainty on values that have been used to develop this algorithm is ±0.0107 g/kg and the uncertainty introduced by the algorithm is 0.0048 g/kg. For coastal environment measurements, everything remains to be defined.

In order to take into account all of the problems involved in defining the concept of absolute salinity, the TEOS-10 separates S_A into different components. First, to integrate the problem of traceability of the SI units, the notion of "salinity – density", S_A^{dens}, was introduced. By definition, it corresponds to *the mass fraction of the solution that has the same density as the sample from which it came, at a temperature of 25°C and a pressure of 101,325 Pa.* It is this definition that is the closest to the notion of absolute salinity S_A, defined by relationship [1.41], but does not reflect the notions of pure water and dissolved substances.

"Solution salinity" S_A^{soln} was defined as the mass fraction of components other than water that can be found when a sample of sea water is brought to a temperature of 25°C and a pressure of 101,325 Pa. As chemical equilibriums are dependent on temperature and pressure, so is absolute salinity. When required to reconstitute sea water in the laboratory or to study the mass flows of rivers or hydrothermal sources, it is necessary to use the concept of "mass added to salinity", S_a^{add}. S_A^{add} is equal to S_R plus the concentration of material mass that must be added to obtain the concentration of all components of the analyzed sample after chemical equilibrium has been reached, at a temperature of 25°C and a pressure of 101,325 Pa.

With these definitions, the notion of a conservative tracer of water masses is somewhat lost. Then, a "preformed salinity", S_*, has been defined as being *an absolute salinity independent of bio-geo-chemical interactions, i.e. which mass fraction in solution that has the same chlorinity as the collected sample*, the bio-geo-chemical reactions being able to locally vary the chemical equilibrium reactions of sea water. The point in common between all of these salinities is standard sea water, for which: $S_R = S_A = S_A^{soln} = S_A^{add} = S_*$.

These definitions are new, are rarely used and are still being studied. For these reasons, the TEOS-10 recommends practical salinity measurements continue to be measured and stored in databases.

1.2.3.2. *Definition and characteristics of the conductivity of sea water*

Sea water is an electrically neutral solution in which the movements of anions and cations participate in thermal agitation, but are statistically compensated for. It is only when we apply an electric field, *E*, for example to two distant electrodes subject to different potentials, that the different ions move in an orderly way, under the influence of the force exerted by this field. Within the field we can then see the circulation of a current for which the intensity corresponds to the algebraic sum of the charge fluxes transported simultaneously by anions and cations in the water. The speeds of movement of the ions, however, are reduced by the viscosity of the environment that creates friction forces and by the presence of different types of ions that exert attraction forces on one another. The tube field that is generated has a certain resistance, or conversely a certain electrical conductivity.

Viscosity is a property essentially linked to the solvent, i.e. water molecules. It depends very little on the types of ions in the solution. Despite the effects of viscosity and attraction forces, it is assumed that the speeds at which ions move are the product of the amplitude of E by ion mobility. From this, we can show that conductivity is essentially the sum of the products of electrical charges of present ions, by their volume concentration and ionic mobility.

The composition of sea water, being relatively constant at fixed temperatures and pressures, we can show that the conductivity–chlorinity relationship is linear. We can also show that at constant pressure and chlorinity, an increase in temperature results in a decrease in water density. This results in a decrease in the viscosity of the environment and then a decrease in the number of ions per unit of volume. Viscosity varies inversely with conductivity. If temperature rises, there is an increase in ion mobility and in the conductivity of water. It was therefore determined that in most seas the contribution of salinity to the variation in conductivity is very small relative to the contribution of temperature. For example, in the North Pacific at a depth of between 100 and 150 m, salinity contributes only 6.6% of the variations of conductivity measured.

At constant temperature and chlorinity, an increase in pressure also results in an increase in conductivity. Thus, if pressure increases from 0 to 10,000 dbar, the relative conductivity $\Delta C/C$ of water of salinity 35 at 0°C increases by 11.5%. This variation is due to the increase in ion volume concentration, the decrease in viscosity under the effects of pressure (contrary to what we might think), and the dissociation of certain electrolytes under the effect of pressure.

At ambient pressure, the effect of a variation of 0.01°C on a water of salinity 35 results in a variation of conductivity of 0.01 mS/cm. However, the WHP preconizes an accuracy of 0.002 on salinity calculations based on the PSS-78, and a repeatability of 0.001.

To address these constraints, it is necessary to determine the absolute temperature of the sample of sea water for which we wish to know the salinity to at least 0.002°C, which is not easy to achieve. However, the conductivity ratio of two samples of sea water of different salinities is relatively insensitive to variations in temperature. It is this property that is used in instruments that measure salinity in laboratories. Temperature primarily acts on the viscosity of water. Two samples of differing salinities will have almost identical viscosities if their temperatures are the same. If we establish a ratio of conductivities to constant temperature, we can show that the ratio is equal to that of salinity.

1.2.3.3. *Standard sea water bottles*

To make salinity measurements based on the principle outlined in section 1.2.3.1, and to cling to the definition of salinity, a "standard" is required for reference conductivity measurements. As it is not easy to prepare a *KCl* solution as defined, the standard used since 1900 is a standard sea water or normal water. To create normal water, a certain amount of sea water is sampled periodically at depth in the North Atlantic between the longitudes 40°W and 50°W. The salinity, always being slightly above 35, is adjusted by dilution to obtain a chlorinity close to 19.374 $\times 10^{-3}$. *Chlorinity* is measured by the Mohr-Knudsen method based on the following definition that was adopted in 1985:

> *"the chlorinity of a sample of sea water (Cl) represents 0.3285234 times the ratio of the mass of pure silver required to precipitate all dissolved chloride, bromide and iodide contained in the sea water sample, over the mass of the sample."*

Here, chlorinity is defined from a mass ratio. It is therefore a dimensionless quantity, similarly to practical salinity. It is this standard sea water that is used to define a reference composition and a standard salinity, S_R, by relationship [1.40]. From this relationship, we can determine that a sea water sample that meets the reference composition and has a practical salinity of 35 has a reference salinity of 35.16504 g kg^{-1}. The definition of a reference composition is discussed in section 2.11.1.

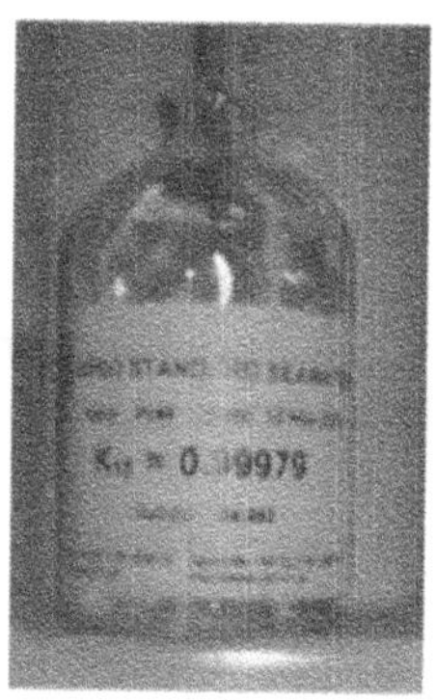

Figure 1.8. *Standard sea water bottle with salinity close to 35*

Standard sea water is packaged in bottles or sealed vials of 275 ml. Each label gives the exact chlorinity of the sample that was used to fill one lot of vials and the ratio K_{15} measured by a salinometer. For one lot of vials, the spread of K_{15} can be guaranteed to $\pm 1.10^{-4}$, which corresponds to a salinity of $\pm 3.10^{-4}$. The tolerance specified by the manufacturer on the vials is ±0.002 salinity, however, which corresponds to $K_{15} = \pm 7.10^{-4}$. These tolerances are guaranteed for two years. To improve the link between salinity measurements and SI and to stick to the notion of absolute salinity within the meaning of S_A^{dens}, we currently aim to include a density value on each bottle. In recent years, other salinity vials have started to exist: 38, 30 and 10.

Initially the standard sea water was prepared by the IAPSO in Copenhagen, which is why it was formerly known as Copenhagen normal water. At present, this work is ensured by the *IAPSO Standard Sea Water Service* at the Oceanographic Institute of Wormley in the United Kingdom.

1.2.3.4. *Laboratory salinometers*

Laboratory *salinometers* are reference instruments that allow the salinity values measured from sea water samples to be linked to the PSS-78. The measuring principle is based on the comparison of a reference sea water sample with a sea water sample to be analyzed. As explained in section 1.2.3.2, the ratio of conductivity of two sea waters with differing salinity is insensitive to variations in temperature. The samples were therefore previously at the same temperature and the comparison that follows is extremely stable.

The first laboratory salinometers appeared in 1930. They were the work of Franck Werner, who worked for the Bureau of Coast Guards. The Bureau provided accurate measurements that were equivalent to those that could be obtained by the Knudsen method, or to within ±0.03 salinity units. In 1956, Alvin Bradshaw, Karl Schleicher (of the Woods Hole Oceanographic Institution, WHOI) and Roland Cox (of the National Oceanographic Institution, NOI) developed a more effective instrument that neared the accuracy characteristics of current salinometers (or ±0.003 salinity units), but this was cumbersome in weight and size, which reduced its usage in laboratories. In 1961, an Australian named Bruce Hamon developed a lighter and more transportable device that sought to establish an equivalent accuracy with an inductive cell (section 2.2.2.3).

The most common and successful salinometers to date use an operation principle and a measurement cell developed by Dauphinee and Klein in 1975. They are marketed under the names AUTOSAL and PORTASAL by the GUILDLINE Society. Conductivity is measured by means of a four-electrode cell that is kept at constant temperature by a thermostatted bath. The two internal electrodes are traversed by an excitation signal of low amplitude and a frequency of 250 Hz, in order to minimize polarization. The two external electrodes are linked to a voltage amplifier with large input impedance. The electrodes are therefore traversed by a low current, which once again limits the polarization risk.

Before entering the cell, the water to be analyzed flows through a capillary that passes through a heat exchanger in order to produce the bath temperature t_b, which can be set between 18°C and 33°C using a switch. When measurements are made, it must be set so that:

$$t_{ambient} - 2°C < t_b < t_{ambient} + 4°C.$$

The accuracy of this temperature is in the order of 0.01°C, but its stability is huge: it is ±0.001°C/day for the AUTOSAL, which in the best conditions produces an uncertainty of ±0.0001 over the conductivity ratio. The uncertainty in salinity, which is dependent on the standard sea water vials used, is guaranteed to ±0.002 according to the manufacturer's documentation. It does in fact vary depending on the value of salinity measured, as shown in a 2009 study, and the accuracy of the instruments is limited by linearity errors that occur when salinity is above 38 or below 30 [LEM 09].

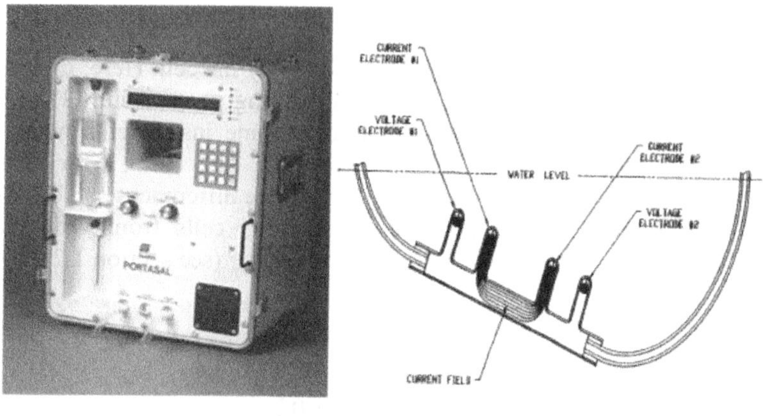

Figure 1.9. *The PORTASAL salinometer is the transportable version of the AUTOSAL. It is found onboard many oceanographic research vessels, but it is also an excellent reference tool in the laboratory. Here it is equipped with a standard sea water vial that is used for its calibration (courtesy of Guildline Instruments Limited). The water to be analyzed flows into a capillary that passes through a heat exchanger in order to produce the temperature of a thermo-regulated bath into which the conductivity cell is also immersed. It is a four spiral-electrode cell. The two internal electrodes are linked to a current generator and the two external electrodes are linked to an electronic circuit voltage measurer (courtesy of Sea Bird Electronics, Inc.)*

1.2.3.5. In situ *salinity measurements*

The measurement of the conductivity ratio is not necessarily carried out at 15°C, so we cannot directly use the definition equation [1.39] giving S_P as a function of K_{15}. Corrections have therefore been introduced. For such corrections, a term ΔS was added and R_t was appointed the conductivity ratio measured at temperature t. After these changes, equation [1.39] becomes:

$$S_P = 0.0080 - 0.1692\text{x } R_t^{1/2} + 25.3851\text{x } R_t + 14.0941\text{x}.\ R_t^{3/2} - 7.0261\text{x } R_t^2 + 2.7081x\, R_t^{5/2} + \Delta S \quad [1.42]$$

with:

$$\Delta S = \frac{(t-15)}{1+0{,}0162.(t-15)}.\left(0{,}0005 - 0{,}0056.R_t^{1/2} - 0{,}0066.R_t - 0{,}0375.R_t^{3/2} + 0{,}0636.R_t^2 - 0{,}0144.R_t^{5/2}\right) \quad [1.43]$$

Laboratory salinometers used at atmospheric pressure do not require any corrections for pressure effects. Equations [1.42] and [1.43] give the salinity value from the measurement of R_t. When measurements are made *in situ* with conductivity cells, however, it is necessary to take the effects of pressure into account. R is the conductivity ratio of cells used *in situ*:

$$R = C(S_P, t, p) \,/\, C(35{,}15{,}0) \quad [1.44]$$

where $C(S_P, t, p)$ is the conductivity value measured by the cell. Here, $C(35{,}15{,}0)$ is a digital reference programmed into the measuring instrument. The PSS-78 recommends using $C(35{,}15{,}0)$ = 42.914 mS/cm. Some instruments made before 1978 use a different value. It is possible to find, for example, $C(35{,}15{,}0)$ = 42.896 mS/cm. Shifts resulting from these differences can be corrected for during the calibration of instrument conductivity cells from water samples analyzed by a correctly referenced laboratory salinometer (see section 1.2.3.4).

The ratio R [1.44] can be split into three factors:

$$R = \frac{C(S_P,t,p)}{C(S_P,t,0)} \frac{C(S_P,t,0)}{C(35,t,0)} \frac{C(35,t,0)}{C(35,15,0)} \quad [1.45]$$

also written:

$$R = R_p R_t r_t \quad [1.46]$$

In this equation, the term R_t is equivalent to that of relationship [1.42]. To calculate salinity from conductivity measurements, we must write:

$$R_t = R / (R_p r_t) \qquad [1.47]$$

In order to determine R_t, R_p, which is the correction ratio for the effect of pressure, must be calculated. R_p can be expressed as a function of the conductivity ratio R of the temperature and pressure that have been measured:

$$R_p = 1 + \frac{p\left(2.07\text{x}10^{-5} - 6.370\text{x}10^{-10}p + 3.989\text{x}10^{-15}p^2\right)}{1 + 3.426\text{x}10^{-2}t + 4.464\text{x}10^{-4}t^2 + R\left(4.215\text{x}10^{-1}t - 3.107\text{x}10^{-3}t\right)} \qquad [1.48]$$

The value of r_t, which is the correction ratio for the effect of temperature at normal pressure and salinity, must also be calculated. Thus, it only depends on measured temperature:

$$R_t = 0.6766097 + 2.00564\text{x}10^{-2}t + 1.104259\text{x}10^{-4}t^2 - 6.9698\text{x}10^{-7}t^3 + 1.0031\text{x}10^{-9}t^4 \qquad [1.49]$$

Once these calculations are done, the value of R_t can be placed in equations [1.42] and [1.43], which will then give a salinity value that is corrected for the effects of pressure and temperature and conform to the 1978 Practical Salinity Scale.

1.2.3.6. *Calibration of conductivity sensors*

It is relatively difficult and time consuming to generate variations in salinity. We must therefore have a system that allows the variation of salinity in a calibration bath with a large capacity or we must have several calibration chambers containing waters at different salinities. In addition, the sensitivities of variation in the conductivity of sea water as a function of temperature and salinity are roughly equivalent. At 20°C, for $32 < S_P < 39$, conductivity varies by around 2.1% of its value per degree Celsius for $0 < t < 18$°C. Then, the points necessary to establish a calibration curve are generally obtained from variations in temperature. This procedure usually corresponds fairly well to the conditions met *in situ* where temperature is the major contributor to variations in conductivity.

For the calibration of conductivity sensors, we proceed in the same manner as for the calibration of temperature sensors (section 1.2.1.2). We must, however, have sea water with salinity close to that of the environment in which the instrument will be used. If this sea water cannot be attained, it is possible to approximately reconstitute the composition of such water by using a formula validated by Millero in 1996 [MIL 96] based on an analysis of fundamental ionic components carried out by Khoo. Millero's formula uses the four major salts in sea water and an equivalent ionic force can be found in terms of salinity. With the correct mix of these salts, it is possible to accurately acquire the desired salinity. Thus, to get salinity close to 35, the following must be added per liter of pure water:

– 30 g $NaCl$;

– 5.06 g $MgCl_2$;

– 1.2 g $CaCl_2$; and

– 0.76 g KCl.

This reconstituted sea water can be used to calibrate conductivity cells because these cells measure an overall conductivity, which is equivalent to a total ionic force.

We try our best to calibrate an instrument by placing it completely (without dismantling its sensors) into a tank of thermo-regulated sea water. At each stage where the temperature of the bath is stabilized, the values of conductivity generated by the sensor to be calibrated are recorded. During each stage, samples of sea water are taken using small bottles that are then kept hermetically sealed. After calibration, the contents of each bottle are analyzed by a laboratory salinometer that was previously calibrated using standard sea water vials. To obtain reference conductivity values, salinity values provided by the salinometer are converted to conductivity values, knowing the average temperature of the bath, measured using an ITS-90 referenced probe. These values must nonetheless be converted to IPTS-68 (section 1.2.1.2) before being used in relations found within the PSS-78.

The salinity–conductivity conversion is achieved by inverting the salinity calculation relationships found in PSS-78. As equations [1.42] and [1.43] are polynomials with degrees superior to two, they can only be inverted by proceeding iteratively through, for example, a Newton-Raphson type algorithm. In these relationships, S_P is only a function of R_t and t:

$$S_P = S(R_t, t) \qquad [1.50]$$

To initiate the algorithm, having measured the value of t, it is sufficient to calculate $R_t = S_P/35$, where S_P is the salinity value obtained from the salinometer, and to insert these values into equations [1.42] and [1.43]. This gives a first estimate of S_n. A Taylor expansion of relationship [1.50] allows us to write,

$$\left(S_P - S_n\right) = \left[\frac{\partial S_P}{\partial R_t}\right]\left(R_{t_{n+1}} - R_{t_n}\right) \qquad [1.51]$$

or:

$$R_{t_{n+1}} = R_{t_n} + (S_P - S_n)\left[\frac{\partial S_P}{\partial R_t}\right]^{-1} \qquad [1.52]$$

The last term in his expression is obtained by differentiating the relationships [1.42] and [1.43] around the initial value of R_t. Given that in these equations R_t is raised to non-integer powers, it is recommended that numerical calculations are performed with the value $\sqrt{R_t}$. This calculation must be repeated as long as the value $S_P - S_n$ is not lower than a fixed threshold (for example 1.10^{-4}). The $R_{t_{n+1}}$ value obtained can then be inserted into equation [1.48], having first determined the value of r_t in relationship [1.49] and R_p in equation [1.48], in which variable R is approximated by the product $r_t R^2_{t_{n+1}}$. From equation [1.44], we can finally extract the reference value $C(S_P, t, p)$ and calculate its deviation from the value given by the sensor to calibrate.

1.2.4. *Velocity*

For many applications, it is sufficient to say that the propagation speed of sound underwater is 1,500 m/s. This value varies at most by 5% or 60–70 m/s, if working at pressures below 1,000 dbar. The propagation of sound underwater is very sensitive to small variations in speed, however, and the variation of this speed increases considerably with increasing pressure. Moreover, if we carry out acoustic propagation calculations, it is necessary to approach an uncertainty of 0.1 m/s based on our knowledge of the propagation speed. This implies the ability to precisely calculate the propagation speed of sound. To do this, there are currently over 30 formulas! Only six of them are explained in this section.

Velocity is defined as the square root of the derivative of pressure generated by an acoustic wave relative to the density of the environment. The density of the water depends on its temperature, its salinity and its pressure. However, before 1949 the acoustic propagation speed c measurements were made directly from the measurements of isothermal compressibility κ and density ρ, by relationship [1.53] which we owe to Laplace:

$$c = \sqrt{\frac{\gamma}{\kappa\rho}} \qquad [1.53]$$

with:

$$\gamma = \frac{c_p}{c_v}$$

where c_p is the specific heat of the water at constant pressure and c_v is its specific heat at constant volume.

In 1949, the first relationship that was able to express velocity as a function of temperature, salinity and depth was developed by Wood. It was used until 1962 by the US Navy, when it was replaced by another relationship published by Wilson in 1960, of which, the general formula is:

$$c(t,S,p) = c_0(0, 35, 0) + c(t) + c_S(S_P - 35, t) + c_P(p, t, S_P - 35) \qquad [1.54]$$

The variables in brackets represent temperature t, salinity S and pressure p. This formula actually contains 20 terms (c_0, c_S, c_P, broken down into polynomials). Its calculation requires a lot of time, bearing in mind computing resources at the time, and so several approximate formulas were developed (the Leroy, Kinsley and Frey formulas).

Towards the end of the 1960s, Del Grosso then Millero managed to show that the Wilson formula was not accurate, particularly at low temperatures. Both published new polynomial formulas. The speed of sound velocity values that we obtain from both of these formulas are similar if p is close to atmospheric pressure, but they differ from each other and relative Wilson's calculations beyond 1,000 dbar.

During the 1970s Donald Ross, who was working at the Saclantcenter (a NATO center involved in underwater research), tried to reduce this bias by taking the values measured by Millero and Del Grosso and publishing new equations. For salinities close to 35, he found that:

$$C_a = 1449.10 + 4.565\,t - 0.0517\,t^2 + 2.21 \times 10^{-4} t^3 + 1.338(S_P - 35) - 0.013\,t\,(S_P - 35) + 1.0 \times 10^{-4} t^2 (S_P - 35) \qquad [1.55]$$

In this equation, t must be expressed in IPTS-68. According to the author, this formula gives speeds to 0.1 m/s for $0 < t < 40$°C and is in better agreement with Del Grosso's results than those of Millero.

For measurements in the Mediterranean, where S_P is close to 38, he developed another formula:

$$C_a = 1510.18 + 3.133\ (t - 15) - 0.0414\ (t - 15)^2 + 2.20 \times 10^{-4}\ (t - 15)^3 + 1.166\ (S_P - 38) - 0.010\ (t - 15)\ (S_P - 38) \qquad [1.56]$$

Finally, pressure dependence is obtained by two other equations. The first is used for the waters of the Atlantic, for which salinity is close to 35:

$$C_p = 0.1592\ p + 1.25 \times 10^{-5}p^2 + 2.0 \times 10^{-4}t\ p - 7.5 \times 10^{-7}t\ p^2 + 2.0.10^{-4}(S_P - 35)\ p - 2.4 \times 10^{-7}\ (S_P - 35)p^2 \qquad [1.57]$$

and the second is for Mediterranean waters, for which salinity is close to 38:

$$C_p = 0.1630\ p + 2.0 \times 10^{-4}(t\text{-}15)\ p - 7.5 \times 10^{-7}(t - 15)\ p^2 \qquad [1.58]$$

where p has to be expressed in kg/cm^2. (1 kg/cm^2 = 1 dbar × $g(\varphi)$, where $g(\varphi)$ is given by equation [1.17]). Speed corrected for the effect of pressure is given by:

$$c = c_a + c_p \qquad [1.59]$$

According to the author, this time the equation is in closer agreement with Millero's results. Disagreements between these formulas are in the order of 0.1 m/s at 300 m, 0.3 m/s at 1,000 m and 0.6 m/s beyond this distance.

In 1974, Del Grosso [DEL 74], who worked at the Naval Research Laboratory in Washington DC, also published a formula that allowed us to calculate c for $0 < t_{68} <$ 35°C, $0 < S_P < 43$ and $0 < P < 1{,}000$ kg/cm². It is presented as:

$$c(t, S_P, p) = 1402.392 + \Delta c(t) + \Delta c(S_P) + \Delta c(P) + \Delta c(S_P, t, p) \qquad [1.60]$$

with:

$$\Delta c(t) = 5.01109398873t - 0.0550946843172t^2 + 0.22153596924 \times 10^{-3}t^3$$

$$\Delta c(S_P) = 1.32952290781\ S_P + 0.12895575684 \times 10^{-3}S_P{}^2$$

$$\Delta c(P) = 0.156059257041\ p + 0.244998688441 \times 10^{-4}p^2 - 8.83392332513 \times 10^{-09}p^3$$

$$\Delta c(S_P, t, p) = -0.0127562783426\ t\ S_P + 0.00635191613389\ t\ p + 0.265484716608 \times 10^{-7} t^2 p^2 - 0.159349479045 \times 10^{-5} t\ p^2 + 5.22116437235 \times 10^{-10} t\ p^3 - 0.438031096213 \times 10^{-6} t^3\ p - 1.61674495909 \times 10^{-09} S_P^2\ p^2 + 0.96840315641 \times 10^{-4} t^2 S_P + 0.485639620015 \times 10^{-5} t\ S_P^2\ p - 0.340597039004 \times 10^{-3} t\ S_P\ p$$

It was shown that the validity of relationship [1.60] is excellent until 1,034 bar, for $t < 5°C$ and $33 < S_P < 36$. For $5 < t < 15$ and the same salinity range, it is advisable to restrict use to $p = 207$ bar. For $S_P = 38$, the accuracy of the values obtained has been validated to 414 bar for $5 < t < 15$ and 69 bar for $t < 5°C$. According to Del Grosso, the standard deviation of values obtained with relationship [1.60] is 0.05 m/s.

Despite the recognized reliability and ease of implementing the Ross and Del Grosso relationships, the formulas officially recommended by UNESCO since 1977, and which must be used, are those put forward by Chen and Millero [MIL 80]. Their formulas are complex. They contain no less than 42 coefficients to program:

$$c(t, S_P, p) = C_W(t, p) + A(t, p)S_P + B(t, p)S_P^{3/2} + D(t, p)S_P^2 \qquad [1.61]$$

with:

$$C_W(t, p) = 1402.388 + 5.03711\ t - 5.80852 \times 10^{-2} t^2 + 3.342 \times 10^{-4} t^3 - 1.478 \times 10^{-6} t^4 + 3.1464 \times 10^{-9} t^5 + (0.153563 + 6.8982 \times 10^{-4} t - 8.1788 \times 10^{-6} t^2 + 1.3621 \times 10^{-7} t^3 - 6.1185 \times 10^{-10} t^4)\ p + (3.126 \times 10^{-5} - 1.7107 \times 10^{-6} t + 2.5974 \times 10^{-8} t^2 - 2.5335 \times 10^{-10} t^3 + 1.0405 \times 10^{-12} t^4)\ p^2 + (-9.7729 \times 10^{-9} + 3.8504 \times 10^{-10} t - 2.3643 \times 10^{-12} t^2)\ p^3$$

$$A(t, p) = 1.389 - 0.01262\ t + 7.164 \times 10^{-5} t^2 + 2.006 \times 10^{-6} t^3 - 3.21 \times 10^{-8} t^4 + (9.4742 \times 10^{-5} - 1.258 \times 10^{-5} t - 6.4885 \times 10^{-8} t^2 + 1.0507 \times 10^{-8} t^3 - 2.0122 \times 10^{-10} t^4)\ p + (-3.9064 \times 10^{-7} + 9.1041.10^{-9} t - 1.6002 \times 10^{-10} t^2 + 7.988 \times 10^{-12} t^3)\ p^2 + (1.1 \times 10^{-10} + 6.649 \times 10^{-12} t - 3.389 \times 10^{-13} t^2)\ p^3$$

$$B(t, p) = -0.01922 - 4.42 \times 10^{-5} t + (7.3637 \times 10^{-5} + 1.7945 \times 10^{-7} t)\ p$$

$$D(t, p) = 1.727.10^{-3} - 7.9836.10^{-6}.p$$

In theory, this formula is valid for:

$0 < t_{68} < 40°C$, $0 < S_P < 40$, $0 < p < 10{,}000$ dbar.

It has a standard deviation of 0.19 m/s, which is reduced to 0.05 m/s at zero pressure. It should be noted that in 1994 Millero published a correction to these formulas that is valid for low temperatures and high pressures. A publication by Meinen and Watts [MEI 97]), however, attempts to show that despite the corrections, Del Grosso's 1974 formula provides more accurate speeds at pressures above 1,000 dbar. The debate regarding the use of these relationships is thus not completely settled. Recently, other authors have found simpler equations approaching the official one by Chen and Millero, in order to calculate velocities in digital models of ocean behavior or for calculations of acoustic tomography [BRY 95]. One of the simplified relationships, used when a large speed calculation is necessary and achieves a velocity value corrected for temperature, salinity and depth, was devised by Medwin [MED 75]:

$$c(t, S, p) = 1449.2 + 4.6\,t - 0.055\,t^2 + 0.00029\,t^3 + (1.34 - 0.01\,t)(S_P - 35) + 0.016\,p \qquad [1.62]$$

where p is expressed in m. It is valid for 0°C < t < 35°C, 0 < S_P < 45 and 0 < p < 1,000 m; however use for t > 20°C is not advised as it diverges considerably from the results obtained by the other formulas. It is based on the Del Grosso relationships and shows that velocity varies in an almost linear manner with changing salinity; whereas it varies in a quadratic, even cubic, manner with changing temperature.

p (bar)	t °C (IPTS-68)	S_P (PSS-78)	c_{CM} (m/s)	c_R (m/s)	c_{DG} (m/s)	$c_{CM} - c_r$	$c_{CM} - c_{DG}$
0	0	35	1449.14	1449.10	1449.08	0.04	0.06
0	15	35	1506.66	1506.69	1506.67	-0.03	0.00
0	35	35	1555.18	1555.02	1555.00	0.16	0.17
300	10	35	1539.69	1539.58	1539.31	0.10	0.37
600	2	35	1560.16	1559.79	1559.44	0.36	0.71
0	0	38	1453.16	1453.13	1453.10	0.03	0.06
0	15	38	1510.16	1510.18	1510.17	-0.02	-0.01
0	35	38	1558.21	1558.04	1558.04	0.17	0.17
300	10	38	1543.34	1543.36	1543.07	-0.02	0.27

Table 1.1. *Comparison of velocity values obtained using the Chen and Millero formula (c_{CM}; relationship [1.61]), the Ross formula (c_R; relationship [1.59]) and the Del Grosso formula (relationship [1.60]). At zero pressure, the three formulas are in agreement except at high temperatures. At high pressure, divergences appear. The Del Grosso relationship is then preferred*

It should be noted, finally, that the TEOS-10 also redefines the notion of velocity by linking it to absolute salinity:

$$c = c(S_A, t, p) = \sqrt{\frac{\partial p}{\partial \rho}}\Big|_{S_A,\eta} \qquad [1.63]$$

where pressure p is expressed in Pa, ρ in kg/m^3 and the specific entropy η in J kg^{-1} K^{-1}.

Velocity is a calculable quantity but can also be measured. Velocity measurement instruments used *in situ*, however, require a calibration that involves calculation of the deviation between the approximate value of velocity provided and a reference value calculated based on relationship [1.61]. In the same way as for temperature calibrations, a temperature-regulated tank is required (see section 1.2.1.2). This tank must be filled with sea water. At each stage where the temperature of the tank is stabilized, the velocity values generated by the sensor to be calibrated are recorded. During these stages, a salinity sample is created using small sample bottles that are then hermetically sealed. After calibration, the contents of these bottles are analyzed by the salinometer that was previously calibrated using standard sea water vials. Water temperature is measured using a reference probe linked to the IPTS-90, then transformed into t_{68} (the influence of this transformation is 0.05 m/s at 35°C for S = 35). The salinities and reference temperatures t_{68} are introduced into relationship [1.61] to obtain reference velocity values, from which it is easy to calculate corrections.

1.2.5. *Time*

The measurement of time is necessary for referencing data collected in oceanography. It is also relevant in the function of a number of instruments in which variations in quantity result in variations in frequency or phase. If this measurement does not pose any fundamental problems to the electronic metering plan, it is the signal generators used that have the greatest influence on the quality of the dating or positioning. It is the long-term and temperature-dependent stability, as well as the frequencies generated, that is mainly used to distinguish between the generators. To appreciate these qualities, references are necessary. These are given by the time scales.

A practical time scale destined to be used all over the world needs two fundamental elements: a realization of the unit of this quantity and a continuous time reference. The first reference used dates back to 1736. This was universal time, denoted UT_1, defined by the International Astronomic Union. It is based on works by Euler, who showed the invariability of the angular rotation speed of the Earth in space. This work led to the choosing of a coefficient of proportionality so that the UT_1 moments, UT_1=12 hours, correspond each day on average to the passing of the sun at the first meridian. Universal time or solar time is also called GMT, which stands for *Greenwich Mean Time*. For synchronization purposes, however, each country has a practical reference for time that is usually acquired from a cesium clock that operates in connection with the definition of the second:

The second is the duration of 9,192,631,770 periods of the radiation corresponding to the transition between two hyperfine levels of ground state of the cesium 133 atom.

In principle, this clock irradiates cesium atoms using microwave rays whose frequency has been adjusted in such a manner as to provoke the desired atomic transition. When the phenomenon reaches its maximum level, the frequency of the incident radiation is in coincidence with that of cesium and is extremely stable. The microwave generator is controlled by a quartz oscillator that was itself stabilized with respect to the chosen transition of hyperfine cesium levels. At the end the feedback produces a signal whose stability is difficult to match.

These clocks are able to maintain a frequency with a relative stability of greater than 10^{-14} over a period of several months. In order to demonstrate the significance of this, let us take, for example, the case of a nominal clock frequency f = 10 MHz. With a relative stability of 10^{-14}, the variation of this frequency value will be less than δf = 0.1 µHz over several months. The relative uncertainty of the knowledge of these clocks in the long run is in the order of 10^{-12}. Some experiments included the use of "atom fountains" or "comb lasers" and some newer fixtures have been used with the aim of replacing cesium clocks that have obtained stabilities in the order of 10^{-15} to 10^{-16}.

The time references of countries that have cesium clocks are called Coordinated Universal Time UTC(i), where i designates the responsible laboratory. By international agreement, the UTC(i) are held close to Coordinated Universal Time (*UTC*) established under the authority of the BIPM. UTC and UTC(i) times have regular leaps of intercalated seconds in order to remain synchronized with the rotation of the Earth, to better than 0.9 s over the span of one year:

$$(UTC - UT_1) < 0.9\ s$$

The choice of dates and the announcement of corrections are made by the *International Earth Rotation and Reference System Service*, which is responsible for the determination of parameters of the Earth's rotation and conservation of the associated terrestrial and celestial reference systems.

These timing constraints are contradictory to the qualities of stability and accuracy that we seek; these are "pure" atomic times that approach ideal physical time. These are the International Atomic Time (*IAT*) and at a national level, the national Atomic Time (TA(*n*)), where *n* represents the initial of the country. They use a set of measurements from some 230 clocks maintained by 65 laboratories around the world. It is important that these clocks can be compared with each other regularly to the highest level of accuracy. This comparison is carried out using satellite or Hertzian transmitters. By this means, after various corrections including propagation and relativist corrections (which can lead to a difference of 39 ms/d between a terrestrial clock and one on a GPS satellite because of the difference in gravitational potential), it is possible to determine the difference between clocks that drive transmitter–receivers.

These are of diverse origins. They can be:

– television transmitters, frame signals have been used for 30 years to compare atomic clocks;

– radio transmitters emitting coded times; and

– transmitters located on telecommunication satellites (linked *Two-Way Satellite Time and Frequency Transfer* microwave) or on radio-navigation systems, such as those on the GPS or the *Global Navigation Satellite System* (GLONASS), see sections 2.6 and 2.7.

By these means, it is possible, for example, to connect the French reference UTC(*OP*) that is given by the Observatory of Paris to the UTC(*USNO*) that is given by the United States Naval Observatory. This connection can be achieved by time or by frequency. To compare signal dating, the user can acquire all of the necessary information from the primary laboratory in charge of the time in their country. The latter provides information reports. Connection to international references is also possible. The user must then refer to www.bipm.org [BIP 05], which also issues a monthly newsletter (*Circular T*) in which an evaluation of daily time differences can be found, notably: (*UTC – GPS Time*) and (*UTC – GLONASS Time*). On January 1, 2006 at 0 h UTC, for example, it was:

$$(\text{UTC} - \text{GPS Time}) = -14 \text{ s} + C_0$$

where C_0 is given at 0 h UTC each day, to 10 ns close. On this same date:

$$(\text{UTC} - \text{GLONASS Time}) = 0 \text{ s} + C_1$$

where C_1 is in the order of several hundred nanoseconds.

Transmitter	Accuracy of linkage to UTC(OP)	Uncertainty of measurement	Accessibility
Radio	± 1 ms	± 0.3 to ± 2 ms	Any point where long-wave radio reception is possible
Television	± 10 ns	± 5 ns	Any point of visibility of a television transmitter in a country
GPS satellites	± 0.1 to ± 1 µs	± 50 to 1 ns	Any point in the globe where a frequency of 1.5 GHz can be received

Table 1.2. *Characteristics of some of the means enabling time linkage to UTC(OP). The accuracy and uncertainty values given depend on the propagation conditions, visibility of transmitters, corrections and stability of receptor instrumental delays*

Transmitter	Frequency	Accuracy of linkage to UTC(OP)	Stability over 1 day
Radio	162 kHz	3.10^{-13}	2.10^{-13}
LORAN-C	100 kHz	5.10^{-12}	5.10^{-12}
GPS satellites	1.475 GHz	20 to 5.10^{-13}	20 to 5.10^{-13}

Table 1.3. *Characteristics of some of the means enabling frequency linkage to UTC(OP)*

On January 1, 1958, the IAT was synchronized to the UT_1 and a unit interval scale was established so that the IAT does not deviate too much from the UT_1. Today, IAT is acquired by comparing clocks over long periods to ensure long-term stability. In 1997, its relative stability was estimated at 2.10^{-15} over two months. Consequentially, the IAT is only accessible in delayed times of a few weeks. The UTC was defined in such a manner that it differed from the IAT by an integer number of seconds: IAT – UTC = 34 s on January 1, 2009. The differences between the UTC(*i*) and the UTC are generally reduced to a few hundred nanoseconds.

As with the UTC, on January 1, 2006 at 0 h UTC, we had:

$$(IAT - GPS\ Time) = 19\ \text{s} + C_0$$

where C_0 is in the order of several hundred nanoseconds. On this same date:

$(IAT - GLONASS\ Time) = 33\ \text{s} + C_1$

The values of C_0 and C_1 are provided by the BIPM at 0 h UTC every day.

1.3. Calculation of density

There is often confusion between the terms *density* and *relative density* or *specific gravity*. Density is expressed in kg m^{-3} as it is the ratio of the mass of a body to its volume; while relative density is a dimensionless quantity as it expresses the ratio of the density of a substance on distilled water measured under standard conditions.

In the field of oceanography, density is a quantity that cannot currently be directly measured *in situ*. Its range of variation for sea water is narrow compared to its standard value. It extends from around 995 to 1,070 kg/m^3. It is therefore necessary to approach a low relative uncertainty ($\pm$ 1 $\times$ 10^{-6}) for its estimation. Density ρ depends on water temperature t, absolute salinity S_A and, at depth, on the pressure that is exerted p. It is thus a calculable quantity and its calculation was carried out until 2010 by using several equations published in 1980 called the *equations of state of sea water* or *EOS-80*. These equations were redefined by TEOS-10 and are now based on equation [1.5]. For theoretical reasons linked to model making, however, temperature t is replaced by potential temperature θ [1.10] or conservative temperature Θ [1.9], both referenced to the reference pressure p_r = 0 dbar. In the TEOS-10 annex K, the coefficients of the 25 terms of the expressions of $\rho = \tilde{\rho}(S_A, \theta, p) = \hat{\rho}(S_A, \Theta, p)$ are given. The residual quadratic error of this polynomial approximation is 0.0015 kg m^{-3}.

1.3.1. *Density and EOS-80*

When it is necessary to calculate density values from values of S, t, and p, it is always possible to use the EOS-80 equations if S designates practical salinity S_P. In these equations, ρ is not defined in relation to mass and length calibrations, but by measuring differences to distilled water, of which the isotopic composition is that of oceans. Its calculation is split into several equations. The first, $\rho(S_P, t, 0)$, defines its value under standard atmosphere and is expressed in kg m^{-3}. One standard atmosphere is equivalent to 101,325 Pa or 1.01325 bar. The definition of $\rho(S_P, t, 0)$ appeals to the notion of the density of *standard mean ocean water* or SMOW. In 1961, Graig revealed that while sampled sea water has a constant isotopic composition, inland fresh water does not. It then became a benchmark for isotopic composition and has been used to create the SMOW by distillation. The maximum density of the SMOW was determined in 1972 by Girard and Ménaché [MÉN 73].

This value, recommended by the *International Union of Pure and Applied Chemistry* is 999.975 kg m^{-3}. Recently, following several experimental works over the past few years, the relationships used to calculate SMOW density have been redefined. The BIPM now recommends using relationship [1.64], which is valid for 0°C < t < 40°C (t is expressed in ITS-90) and allows SMOW value calculations with an uncertainty at two standard deviations less than 9×10^{-4} kg m^{-3}. This relationship can be corrected for the isotopic abundance of the water used, for dissolved water and for compressibility (see [TAN 01]).

$$\rho_W = 999.97495\left[1-\frac{(t-3.983035)^2(t+301.797)}{522528.9(t+69.34881)}\right] \qquad [1.64]$$

Relationships [1.65] and [1.66] published by UNESCO and based on work by Millero and Poisson [MIL 81], are in agreement with this value.

$$\rho(S_P, t, 0) = \rho_w + (0.824493 - 4.0899 \times 10^{-3}.t + 7.6438 \times 10^{-5}t^2 - 8.2467 \times 10^{-7}.t^3 + 5.3875 \times 10^{-9}t^4)S_P + (-5.724\,66 \times 10^{-3} + 1.0227 \times 10^{-4}t - 1.6546 \times 10^{-6}t^2)S_P^{3/2} + 4.8314 \times 10^{-4}S_P^2 \qquad [1.65]$$

with:

$$\rho_w = 999.842594 + 0.06793952t - 9.09529 \times 10^{-3}t^2 + 1.001685 \times 10^{-4}t^3 - 1.120083 \times 10^{-6}t^4 + 6.536332 \times 10^{-9}t^5 \qquad [1.66]$$

They are valid for $0 < S_P < 40$ and $-2 < t < 40$°C. They present an uncertainty value of 0.0036 kg m^{-3} to one standard deviation. This uncertainty value was recently confirmed in a study led by Millero and Huang [MIL 09b]. In 1993, Poisson and Gadhoumi [POI 93] extended these equations to $S_P = 50$ and $15 < t < 30$°C. It should be noted that in order to use these relationships, it is necessary to reference measured temperatures to the IPTS-68.

$$\rho(S,t,p) = \frac{\rho(S,t,0)}{1-\dfrac{p}{K(S,t,p)}} \qquad [1.67]$$

with:

$$K(S_P, t, p) = K(S_P, t, 0) + Ap + Bp^2 \qquad [1.68]$$

where:

$$K(S_P, t, 0) = K_w + (54.6746 - 0.603\,459t + 0.010\,9987t^2 - 6.167 \times 10^{-5}t^3)S_P + (0.079\,44 + 0.016\,483t - 5.3009 \times 10^{-4}t^2)S_P^{3/2} \quad [1.69]$$

$$A = A_w + (2.2838 \times 10^{-3} - 1.0981 \times 10^{-5}t - 1.6078 \times 10^{-6}t^2)S_P + 1.910\,75 \times 10^{-4}S_P^{3/2} \quad [1.70]$$

$$B = B_w + (-9.9348 \times 10^{-7} + 2.0816 \times 10^{-8}t + 9.1697 \times 10^{-10}t^2)S_P \quad [1.71]$$

Millero *et al.* [MIL 81] also worked out relationships for calculating density as a function of depth. They involved $\rho(S_P, t, 0)$ and measured pressure p, which is expressed in bar. They are valid for $0 < p < 1{,}000$ bar and, over this range, they show an estimation uncertainty of 0.009 kg m^{-3}. This calculation is done using relationships [1.67] to [1.71].

$K(S_P, t, p)$ is also known as the cubic elasticity module at pressure p. In expressions [1.69] to [1.71], the terms K_w, A_w and B_w correspond to values established with pure distilled water. They themselves are calculated from polynomials:

$$K_w = 19{,}652.21 + 148.4206t - 2.327105t^2 + 1.360477 \times 10^{-2}t^3 - 5.155288 \times 10^{-5}t^4 \quad [1.72]$$

$$A_w = 3.239908 + 1.43713 \times 10^{-3}t + 1.16092 \times 10^{-4}t^2 - 5.77905 \times 10^{-7}t^3 \quad [1.73]$$

$$B_w = 8.50935 \times 10^{-5} - 6.12293 \times 10^{-6}t + 5.2787 \times 10^{-8}t^2 \quad [1.74]$$

It should be noted that, according to Fofonoff [FOF 85] from the Woods Hole Oceanographic Institution, differences of up to 0.05 kg m^{-3} can exist between the density of natural sea water and the density calculated using these equations because the conductivity ratio necessary to calculate salinity does not reflect the proportions of different dissolved salts. The variations in the concentration of silicates (SiO_2), carbonates and nitrates (NO_3) are not accounted for in the measurement of conductivity as well as those of dissolved biological materials. Also, in 2000, Millero calculated corrections $\Delta\rho$ that needed to be inserted into the calculation of ρ, by using equations of the type:

$$\underline{\Delta\rho} = a\,\Delta TCO_2 + b\,\Delta SiO_2 + c\,\Delta NO_3 \quad [1.75]$$

where a, b and c are empirically determined constants. These constants were evaluated for the Pacific Ocean ([MIL 00] and [MIL 09b]), the Indian Ocean and for the two oceans together [MIL 08]. Other works of this kind have been carried out in the Baltic Sea, but this time linking to the notion of absolute salinity [FEI 10].

1.3.2. *Laboratory densitometers*

If *in situ* measurements of density were not possible today, there are laboratory instruments that exist that would allow measurements from sea water samples. The most widely used for studies in oceanography rest on the principle of the vibrating tube. The measurement cell is a U-shaped borosilicate glass tube. It is filled with 0.7 ml of the sample to be analyzed and vibrated by a coil and magnet. Another coil measures the oscillation frequency of the tube for which the period τ is given by the relationship:

$$\tau = 2\pi\sqrt{\frac{\rho v + m}{c}} \qquad [1.76]$$

where v is the cell volume, m its mass when empty and c a spring constant. From this relationship, it is possible to extract the value of density ρ of the sample. This is a function of its measured oscillation period and can be expressed in certain instruments relative to the vibration of a reference tube in the form of a quotient Q. However, as v, m and c cannot be measured with sufficient uncertainty, we must proceed by calibration to obtain ρ. Bearing in mind relationship [1.76], ρ is a function of Q or τ squared multiplied by a constant A, which can only be obtained by a first standard substance. Given the form of equation [1.76], we must also subtract constant B from the previous product. The value of B is acquired by calibration with a second standard substance. The substances in question are usually air and distilled water. With distilled water, it is easy to obtain reference values using formula [1.64] and some eventual corrections. With humid air, a simple formula also exists (see [PIC 08]) that can be used to calculate reference values.

The viscosity of the fluid influences the damping of the vibration of the tube so it is necessary to introduce a third constant, C, and a damping value provided by the device. For substances for which viscosity is less than 30 mPa s, ρ is obtained using the following equation:

$$\rho = A - BQ^2 + \frac{C}{Damping} \qquad [1.77]$$

Constants A, B and C are independent of the properties of the fluid to be measured. It is then possible to obtain an expanded uncertainty of less than ± 0.015 kg/m^3 on the measurements.

The most widely used densitometer in the field of oceanography is the DMA 5000 produced by Anton Paar GmbH. Anton Paar has sold this type of instrument since 1967, but the DMA 5000 is the result of work carried out in 1998 by Hans Stabinger [STA 94]. Its linearity error was evaluated at 0.0013 kg/m^3 (the maximal value) and the stability of values displayed is at worst 0.002 kg/m^3 with distilled water. The most difficult source of error to master on this instrument comes from the measurement of temperature. This is done using a probe (that is not easily removable) inserted close to the measurement zone. When the measurement is taken at a temperature that is different to room temperature, the wires of the probe can conduct heat and bias the measurement. Similarly, it is necessary to wait for thermal equilibrium on the whole assembly (at least 10 min) before making density measurements. The existence of thermal gradients between the probe and the tube and the stability of measurements was evaluated to be 0.002°C at thermal equilibrium. Finally, calibration of this probe is difficult to achieve to the required accuracy, and it is simpler to correct its drifts by regularly recalculating coefficients A, B and C from relations [1.77].

1.3.3. *Density and absolute salinity*

In a study published in 2009, Millero and Huang [MIL 09b] used a Paar 500 vibrating tube to establish relationships between ρ and S_A at atmospheric pressure, and confirmed the standard uncertainty for relationships between ρ and S_P (0.0036 kg/m^3) in this study. This uncertainty is of the same magnitude as the residual standard deviation of the polynomial they had developed to connect ρ and S_A in the range 0–40°C (0.0037 kg/m^3). This polynomial has the following form:

$$\rho - \rho_0 = AS_A + BS_A^{1,5} + CS_A^2 \qquad [1.78]$$

where ρ_0 is the density of distilled water and A, B and C are functions of absolute temperature T (expressed in Kelvin).

$$A = a_0 + a_1T + a_2T^2 + a_3T^3 + a_4T^4 + a_5T^5$$

$$B = b_0 + b_1T + b_2T^2 \qquad [1.79]$$

$$C = c_0$$

In the 0–40°C range, the coefficients of these functions have the values:

$a_0 = 8.211458\times10^{-01}$ $\quad$ $b_0 = -6.058307\times10^{-03}$ $\quad$ $c_0 = 5.265280\times10^{-4}$

$a_1 = -3.959680\times10^{-03}$ $\quad$ $b_1 = 8.265457\times10^{-05}$

$a_2 = 7.305182\times10^{-05}$ $b_2 = -1.077747\times10^{-06}$

$a_3 = -8.282446\times10^{-07}$

$a_4 = 5.386657\times10^{-09}$

$a_5 = 0$

In these expressions, S_A is expressed in g/kg and ρ is in kg/m^3.

Using a specially adapted (steel) vibrating tube, this study was extended by Safarov *et al.* [SAF 09], in the same year, to high temperatures and pressures but only for practical salinity $S_P = 35$. They established a relationship between pressure p expressed in MPa and ρ expressed in g/cm^3 via coefficients $A(T)$, $B(T)$ and $C(T)$ as functions of absolute temperature T (expressed in Kelvin).

$$p = A\rho^2 + B\rho^8 + C\rho^{12} \quad [1.80]$$

with:

$$A = a_1T + a_2T^2 + a_3T^3 + a_4T^4$$

$$B = b_0 + b_1T + b_2T^2 + b_3T^3 \quad [1.81]$$

$$C = c_0 + c_1T + c_2T^2 + c_3T^3$$

In the 0–195°C range and for pressures from 0 to 140 MPa, the coefficients of these functions have the following values:

$a_1 = -2.157761589$ $b_0 = 3638.11368199$ $c_0 = -2235.76774015$

$a_2 = -0.10341365\times10^{-02}$ $b_1 = -27.97107636$ $c_1 = 17.6361364636$

$a_3 = 0.103809737\times10^{-04}$ $b_2 = 0.0815083395$ $c_2 = -0.04771579895$

$a_4 = -0.305662763\times10^{-08}$ $b_3 = -0.74604739\times10^{-04}$ $c_3 = 0.4100691661\times10^{-04}$

The standard uncertainty for these density values is 0.084 kg/m^3.

1.4. Bibliography

1.4.1. *Quantities that we want to know*

[APE 87] APEL J.R., *Principles of Ocean Physics*, Vol. 38, International Geophysics Series, Academic Press, 1987.

[FES 08] FEISTEL R., "A Gibbs function for seawater thermodynamics for – 6 and 80°C and salinity up to 120 g/kg", *Deep-Sea Research* I, vol. 55, pp.1639-1671, 2008.

[FOF 91] FOFONOFF N.P., MILLARD R.C., "Calculation of physical properties of seawater", *WHP Operations and Methods*, WHP (WOCE Hydrographic Program Office), 1991.

[FOR 65] FORCHHAMMER G, "On the composition of seawater in different parts of the ocean", *Philos. Trans. R. Soc. London*, vol. 155, pp. 203-262, 1865.

[IAP 96] THE INTERNATIONAL ASSOCIATION FOR THE PROPERTIES OF WATER AND STEAM (IAPWS), "The IAPWS formulation 1995 for the thermodynamic properties of ordinary water substance for general and scientific use", *The International Association for the Properties of Water and Steam*, Fredericia, Denmark, September 1996.

[IAP 09] The INTERNATIONAL ASSOCIATION FOR THE PROPERTIES OF WATER AND STEAM (IAPWS), "Computational efficient thermodynamic formulation for liquid water for oceanographic use", Presented at: *The International Association for the Properties of Water and Steam*, Arnhem, Netherlands, 2009.

[IOC 10] IOC, SCOR and IAPSO, The International Thermodynamic Equation of Seawater 2010 (TEOS-2010): Calculation and Use of Thermodynamic Properties, Intergovernmental Oceanographic Commission, Manuals and Guides no. 56, UNESCO, 2010. (www.TEOS-10.org)

[IVA 72] IVANOFF A, "Introduction à l'océanographie, propriétés physiques et chimiques des eaux de mer", *Vuibert,* vols 1 and 2, 1972.

[JOY 88] JOYCE T.M., "The WOCE hydrographic program: a status report", *Woce Newsletter,* no. 6, pp. 7-10, October 1988.

[JOY 91] JOYCE T.M., "Introduction to the collection of expert reports compiled for the WHP programme", *WHP Operations and Methods,* 1991.

[PER 80] PERKIN R.G., LEWIS E.L., "The practical salinity scale 1978: Fitting the data", *IEEE J. Oceanic Eng.*, OE-5, vol. 1, pp. 9-16, 1980.

[UNE 83] UNESCO, Algorithm for computation of fundamental properties of seawater, UNESCO technical papers in Marine Sciences, no. 44, 53p., 1983.

[UNE 88] UNESCO, The acquisition, calibration, and analysis of CTD data. A report of SCOR Working Group 51, Unesco Technical Papers in Marine Science, no. 54, 92p., 1988.

1.4.2. *Linking of essential quantities in oceanography*

1.4.2.1. *Temperature*

[ECH 90] ECHELLE INTERNATIONALE DE TEMPÉRATURE de 1990, *Bulletin du Bureau National de Métrologie* no. 79, 1990.

[PRE 90] PRESTON-THOMAS H., "The International Temperature Scale of 1990 (ITS-90)", *Metrologia*, vol. 27, pp. 3-10, 1990.

[SAU 91] SAUNDERS P.M., MAHRT K.H., WILLIAMS R.T., Standards and laboratory calibration, WHP Operations and Methods, WHP (WOCE Hydrographic Program Office), 1991.

1.4.2.2. *Pressure*

[DAD 82] DADSON R.S., LEWIS S.L., PEGGS G.N., *The Pressure Balance, Theory and Practice*, National Physical Laboratory, HMSO, London, 1982.

[KAJ 05] KAJASTIE H., RANTANEN M., RISKI K., "An absolute pressure balance with a removable weight exchange system", *Measurement Science and Technology*, vol. 16, pp. 33-37, 2005.

1.4.2.3 *Salinity*

[BAC 07] BACON S., CULKIN F., HIGGS N., RIDOUT P., "IAPSO standard seawater: definition of the uncertainty in the calibration procedure and stability of recent batches", *Journal of Atmospheric and Oceanic Technolology*, vol. 24, pp. 1785-1799, 2007.

[CUL 79] CULKIN F., and SMED J., "The history of standard seawater", *Oceanologica Acta*, vol. 2, no. 3, pp. 355-364, 1979.

[CUL 80] CULKIN F., SMITH N.D., "Determination of the concentration of potassium chloride solution having the same electrical conductivity at 15°C and infinite frequency, as standard seawater of salinity 35,000‰ (chlorinity 19,37394‰)", *IEEE Journal of Oceanic Engineering*, OE-5, vol. 1, pp. 22-23, 1980.

[CUL 97] CULKIN F., RIDOUT P.S., "Stability of IAPSO standard seawater", *Journal of Atmospheric and Oceanic Technolology*, vol. 15, pp. 1072-1075, 1997.

[DAU 75] DAUPHINEE T.M., KLEIN H.P., "A new automated laboratory salinometer", *Sea Technology*, vol. 16, pp. 23-25, 1975.

[DIT 84] DITTMAR W., "Physics and Chemistry", *Rept. Scient. Res. H.M.S. Challenger*, 1873-76, vol. 1, p. 251, 1884.

[FEI 10] FEISTEL R., WEINREBEN S., WOLF H., SEITZ S., SPITZER, P., ADEL B., NAUSCH G., SCHNEIDER B., WRIGHT D.G., "Density and absolute salinity of the Baltic Sea 2006-2009", *Ocean Science*, vol. 6, pp. 3-24, 2010.

[FOF 83] FOFONOFF N.P., MILLARD R.C., Algorithm for computation of fundamental properties of seawater, UNESCO Technical Papers in Marine Sciences, No. 44, 1983.

[FOF 85] FOFONOFF N.P., "Physical properties of seawater: a new salinity scale and equation of state for seawater", *Journal of Geophysical Research*, vol. 90, no. C2, pp. 3332-3342, 1985.

[JAC 06] JACKETT D.R., MCDOUGALL T.J., FEISTEL R, WRIGHT D.G., GRIFFIES S.M., "Algorithms for density, potential temperature, conservative temperature, and the freezing temperature of seawater", *Journal of Atmospheric and Oceanic Technology*, vol. 23, pp. 1709-1728, 2006.

[KAW 05] KAWANO T., AOYAMA M., TAKATSUKI Y., "Inconsistency in the conductivity of standard potassium chloride solutions made from different high-quality reagents", *Deep-Sea Research*, I, vol. 52, pp. 389-396, 2005.

[LEM 09] LE MENN M., "About uncertainties in practical salinity calculations", *Ocean Science Discussion*, vol. 6, pp. 2461-2485, 2009.

[MAC 09] MCDOUGALL T.J., JACKETT D.R., MILLERO F.J., "An algorithm for estimating Absolute Salinity in the global ocean", *Ocean Sci. Discuss.*, vol. 6, pp. 215-242, 2009.

[MIL 96] MILLERO F.J., *Chemical Oceanography*, 2nd edition, CRC Marine Science Series, 1996.

[MIL 08] MILLERO F.J., FEISTEL R, WRIGHT D.G., MCDOUGALL T.J., "The composition of standard seawater and the definition of the reference-composition salinity scale", *Deep-Sea Research* I, vol. 55, pp. 50-72, 2008.

[POI 80] POISSON A., "Conductivity/salinity/temperature relationship of diluted and concentrated standard Seawater", *IEEE Journal of Oceanic Engineering*, vol. OE-5,1, pp. 41-50, 1980.

[SEI 10a] SEITZ S., SPITZER P., BROWN R.J.C., "CCQM-P111 study on traceable determination of practical salinity and mass fraction of major seawater components", *Accred. Qual. Assur.*, vol. 15, pp. 9-17, 2010.

[SEI 10b] SEITZ S., FEISTEL R., WRIGHT D.G., WEINREBEN S., SPITZER P., DE BIÈVRE P., "Metrological traceability of oceanographic salinity measurement results", *Ocean Sci. Discuss.*, vol. 7, pp.1303-1346, 2010.

[WRI 11] WRIGHT D.G., R. PAWLAOWICZ R., MACDOUGALL T.J., FEISTEL F, MARION G.M., "Absolute salinity, density salinity, and the reference composition salinity scale: present and future use in the seawater standard TEOS-10", *Ocean Science,* vol. 7, pp. 1-26, 2011.

[UNE 79] UNESCO, Ninth report of the joint panel on oceanographic tables and standards, UNESCO Technical Papers in Marine Science, No 30, 33p., 1979.

[UNE 81] UNESCO, Background papers and supporting data on the Practical Salinity Scale, 1978, UNESCO Technical Papers in Marine Science, No. 37, 193p., 1981.

1.4.2.4. *Velocity*

[BRY 99] BRYDON D., SUN S., BLECK R., "A new approximation of the equation of state for seawater, suitable for numerical ocean models", *Journal of Geophysical Research*, vol. 104, no. C1, pp. 1537-1540, 1999.

[CHE 77] CHEN C.T., MILLERO F.J., "Speed of sound in seawater at high pressures", *Journal of the Acoustical Society of America*, vol. 62, no. 5, pp. 1129-1135, 1977.

[DEL 74] DEL GROSSO V.A., "New equation for the speed of sound in natural waters (with comparison to other equations)", *Journal of the Acoustical Society of America*, vol. 56, no. 4, pp. 1084-1091, 1974.

[MED 75] MEDWIN H., "Speed of sound in water: a simple equation for realistic parameters", *Journal of the Acoustical Society of America*, vol. 58, no. 6, pp. 1318-1319, 1975.

[MEI 97] MEINEN C.S., WATTS D.R., "Further evidence that the sound-speed algorithm of Del Grosso is more acurate than that of Chen and Milero", *Journal of the Acoustical Society of America*, vol. 102, no. 4, pp. 2058-2062, 1997.

[ROS 78] ROSS D., *Revised Simplified Formulae for Calculating the Speed of Sound in Sea Water,* Saclant ASW Research Centre Memorendum, La Spezia, Italy, 1978.

[URI 83] URICK R.J., *Principles of Underwater Sound*, 3rd Edition, McGraw-Hill, New York, 1983.

1.4.2.5 *Time*

[ACH 07] ACHKAR J., VALAT D., "Recent development in time metrolog", *13ème Congrès International de Métrologie*, Lille, June 2007.

[BIP 05] BIPM., Annual report of the IBWM time section, Rapport Annuel de la Section du Temps du IBWM, vol. 18, IBWM, 2005.

[DEF 03] DEFRAIGNE P., PETIT G., "Time transfer to TAI using geodetic receivers", *Metrologia*, vol. 40, pp. 326-334, 2003.

[DEL 06] DELPORTE J., MERCIER F., MARECHAL J., JEANNOT M., RICHARD J.-Y., UHRICH P., "Performance assessment of the time difference between EGNOS Network Time and UTC", *Proc. of the ION GNSS*, Fort Worth, 2006.

[LAH 06] LAHAYE F., ORGIAZZI D, TAVELLA P., CERRETTO G., "GPS time transfer", *GPS World*, pp. 44-49, Nov. 2006.

[PET 08] PETIT G., Z. JIANG Z., "GPS All in View time transfer for TAI computation", *Metrologia*, vol. 45, pp. 35-45, 2008.

[PLU 05] PLUMB J., LARSON K.M., WHITE J., POWERS E., "Absolute calibration of a geodetic time transfer system", *IEEE Trans. Ultrason., Ferroelect., Freq. Contr.*, vol. 52, no. 11, pp.1904-1911, 2005.

1.4.2.6. *Density*

[FEI 10] FEISTEL R., WEINREBEN S., WOLF H, SEITZ S, SPITZER P., ADEL B., NAUSCH G., SCHNEIDER B., WRIGHT D.G., "Density and absolute salinity of the Baltic Sea 2006-2009", *Ocean Science*, vol. 6, pp. 3-24, 2010.

[FOF 85] FOFONOFF N.P., "Physical properties of seawater: A new salinity scale and equation of state for seawater", *Journal of Geophysical Research*, vol. 90, no. C2, pp. 3332-3342, 1985.

[HAR 09] HARVEY A.H., SPAN R, FUJII K., TANAKA M, DAVIS R.S., "Density of water: roles of the CIPM and IAPWS standards", *Metrologia*, vol. 46, pp.196-198, 2009.

[H&D 00] H&D FITZGERALD Ltd, Technical Assessment of the Anton Paar DMA5000 Density Meter, www.density.co.uk, 2000.

[MAS 96] MASUI R., FUJII K, TAKENAKA M., "Determination of the absolute density of water at 16 °C, and 0,101 325 Mpa", *Metrologia*, vol. 32, pp. 333-362, 1995-96.

[MÉN 73] MÉNACHÉ M., GIRARD G., "Concerning the different tables of the thermal expansion of water between 0 and 40°C", *Metrologia*, 9, 62-68, 1973.

[MIL 80] MILLERO F.J., CHEN C.T., BRADSHAW A., SCHLEICHER K., "A new height pressure equation of state for seawater", *Deep-Sea Research*, vol. 27A, pp. 255-264, 1980.

[MIL 81] MILLERO F.J., POISSON A., "International one-atmosphere equation of state of seawater", *Deep-Sea Research*, vol. 28A, no. 6, pp. 625-629, 1981.

[MIL 00] MILLERO F.J., "Effect of changes in the composition of saewater on density-salinity relationship", *Deep-Sea Research*. I, vol. 47, pp.1583-1590, 2000.

[MIL 08] MILLERO F.J., WATERS J, WOOSLEY R., HUANG F., CHANSON M., "The effect of composition on the density of Indian Ocean waters", *Deep-Sea Research*, I, vol. 55, pp. 460-470, 2008.

[MIL 09a] MILLERO F.J., HUANG F., WILLIAMS N., WATERS J., WOOSLEY, R. "The effect of composition on the density of South Pacific Ocean waters", *Marine Chemistry*, vol. 114, pp. 56-62, 2009.

[MIL 09b] MILLERO F.J., HUANG F., "The density of seawater as a function of salinity (5 to 70 g kg^{-1}) and temperature (273.15 to 363.15 K)", *Ocean Science*, vol. 5, pp. 91-100, 2009.

[PIC 08] PICARD A, DAVIS R.S., GLÄSER M, FUJII K., "Revised formula for the density of moist air (CIPM-2007)", *Metrologia*, vol. 45, pp. 149-155, 2008.

[POI 93] POISSON A., GAGHOUMI M.H., "An extension of the Practical Salinity Scale 1987 and the Equation of State 1980 to high salinities", *Deep-Sea Research*, vol. 40, pp. 1689-1698, 1993.

[SAF 09] SAFAROV J., MILLERO F.J., FEISTEL R., HEINTZ A, HASSEL E., "Thermodynamic properties of standard seawater: extension to high temperatures and pressures", *Ocean Science*, vol. 5, pp. 235-246, 2009.

[STA 94] STABINGER H., Density measurement using modern oscillating transducers, South Yorkshire Trading Standard Unit, Sheffield, 1994.

[SUN 08] SUN H., FEISTEL R., KOCH M, MARKOE A., "New equations for density, entropy, heat capacity and potential temperature of a saline thermal fluid", *Deep-Sea Research*, I, vol. 55, pp. 1304-1310, 2008.

[TAN 01] TANAKA M., GIRARD G., DAVIS R., PEUTO A., BIGNELL N., "Recommended table for the density of water between 0°C and 40°C based on recent experimental reports", *Metrologia*, vol. 38, pp. 301-309, 2001.

[UNI 83] UNESCO, "Algorithms for computation of fundamental properties of seawater", *UNESCO Technical Papers in Marine Sciences*, No. 44, 1983.

[WAG 04] WAGNER W., KLEINRAHM R., "Densimeters for very accurate density measurements of fluids over large ranges of temperature, pressure and density", *Metrologia*, vol. 41, pp. S24-S39, 2004.

Chapter 2

Measurement Systems in Practice

2.1. Determining temperature

Temperature is a quantity that occurs in the calculation of most variables allowing characterization and modeling of the ocean. The National Oceanographic and Atmospheric Administration (NOAA) launched satellites installed with instruments such as the advanced very high resolution radiometer (AVHRR) in 1981, which has allowed the observation of sea surface temperature for the past 30 years. However, their observations are limited to the first micrometers of the ocean's surface. The data they collect is very useful in the study of temperature evolution over large surfaces; however only 10% of these data are exploitable due to cloud cover, with accuracy limited to ±0.5°C. The sensitivity of these measurements have been improved with the introduction of micro-wave passive radiometers (National Polar-Orbiting Environmental Satellite System (NPOESS) or Geostationary Operational Environmental Satellite-R (GOES-R) programs) or spectroradiometers such as NASA's (Moderate Resolution Imaging Spectroradiometer (MODIS)) or the European Space Agency's MERIS (medium resolution imaging spectrometer), in which accuracy has improved to ±0.3°C. The usage of measuring equipment *in situ* has, however, remained indispensable.

Furthermore, a very high number of oceanographic instruments are specifically equipped with temperature sensors. Each instrument is allotted a particular use, and the accuracies required are generally 10–100 times greater than those seen in classic industrial applications (see section 1.2); however the range of oceanographic temperatures they are able to measure is restricted. Classically, they can measure temperatures from -2°C to +35°C, but occasionally warmer surface water is found in certain parts of the world, with maximum temperatures being 38–40°C. As these required accuracies are important, the sensors are subject to systematic linearization and calibration. The sensors are often mounted onto voluminous instruments, and therefore the process is carried out in high stability, high volume vats that have a regulated temperature (section 1.2.1.2).

2.1.1. *Principal instruments*

2.1.1.1. *Conductivity–temperature–depth profilers*

In order to respond to recommendations dictated by WOCE HPO (section 1.1), oceanographers must have appropriate instruments at their disposal that are fit for *in situ* use. Their performance must be close to the limits that can be reached with the most advanced laboratory instruments. Conductivity–temperature–depth (CTD) profilers have been created to achieve such limits.

The first system to record measurements off a boat in this domain dates from 1948. It was designed by Jacobsen from the Waterbury Bristol Corporation (Connecticut, USA). Suspended by a carrier cable, it could reach depths of up to 400 m. A second multi-conductor cable allowed the transmission of measured data. Ten years later, Hamon and Brown described an instrument that could be used to measure temperature and salinity at depths of up to 1,000 m within ±0.15°C and ±0.05, respectively. The major innovation of this publication was found in the description of a cable, called an "electro-carrier" (section 3.1.1), that not only supported the instrument's weight but also powered it and transmitted data to the boat.

It was only in 1961 that the first *STD* (for salinity–temperature–depth) sensor, or *scientific CTD profiler*, was commercialized. Around 700 of these sensors were sold before the first CTD (for *conductivity–temperature–depth*) profiler was launched in 1970 (see Figure 2.1). According to a report published in 1974 for the WOCE Hydrographic and Oceanographic Institution by Fofonoff, Hayes and Millard, these CTD probes allowed temperature to be obtained to the nearest ±0.0015°C, salinity to the nearest ±0.003 and pressure to the nearest ±1.5 bar when these measurements were taken under the principal thermocline, when *in situ* calibrations were carried out and adequate data processing was performed. Therefore, CTDs became the instrument of reference for oceanographers.

Today, CTDs are regularly used when measuring the profile of water columns. The carrier ship comes to a stop, the CTD descends vertically into the water with the help of a winch and temperature, conductivity and pressure are continually measured at a rate allowing the correct sampling of crossed water layers during phases of descent and ascent. Additional sensors, such as dissolved oxygen and fluorimeter, etc., can be connected to the supplementary entries and are equipped with analog/digital convertors. The temperature sensors that are made up of thermistors (CTDs made by Sea Bird Electronics, Inc.) or platinum sensors (CTDs made by FSI Instruments or General Oceanic, Inc.), are selected and aged in order to limit their drifts (section 2.1.2). These sensors are protected from the effects of pressure by a thin casing made from stainless steel. The diameter of the casing is optimized in order to conserve a sufficient response time (section 2.1.4). This response time is of great importance when the CTD crosses the first few layers of the ocean, as they are generally zones that have high temperature gradients.

Variable	Range of measure	Resolution	Initial accuracy	Drift	Response time
Temperature	-2°C–+35°C	0.0002°C	0.002°C	0.001°C/year	60 ms
Conductivity	0–70 mS/cm	0.0004 mS/cm	0.003 mS/cm	0.002 mS/cm/month	40 ms
Pressure	0–600 bar	0.001% of the full scale	0.015% of the full scale	0.0015% of the full scale	1 ms
Voltage	0–5 V	1.2 mV	5 mV	1 mV/month	Filter at 5.5 Hz

Table 2.1. *Specifications of the CTD SBE 9⁺ profiler produced by Sea Bird Electronics, Inc.*

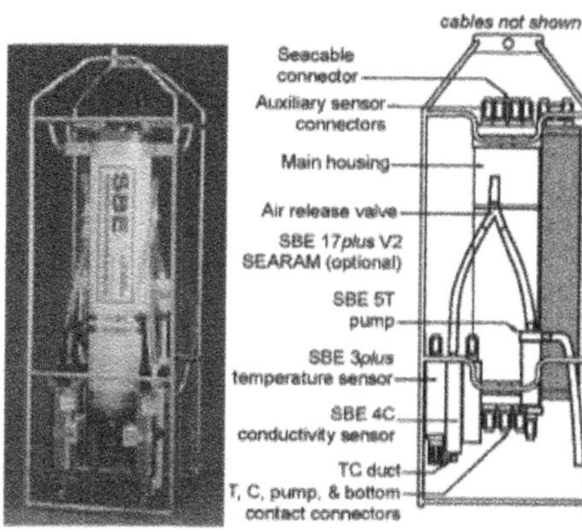

Figure 2.1. *Scientific CTD SBE 9$^+$ by Sea Bird Electronics, Inc. A pumping system that allows sea water to circulate at a constant speed in a pipeline, called the "TC duct", is connected to a temperature sensor SBE 3 and a conductivity sensor SBE 4. This system allows the application of corrections in order to compensate for the differences in the response time of sensors, thanks to the constant flow speed (Courtesy of Sea Bird Electronics, Inc.)*

Theoretically, there is no difference between the performance of these instruments in terms of measurement, even if different technologies are used. Table 2.1 gives us an example of the CTD SBE 9^{+} produced by Sea Bird Electronics, Inc.

2.1.1.2. *Multi-bottle sampling array and reversing thermometers*

Instruments have been developed in order to control punctually the accuracy of CTD measurements *in situ*. A CTD profiler can be rigged on a frame composed of *sampling bottles* and one or more control thermometers. The set is called a *multi-bottle sampling array* (see Figure 2.2). The bottles are tubes made from plastic material. The extremity of each is fitted with valves that can be remotely closed on command by the "sampling array motor". When a bottle closes, the temperature is measured by a control sensor. In order to carry out this process, the uplift is interrupted for several minutes and a measurement is taken on stage. Several stages can be carried out according to the number of bottles that have been rigged. The water sample will then be used for chemical analysis, but also to determine salinity. The salinity will then be compared to the value calculated from the CTD data.

Figure 2.2. *Multi-bottle sampling array before being lowered into water. The bottles are fitted with (white) supports of reversing thermometers. A CTD is horizontally rigged underneath the bottles (Courtesy of © SHOM)*

Control thermometers will also allow verification of the validity of temperature values provided by the CTD. As measurement campaigns at sea can last several weeks, sometimes even months, these controls are realized at regular intervals. They must allow corrections to the sensor drift if necessary. To make such changes possible, thermometers that can be recalibrated and have little drift have been developed.

The principle of reverse thermometers was retained up until the 1990s. This principle, dating from 1874 and developed by Negretti and Zambra, consists of tipping a thermometer to a given depth in order to block the temperature shown. This blockage is difficult to carry out electronically, so in the context of CTD measurements mechanical processes have been developed. A mechanism tips the thermometer frame fixed on a sample bottle when the bottle's valve opens. On the most recent, RPM-type thermometers developed by Sensoren Instrument Systems, an inclinometer detects the tipping and blocks the measurement in progress, which is then memorized on a display. The measurement then just has to be read when the contents of the bottle are sampled. Reversible pressure meters that function on the same principle and register pressure are also available.

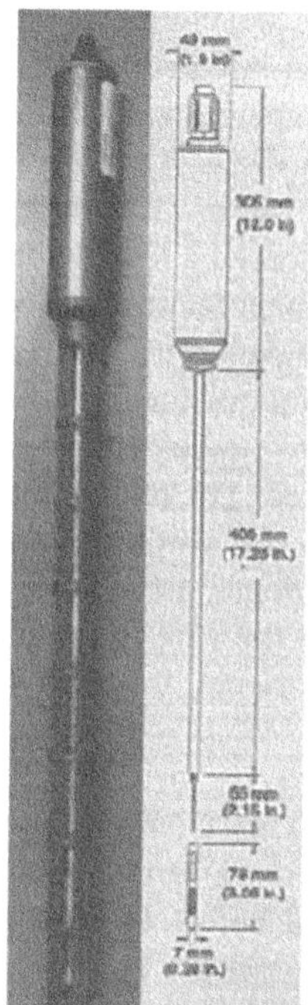

Figure 2.3. *A referenced SBE 35 sensor, made by Sea Bird Electronics, Inc. The sensitive element (thermistor) is protected by a 65 mm long thin tube made from stainless steel. A 70 mm-long Teflon tip can be screwed on to adapt the response time to that of the referenced platinum sensors that are used in laboratories. When used in the sea, the tip is removed and the casing of the sensor is protected by a screwed frame (Courtesy of Sea Bird Electronics, Inc.)*

Due to the geometry of reverse thermometers, their calibration cannot be carried out in the fixed points cells of the ITS (see section 1.2.1.1.), yet temperature measured by the thermometers is supposed to reference those taken by the CTD. As technology has evolved, the accuracy of both of these instruments has become equally impressive. The timing was therefore ideal for the development of a thermometer with even less drift that was able to be directly calibrated to the fixed points of the ITS. The SBE 35 developed by Sea Bird Electronics Inc. allows these things. This thermometer is not tipped during measurement. Instead, average temperature values are memorized when triggered by a signal that leads to close of the sample bottles. This sensor, which is able to be calibrated directly to the fixed points of the ITS, is a real reference for temperature.

2.1.1.3. *Losable probes*

The need to carry out sufficient temperature profiles in both operational and research domains, led to the development of an instrument known as a *bathythermograph* or a *bathythermal probe.* This name was given as it allows measurements of variations in temperature according to depth. The advantage of the bathythermograph as opposed to the CTD is that it can be used while a vessel is moving. These tools were first made in 1938 by Spilhaus. At the end of the 1960s, three American manufacturers released a losable bathythermograph onto the market, named Sippican after one of its manufacturers. Nowadays, there are several models of Sippican probes. The most popular is the XBT (*eXpendable Bathy Thermograph*). It is made up of a (unprotected) thermistor fixed in a zinc ballast. This thermistor is welded to an isolated biconductor wire with a diameter of 0.2 mm, which then unravels when the probe is released. The release mechanism of the probe is linked to a microcomputer that translates the variations in resistance of the thermistor in temperature and calculates the depth using knowledge of the probe's maximum descent speed in the water. When the wire is completely unraveled, it breaks and the probe is lost. Depending on the model, the wire can vary from 200 to 1,800 m in length.

The accuracy of this set can vary from one probe to another. The manufacturer fixes a temperature tolerance at ± 0.1°C (knowing that the probes are not individually laboratory calibrated before their delivery) and in depth at 2%, while the system resolution is fixed at 0.01°C. The accuracy is limited by the length and variation in the resistance of the wire released, but also by the variation in descent speed according to the density of the various different layers that are crossed, resulting in difficulties when lining up temperature and immersion. The variations in resistance can be provided by poor ground contact (as the return signal is transmitted by the boat ground) or mechanical constraints applied by the wire, which can be stretched during the probe's descent, there could be bad tracking of the wire linking it to the interface board, etc.

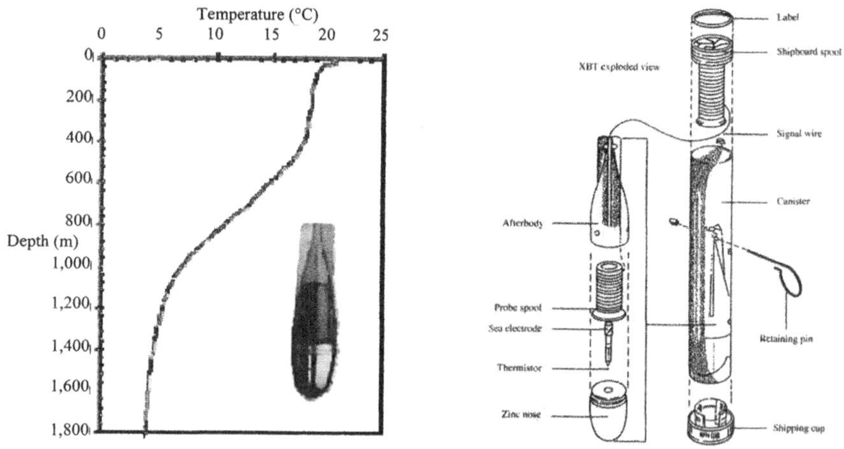

Figure 2.4. *Left: The XBT sensor made by Lockheed Martin Sippican, and an example of the temperature profile that can be obtained between 0 and 1,800 m. Right: diagram of the elements that make up a XBT sensor (Courtesy of Lockheed Martin)*

Taking into account the number of losable probes released by every country in the world in seas across the globe, in 1994 UNESCO published an article in its technical review to standardize the equations used when calculating depth from the descent speed of the XBT probe (and XCTD, the equivalent used to measure conductivity, see section 2.2.1). The depth d is then linked to the time t that starts running as soon as the probe enters the water, by a relation of the form: $d(t) = at - bt^2$, where coefficients a and b are determined for each type of probe. Taking into account the importance of these measurements in the estimation of the thermal content of oceans and thus their warming or cooling, more recent publications have attempted to correct the biases of these data. In 2008, Kizu *et al.* [KIZ 08] estimated that the accuracy of 2% depth was considered over 20 m with XCTD, while Willis *et al.* [WIL 09] showed in 2009 that the XBT assigned temperatures to depths with a bias of + 2%. In 2010, Gouretseki and Reseghetti [GOU 10] estimated that only having one correction of depth by a constant factor did not allow total elimination of the biases on all water columns, and that the descent speed was in relation to the background temperature.

In order to carry out temperature–conductivity profiling at a lower cost, and to replace XBT probes, systems based around a profiled and ballasted CTD and fast unwinding winch have been created [RUD 07], but their usage currently remains rather limited.

2.1.1.4. *Hull thermosalinographs*

Figure 2.5. *Hull thermosalinograph SBE 21 made by Sea Bird Electronics, Inc. A valve system allows water intake to be cut off in order to clean the cell. These valves can also be used to take water samples (Courtesy of Sea Bird Electronics, Inc.)*

In order to measure surface water temperature and its salinity, oceanographic ships are themselves equipped with instruments known as *hull thermosalinographs*. These instruments are constructed from a base with a temperature sensor and a conductivity sensor. Sea water is pumped into this base and circulates around the conductivity cell. These temperature–conductivity values allow the calculation of surface salinity values, with an uncertainty that attempts to reach the 100th for each of the parameters. For convenience, the base can be placed at a large distance from the water intake; in this case, the temperature measured is no longer really representative of that of the ocean. The dispositive is therefore generally completed by another surface temperature sensor that is fixed onto the duct where the water enters so that surface temperature can be measured without error. As the ship is likely to roll and pitch, water intake is not at floating point level, but instead 1 to 6 m below this level. It is therefore preferable to talk about subsurface temperature and salinity. The expected uncertainty of these measurements is approximately equal to 0.01 for salinity, but the remote temperature can be measured with an uncertainty equivalent to that of the CTD, i.e. close to 0.002°C.

2.1.1.5. *Thermistor chains*

The array of oceanographic equipment would not be complete if we did not include instruments capable of working from a fixed position, or rather, at anchor. Tools known as *thermistor chains* are used in such circumstances. The chains are in the form of long cables made from isolation casing in which the thermistors are molded at regular intervals. In the case of the chains made by AANDERA, the cables are often 25–150 m in length. The casing contains the linking wires for the thermistor that have an interface with an electronic module placed in a container that is fixed to the other end of the wire. The electronic board is made up of a system of switches that allow the measuring resistance bridge to be tilted from one thermistor to another.

This instrument is used to measure the variation in the temperature of a water column in successive sections, with an accuracy of approximately ±0.1°C. This accuracy is limited by mechanical strains that the cable is submitted to (due to stretching or twisting), but also by the drifts of the comparison resistances of the electronic board. For longer chains (200–300 m), other assemblies have been devised where the electronic conditioning unit of the sensor is inserted into the mold closest to the thermistor, therefore avoiding problems linked to wire lengths (for example, by NKE Electronics). These fixtures benefit from current progress being made in the miniaturization of components. Temperature information is directly digitalized and transmitted to the surface by RS 422 type serial links.

2.1.2. *Sensor technologies*

Due to the high accuracies researched and the small temperature span to be covered, two types of sensor technology are used: metallic resistive sensors and thermistors. Taking into account the low resolution that they allow us to obtain, thermocouples are rarely used in the field of oceanography.

2.1.2.1. *Metallic resistive sensors*

Conductivity of a metal is assured by the mobility of free electrons placed in the conduction zone (see Figure 2.6). The mobility of these electrons varies in relation to the temperature to which the metal is subjected. The conductivity of metal decreases; however, the temperature increases because the vibrations of its atoms induce the release of phonons that cut down the electrons and create electric resistance. As a consequence of this, the *electrical resistivity* $\rho(t)$ of metal increases with the increase in temperature.

The resistance R of a metallic element with a length of l and section of ϕ is proportional to its resistivity. This can be shown by:

$$R(t) = \rho_e(t)\frac{l}{\varphi} \qquad [2.1]$$

If R_0 is its resistance at 0°C, the variation of $R(t)$ with t can break down according to relation [2.2] where α and β are constants that we can define.

$$R(t) = R_0 (1 + \alpha t + \beta t^2) \qquad [2.2]$$

If variations in temperature are low:

$$\frac{dR(t)}{dt} = R_0 \ \alpha \qquad [2.3]$$

where,

$$\alpha = \frac{1}{R_0}\frac{dR(t)}{dt} \qquad [2.4]$$

α is known as the temperature coefficient of the resistance or thermal sensitivity to the temperature t. Its value varies according to the temperature range where resistance is used. It depends equally on the nature of the conductor, which allows its usage to characterize the purity of the metals. For this, we calculate the ratio:

$$\alpha_0^{100} = \frac{(R_{100} - R_0)}{100R_0} \qquad [2.5]$$

The choice of metal for the assembly of a temperature sensor depends on:

– the purity with which we can create and conserve it;

– its sensitivity, i.e. the value of α; and

– the effect of temperature variations and mechanical constraints on its resistivity, and so on the stability of the value of α.

A certain number of metals are able to produce these sensors (see Table 2.2).

Due to its specific resistance to oxidation, platinum is the most frequently used metal, despite its cost and low sensitivity, approximately 0.4 $\Omega.K^{-1}$ for a resistance of 100 Ω or 4.10^{-3} K^{-1} is at 0°C. Platinum sensors have been chosen as reference instruments by the ITS. Those that meet the criteria fixed by the ITS are known as standard platinum resistance thermometers or SPRTs (section 1.2.1.1). Other types of thermometers are considered to be industrial platinum resistance thermometers or IPRTs.

Metal	**Resistivity at 0°C in μΩ cm² cm⁻¹**	**Coefficient α_0^{100}**	**Reason usage is limited**
copper	1.56	0.00425	Weak resistivity, alterable
nickel	6.38	0.00660	Alterable
platinum	9.81	0.00392	Price

Table 2.2. *Characteristics of several metals allowing the production of temperature sensors. Platinum shows the strongest resistivity, but also the lowest sensitivity; however, it is not alterable*

In a SPRT, the platinum wire is mounted in a sealed rod containing dry gas that stabilizes the oxidation of platinum. These instruments are therefore highly sensitive to shocks, which excludes them from all oceanographic usage (except for use in calibration laboratories). In IPRTs, the wire is under mechanical strain as it is rolled around cemented refractory materials or immersed in the cement. In this way, it has a higher resistance to mechanical shocks and vibrations, while remaining somewhat sensitive at this level. The other disadvantage of an IPTR is that the platinum follows variations in volume when the support is subject to thermal variations, and is therefore subject to mechanical strains that may induce errors of linearity and reproducibility.

The Callendar-Van Dusen equation allows us to define the temperature of a platinum wire to 0.1°C, from the measurement of resistances ratio $R(t)/R(0)$, or reduced resistance.

For $t > 0°C$: $$R(t)/R(0) = 1 + At + Bt^2 \quad [2.6]$$

For $-200°C < t < 0°C$: $$R(t)/R(0) = 1 + At + Bt^2 + Ct^3(t - 100°C) \quad [2.7]$$

with:

$$A = \alpha^{100}{}_0(1 + \delta/100)°C^{-1}$$

$$B = \alpha^{100}_{0}\delta x10^{-4}\ °C^{-2}$$

$$C = \alpha^{100}_{0}\beta x10^{-8}\ °C^{-4}$$

where α^{100}_{0} is determined by relation [2.5]. δ is linked to the quantum metallic crystal theory and describes the discrepancy in linearity that occurs at high temperatures. β is a coefficient that is determined by the boiling point of oxygen (-182.97°C at normal atmospheric pressure). $\beta = 0.11$ if $t < 0°C$ and $\beta = 0$ if $t > 0°C$.

The need to dispose of interchangeable equipment has pushed the industrial sector to establish a standard that defines two quality ratings relating to platinum thermometer sensors (Standard CEI 751): class A and class B. At 0°C, the deviation tolerated for class A is from ±0.15°C, or ±0.06 Ω. For class B at 0°C, the deviation tolerated is from ±0.3°C, or ±0.12 Ω.

$$\alpha^{100}_{0} = 3.85 \times 10^{-3}\ °C^{-1}$$

$$A = 3.9083 \times 10^{-3}\ °C^{-1}$$

$$B = -\ 5.775 \times 10^{-7}\ °C^{-2}$$

$$C = -\ 4.183 \times 10^{-12}\ °C^{-4}$$

It is therefore possible to use this equation to directly measure temperature, when the accuracy researched is low. For class A, the maximum inaccuracy will be ±0.15°C and it will be ±0.15°C for class B, on the condition that the reading instrument is well calibrated. When the accuracy researched is higher, it becomes necessary to calibrate to the fixed points of the ITS or by comparison (section 1.2.1.2).

2.1.2.2. *Thermistors*

The word "thermistor" means *thermally sensitive resistor*. These elements can be classified in the same category as resistive sensors; however they are formed from metal oxide ceramics of semi-conductors like MgO, $MgAl_2O_4$, etc. The metal oxide powders are grouped together by compression, forming disks, pearls or cylinders of small dimensions (approximately 1 mm² or smaller).

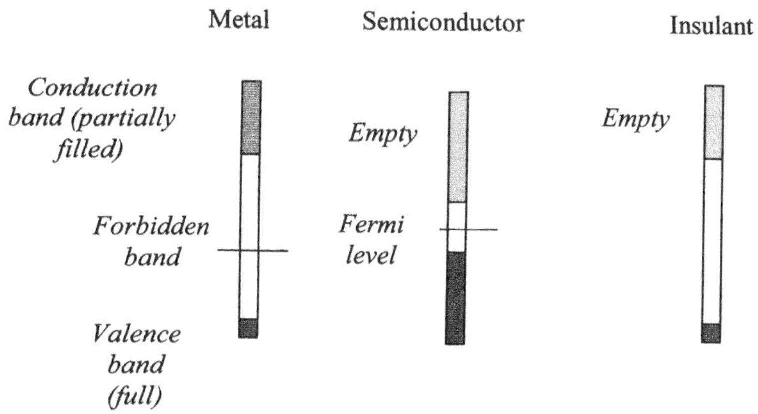

Figure 2.6. *Schematic representation of energy bands from three categories of matter: metal, semiconductor and insulant*

The atomic structure of natural elements can be represented by energy bands that may or may not contain electrons (see Figure 2.6). Semiconducting matter is made of the valence band in which all energy levels are filled, a very weak forbidden band and the conduction band, which is initially empty, as in the case of insulants. However, the Fermi level place in the middle of the forbidden band is not in an eigenstate: when $T = 0°K$, the electrons saturate the valence band. If T increases, the electrons will move into the conduction band and leave holes or positive charges in the valence band.

In the case of conducting matter (metal), the conduction band is partially filled with electrons that will act to circulate the current. The valence band is clearly separated from the conduction band. The Fermi level is in an eigenstate: it is the last to be held at $T = 0°$ K.

The circulation of a current is interpreted by the passage of electrons, from the valence band to the conduction band. Having left one band, the electrons leave room for holes or positive charges. Thermal agitation produced in semiconducting matter provokes the freeing up of pairs of electrons-holes. This release results in the conductivity χ of matter that can be quantified by:

$$\chi = e(n\mu_n + p\,\mu_p) \qquad [2.8]$$

where n and p are the densities of free electrons and holes, μ_n and μ_p their mobility and e the electron charge (1.6×10^{-19} C). It is predominantly free electron densities that vary with temperature. Moving on from this finding, the Maxwell-Boltzman equation shows that electric conductivity χ of a thermistor takes the form of equation [2.9] where A and β are constant characteristics of matter used, a is a coefficient that depends on concentrations in n and p, and T is the thermodynamic temperature of the semiconductor.

$$\chi = A\,T^{a} e^{(-\beta/T)} \qquad [2.9]$$

From this expression we get that of thermistor resistance, where R_0 is the resistance of the absolute maximum temperature:

$$R(T) = R_0 \left(\frac{T}{T_0}\right)^{-b} e^{\left(\beta\left(\frac{1}{T}-\frac{1}{T_0}\right)\right)} \qquad [2.10]$$

Since the influence of the exponential term on the value of R is predominant, we will rewrite this relation in a more condensed way in which coefficient B depends on the temperature:

$$R(T) = R_0 e^{\left(B\left(\frac{1}{T}-\frac{1}{T_0}\right)\right)} \qquad [2.11]$$

By analogy with the ratio giving the sensitivity of resistive sensors, we are able to calculate the reduced sensitivity α_r of a thermistor:

$$\alpha_r = \frac{1}{R(T)} \frac{dR(T)}{dT} \qquad [2.12]$$

Expression [2.11] shows that:

$$\alpha_r = -B / T^2 \qquad [2.13]$$

The value of α_r is generally given at 25°C. Due to the fact that it is negative, thermistors are often termed negative temperature coefficient (NTC). This allows them to be distinguished from silicon resistances that have a positive temperature coefficient (PTC).

Thermistor sensitivity is around 10 times higher than that of platinum resistance of 100 Ω. However, this sensitivity is not constant throughout its range of usage. Thermistor response is therefore nonlinear, but these elements are extremely sensitive; the value of their base resistance R_0 can be chosen from a broad range of values that go from 100 Ω to 5 MΩ. Their ability to interchange is weakened, however, from one element to another. It is therefore necessary to select a high number of components when producing instruments. Their essential assets relating to oceanography are their high resistance to shocks and vibrations and the stability of their characteristics, which is optimal for covered bead thermistors.

For these reasons, studies were undertaken in the 1960s and 1970s to improve the accuracy of thermistors in the 0–30°C range. These studies were mainly carried out by Bennett who, after works carried out by Steinhart and Hart, found that the best polynomial approximation of their temperature response in this field was as follows:

$$\frac{1}{T} = A + B\ln\left(\frac{R}{R_1}\right) + C\left[\ln\left(\frac{R}{R_1}\right)\right]^2 + D\left[\ln\left(\frac{R}{R_1}\right)\right]^3 \qquad [2.14]$$

where R_l is the resistance of the sensor to a reference temperature, and A, B, C and D are constants defined by calibration. He also evaluated the effects of pressure on sensors made up of a thermistor mounted in a protected stainless steel tube 2.5 mm in diameter. The maximum error that he measured at 700 bar was 0.0017°C, its measured uncertainty to one standard deviation being 0.001°C. As the variation observed is linear, he calculated its coefficient to be equal to $(1.3 \pm 1.5)\times 10^{-6}$ °C bar^{-1}. In 1984, Gent [GEN 84] from the US Naval Oceanographic Office measured the effect of pressure on thermistors only protected by a glass bulb. He found a coefficient of 0.050°C/1,000 bar, or 5×10^{-5} °C dbar^{-1}. These works cleared the way for the usage of thermistors in a good number of oceanographic measurements.

2.1.2.3. *Electronic conditioning of temperature sensors*

Conditioning technology used on instruments with limited accuracy (thermistor chains, bathythermographs) is similar to the technology generally used in order to interface with resistive sensors: resistance ratio assembly or Wheatstone bridges that transform resistive variations in voltage measurements. The accuracy and the stability needed for the CTD, however, require fixtures to be tailored.

Figure 2.7 represents the general make up of the acquisition card of a signal from a resistive sensor. In this section, the principal focus is on the "conditioner", being the key point of the measurement principle adopted. This term generally refers to the part of the fixture that makes up the interface between the sensor and the amplifier fixture, which is indivisible. The amplifier fixture increases the tension measured at the fixed interface terminals, and eventually filters the signal. The operational amplifier assemblies' specifications are very nearly perfect (so entry impedance is infinitely larger than the exit impedance of the interface fixture, the frequency band utilization is much bigger than necessary in order to receive the signal and the exit impedance is very low). Their influence on measurement becomes negligible, provided that the passive components surrounding them have sufficiently stable values over time and according to temperature. Errors in measurement introduced by the digitizer, or a digital-to-analog convertor and the voltage reference necessary for its function, are equally negligible for the most part provided that a converter with appropriate dynamic measure and sufficiently stable voltage reference are chosen.

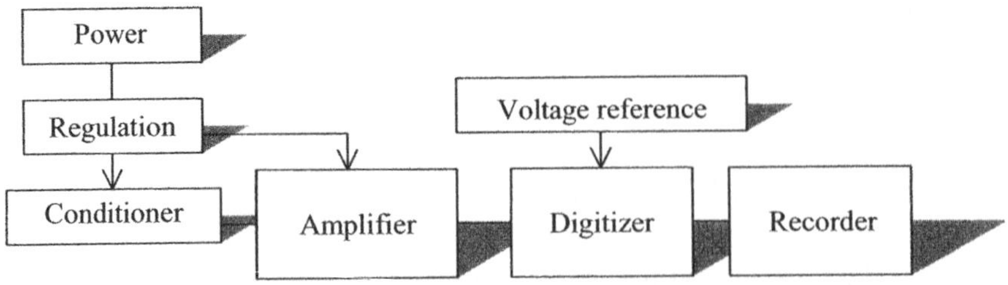

Figure 2.7. *General make up of a resistive signal data capture card*

The most simple and economic sensor interface fixture is without doubt the ratio resistance assembly (see Figure 2.8).

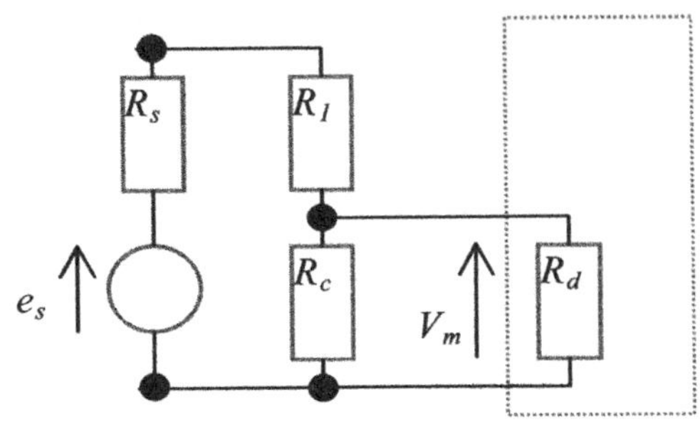

Figure 2.8. *Principal diagram of a ratio resistance assembly*

It is created from two resistances: the sensor, R_c, and R_1, which provides a comparison. They are serially connected to a regulated power source, which is equivalent to that of an electromotive force generator, e_s, and internal resistance, R_s. The voltage of the sensor terminal is predominantly measured by an operational amplifier fixture. If it has an entry resistance R_d, the voltage V_m measured by the fixture is the following:

$$V_m = e_S \frac{R_c R_d}{\left[R_c\left(R_S + R_1\right) + R_d\left(R_S + R_1 + R_c\right)\right]} \qquad [2.15]$$

In order for the terminal voltage of the sensor to be independent of the fixture used, R_d must be considerably larger than R_c, which is normally the case. The equation therefore becomes:

$$V_m = e_S \frac{R_c}{\left(R_S + R_1 + R_c\right)} \qquad [2.16]$$

We therefore have a nonlinear equation of V_m, function of R_c. If we want to find a linear response to the conditioner output, three solutions can be adopted:

i) The fixture can operate with "small signals", either with variations ΔR_c from R_c around a basic value (chosen to be 0°C, for example), which are weak, ahead of the sum: $R_{c0} + R_1 + R_s$. We therefore measure the variations ΔV_m of V_m:

$$V_m + \Delta V_m = e_S \frac{R_c + \Delta R_c}{\left(R_S + R_1 + R_c + \Delta R_c\right)} \qquad [2.17]$$

The variations ΔV_m take the following form, rounded to the second-order:

$$\Delta V_m = e_S \frac{R_{c0}\Delta R_c}{\left(R_S + R_1 + R_{c0}\right)} \qquad [2.18]$$

where ΔV_m is directly proportional to ΔR_c. If the adjustment $R_s + R_1 = R_{c0}$ is chosen, then the sensitivity of the conditioner is maximal, and we have:

$$\Delta V_m = e_S \frac{\Delta R_c}{4R_{c0}} \qquad [2.19]$$

The fixture is linearized. The voltage generators or power voltages that are on the market generally have very low R_s values, as opposed to the usual R_c and R_1 values. To reach maximum sensitivity, $R_1 \approx R_{c0}$ should therefore be chosen. A resistance that has a weak drift according to temperature must be found, however, so that the characteristics of the conditioner are conserved during measurements *in situ*, because electronic packages are often applied to temperatures that are close to the temperatures they are attempting to measure.

ii) The second solution consists of supplying the fixture with a current source, which results in the written equation: $R_s >> R_1 + R_{c0}$. In this case, the condition $\Delta R_c << R_s + R_1 + R_{c0}$ is always verified. In posing: $I_s = e_s / R_s$, relation [2.18] gives:

$$\Delta V_m = I_s \; R_{c0} \; \Delta R_c \qquad [2.20]$$

iii) The third solution involves replacing resistance R_1 with a second sensor that is identical to the first but has opposing variations. This is known as a "push–pull" fixture. We therefore have: $R_1 = R_{c0} - \Delta R_c$. If we replace R_1 by its value in relation [2.17], we get:

$$\Delta V_m = e_S \frac{\Delta R_c}{\left(R_S + 2R_c\right)} \qquad [2.21]$$

ΔV_m is once again directly proportional to ΔR_c.

Ratio resistance assemblies have three major disadvantages:

– they are sensitive to noise, and they do not eliminate direct currents from the signal being measured;

– they are sensitive to fluctuations in voltage supply e_s, or feed-in current I_s, and these fluctuations cannot be separated from variations ΔR_c; and

– they cannot be used directly if there is a long wire between the sensor and the measuring instrument.

Despite this, as they are associated to reference Vishay resistors, and high-resolution (24 bits) analog–digital converters, they allow the production of highly accurate fixtures; they just need to be supplied with an alternative voltage, the temperature of resistance must be monitored and the resistance variations related to temperature must be balanced.

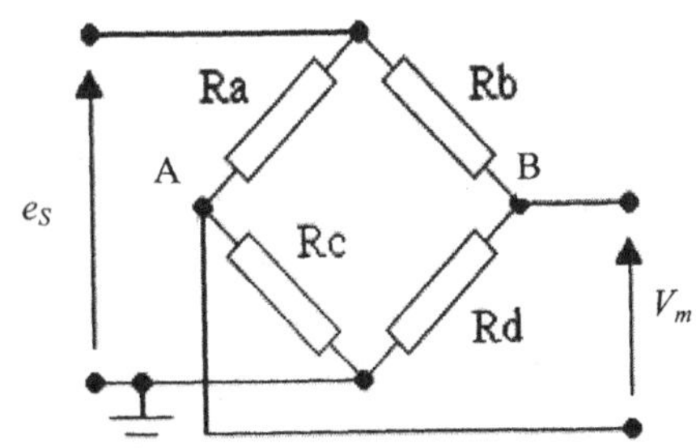

Figure 2.9. *Principle of a Wheatstone bridge. A generator e_s feeds a network of resistors. The difference in voltage V_m generated is measured between points A and B when the resistance value R_c of the sensor varies*

When there is a long wire between the sensor and the conditioner, however, the use of a Wheatstone bridge is preferred (see Figure 2.9). Their advantage is that they allow differential measures to be taken from ratio resistance fixtures, making good use of operational amplifier fixtures for voltage measurement.

For the fixture represented in Figure 2.9, the measurement is taken between points A and B. R_c is the resistance of the sensor. It could be suggested without committing a serious error that the generator has an internally negligible resistance ($R_s = 0$). It could also be suggested that voltage V_m is measured by a fixture whose resistance is infinite compared to the values of R_a, R_c, R_b and R_d. If both of these conditions are gathered, we can establish the expression of voltage $V_m = V_A - V_B$:

$$V_m = e_S \frac{(R_c R_b - R_a R_d)}{[(R_c + R_a)(R_b + R_d)]} \quad [2.22]$$

The bridge is balanced when $V_A = V_B$ or when $R_c R_b = R_a R_d$. In order to create this balance, which is actually the "zero" of the bridge, a potentiometer is sometimes inserted in one of the branches, or values R_a, R_b and R_d are adjusted with high precision in measuring conditions where ΔR_c is zero. From expression [2.22], it can be seen that the sensitivity of the bridge is maximal if $R_a = R_c$ and $R_b = R_d$. Generally, $R_b = R_d = R_{c0}$ is produced, where R_{c0} is the value of R_c at equilibrium; however, R_a, R_b and R_d vary slightly in most cases (with temperature for example) and their variations are different to that of R_c. We therefore have:

$$R_c = R_{c0} + \Delta R_c \qquad R_b = R_{c0} + \Delta R_b$$

$$R_a = R_{c0} + \Delta R_a \qquad R_d = R_{c0} + \Delta R_d \quad [2.23]$$

When combining relations [2.22] and [2.23], we can see that at second order V_m is:

$$V_m = \left(\frac{e_S}{4}\right)\left(\frac{\Delta R_c}{R_{c0}}\right)\left(\frac{1}{1 + \frac{\Delta R_c}{2R_{c0}}}\right) \quad [2.24]$$

This relation is similar to that obtained with a ratio resistance fixture, except for the fact that this time V_m is directly proportional to ΔR_c. V_m is only linear if the variations in ΔR_c are small compared to the value of R_{c0}. It can be shown, however, that the influence of fluctuations in voltage supply e_s on V_m is much smaller than that of a resistance ratio fixture, on the condition that we still have $\Delta R_c << R_{c0}$. It is also possible to linearize a Wheatstone bridge fixture with the help of a current supply or by adopting a "push–pull" fixture. The bridge fixtures are able to be used when there is a long wire between the sensor and the measuring instrument, this being one of their major assets. Different wiring can therefore be used to limit the effects of resistance variations on the linking wires.

The resolution and stability of the measurements that can be obtained with ratio resistance fixtures or Wheatstone bridges have been limited for some time by the stability of supply circuits, and by the signal-to-noise ratio before digital conversion. Furthermore, when direct voltage is used to power the sensors, potential differences are formed at the junctions of the different conductors of the circuit if these junctions are not set to the same temperature. This effect is known as the Peltier effect. In a non-foreseeable way, it introduces inaccuracies into the measured values. To solve this problem, it is necessary to work with alternative voltages. In this way, junctions are positively polarized at each positive alternance and negatively polarized at each negative alternation. The summation of these alternations cancels out parasitic potentials.

These problems of drift and inaccuracy meant that manufacturers in the 1970s and 1980s had to find new solutions in order to interface resistive CTD sensors. Sea Bird Engineering, Inc. submitted a patent in the 1970s to achieve this by interfacing sensors with Wien bridge oscillator circuits. This fixture is a filter made up of two resistance–condensator couples, one in series and the other in parallel, that are made to oscillate when placed in counter-reaction to a fixture whose transfer function is proportionally opposite to that of the filter. The resistance–condensator couples allow the oscillation frequency to be fixed, and the counter-reaction fixture stabilizes this frequency. One of the resistances of the fixture should be replaced by a resistive sensor to dispose of a fixture whose frequency varies with the resistance of the sensor. The resistive variations are therefore transformed to frequency variations, instead of voltage variations. This is of great interest, as on a metrological level frequency is the quantity that is measured with the highest accuracy. Furthermore, signals with modulated frequency are not very sensitive to noise during transmission.

The patent submitted by Sea Bird Engineering, Inc. describes a fixture made up of transistors. Since operational amplifiers have taken the place of transistors in low frequency measuring, counter-reaction fixtures are now produced from these components. In Figure 2.10, couples R_1 C_1 and R_2 C_2 make up the oscillator whose voltage gain A can be calculated. Resistances R_5 and R_6 allow determination of the gain A' of the amplifier. Barkhausen states that such a fixture oscillates if it always responds to the following criterion:

$$AA' = 1 \qquad [2.25]$$

This criterion introduces a condition on the values of gain resistances R_5 and R_6. It is necessary that $R_6 = 2.R_5$ in order to obtain oscillation with a frequency of:

$$f = \frac{1}{2\pi\sqrt{R_1 R_2 C_1 C_2}} \quad [2.26]$$

A characteristic of the Wien bridge oscillator is therefore that the expression of frequency f, does not interfere with the values of the gain components of the fixture, this being an advantage in terms of adjustment. R_2 being the resistance of the sensor, it varies over time, leading to a variation in f. Therefore, to ensure that the relation [2.25] is always filled, it is necessary to control the value of A' in an active way. This is the role of the fixture, made up by the transistor T, the diode D, the resistances R_3 R_4 and the condenser C_3. The diode D realigns the negative alternation of a part of the output V_s signal. The realigned signal is filtered by fixture R_3 C_3 in order to power the base of the field effect transistor (FET) T and to vary the drain-source resistance that just adds itself to the resistance R_5, being a variable gain that allows a constant signal amplitude to be maintained, and therefore permanent oscillation.

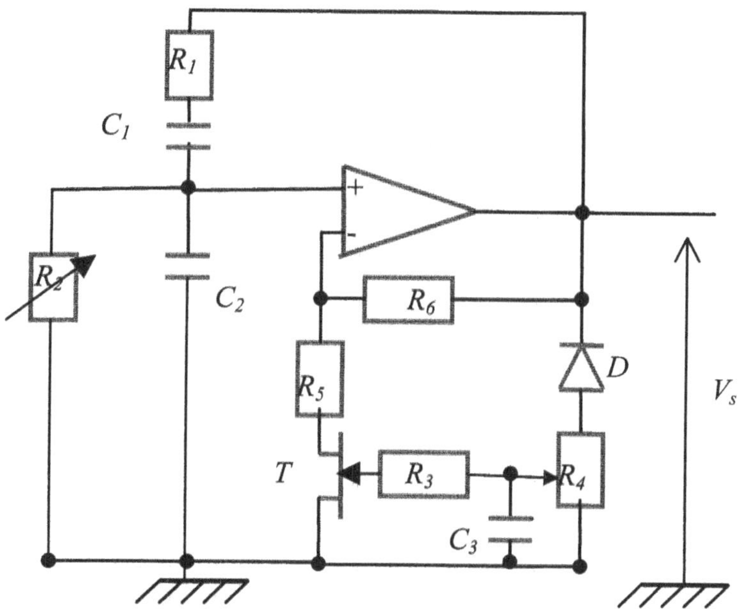

Figure 2.10. *An example of a Wien bridge oscillator fixture where the stability of oscillation is assured by a transistor placed in counter-reaction*

The treatment of the signal that follows this conditioner is different from that represented in Figure 2.7, as the useful information is contained in the signal frequency and not in its amplitude. The sinusoidal signal can easily be transformed into a square signal in which period can be measured by logic components as "binary counter". The digitalization is therefore immediate (Figure 2.11). To function without introducing error, the counter must have a highly stable clock built around a frequency-stabilized oscillator. As the frequency modulation is less sensitive to measurement noise than conventional fixtures, such as those represented in Figure 2.8, it is possible to obtain more sizeable measurement ranges, and therefore higher sensitivity. This sensitivity is not constant across the entire range, however, which results in nonlinearity. The digitalization should therefore be followed by digital treatment that allows linearization of the sensor's response after calibration of the outfit.

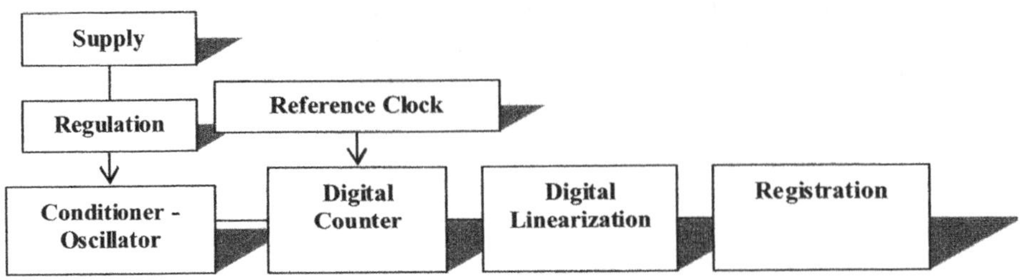

Figure 2.11. *General make-up of a fixture where the conditioner is an oscillator*

By way of an example, the SBE 3 Sea Bird sensor, installed on a CTD SBE 25, delivers a frequency of 6 kHz at –1°C. The sensitivity of the fixture is therefore 146 Hz °C^{-1}. At 31°C, its frequency is 12 kHz and its sensitivity is 233 Hz °C^{-1}. The digitalization system is made of two timers of 12 bits. One is used to measure the integer number of signal periods emitted by the sensor during intervals of 1/36 s, and the second measures the fractional share, i.e. the difference between the clock's signal and the sensor's signal. This system allows a minimum resolution of 2×10^{-4}°C to 31°C. The reference clock, working at 9.4 MHz, has a stability of 5 ppm in temperatures from 0 to 50°C, signifying that the maximum error of measurement brought about by the clock at 31°C is about 4×10^{-4} °C.

Other processes exist that can attain such drift and resolution performances. In this way, Neil Brown of Woods Hole Oceanographic Institution invented a system based on comparison of the resistance value of the measuring sensor in a network of three standard resistances every 1.5 s, the aim of this fixture being to expel electronic instability (see Figure 2.12). This patented system is used on instruments made by Falmouth Scientific, Inc (FSI). The sensor and the network of resistances are fed by a square signal with a frequency of 1,024 Hz. They are then multiplexed and their amplitudes E_s are compared to that of the counter-reaction signal E_s, which is obtained by removing part of the sinusoidal signal E_{ac} emitted from the comparator, in which the phase is detected and integrated. An FET interrupter allows the recreation of the square signal E_c with a frequency 1,024 Hz and of exactly equal amplitude to that of the E_i signal emitted from the integrator. The counter-reaction signal E_{cr} is a square signal whose amplitude is the same as E_s. The amplifier comparator only conserves the fundamental component frequency of signal E_s, it does not take into account the parasitical effects linked to harmonic components from the square signal. The amplitude of this fundamental component E_i which is, in effect, strictly proportional to the amplitude of the signal emitted from the sensor or to the signals emitted from the standard resistances, is measured by an analog-to-digital convertor. A microprocessor can then compare these amplitudes in order to correct drift and to give a temperature value after calibration of the outfit. This system has the advantage of being linear during output, and simultaneously being able to interface several resistive signals. The multiplexer, for example, is able to link up a pressure sensor in addition to a temperature sensor, and in the same way correct the values they provide.

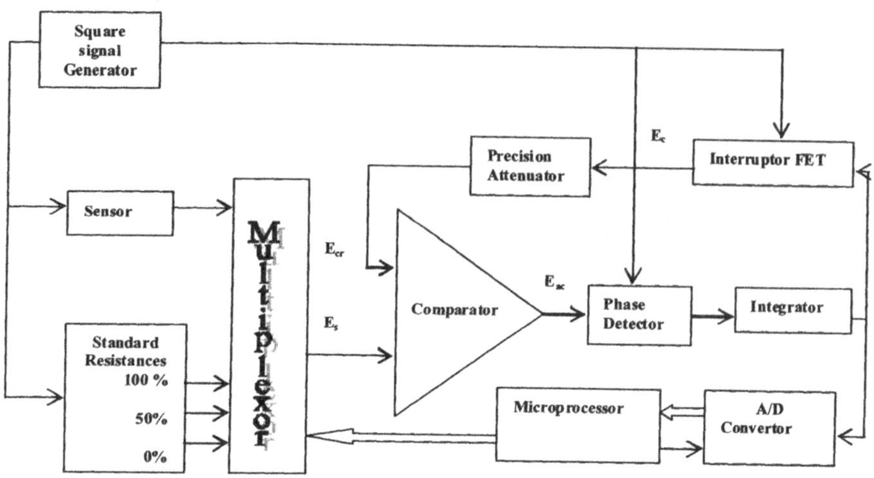

Figure 2.12. *Diagram of a high precision conditioner developed by Brown*

2.1.3. *Thermal transfers*

Temperature is a quantity that characterizes thermal balance. In order for thermal balance to occur, there must previously have been temperature exchanges. All instruments used to measure temperature provide incorrect data if the time of this exchange is not respected. It varies in relation to thermal transfer conditions that are governed in three modes:

– the conductive mode, where heat transfers are made in matter;

– the convective mode, where heat transfers are made with transfer of matter; and

– the radiation mode, where heat transfers are made without support from matter.

The aim of this section is not to explain a complete and rigorous theory of these exchanges, which are the subject of an abundance of specialized literature, but to give several bases that allow understanding of the subject in the framework of oceanographic measurements. More specifically, we aim to explain the problems that arise due to the response time of temperature and conductivity sensors.

When temperature measurement instruments are submerged, they thermally balance themselves with their surroundings, essentially by conduction and convection. Heat transfer by radiation makes very little contribution to these exchanges as soon as the temperature measurements are taken below the surface. This is because the electromagnetic waves are absorbed and reflected very quickly by the marine environment. This mode of exchange will therefore not be tackled.

2.1.3.1. *Conduction exchanges*

In order to have *conduction*, there must be matter. Heat flow ϕ that crosses a thickness dx of mass during 1 s is proportional to the surface S of this mass and the temperature gradient dT. This is the expression of Fourier's law for constant flow, relation [2.27], where λ is the thermal conductivity of the mass (in W m^{-1} °C^{-1}).

$$\varphi = -\lambda \frac{dT}{dx} S \qquad [2.27]$$

This law when applied to different types of geometries shows that the repartition of the quantities of heat exchanged is not the same according to the mass form, for the same exchange surface. For example, a cylinder and a flat wall do not exchange heat in the same way. A case often examined in the case of thermal measure is that of the hollow cylinder, as it allows us to model the transfer mode that rules over temperature probes.

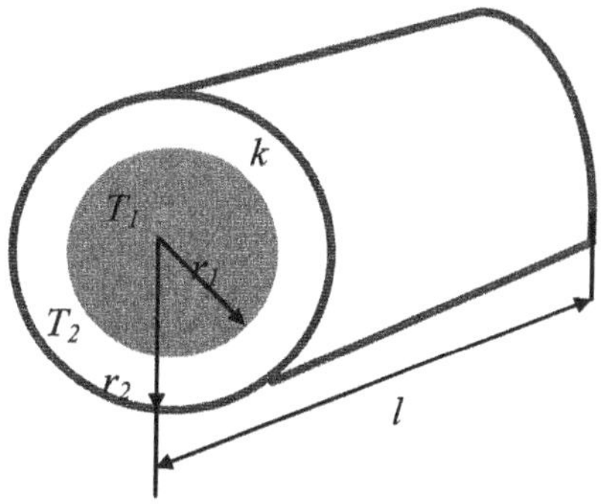

Figure 2.13. *Representation of a hollow cylinder allowing the modeling of conduction exchange*

The application of Fourier's law to the cylinder shown in Figure 2.13 gives:

$$\varphi = -\lambda 2\pi r l \frac{dT}{dr} \qquad [2.28]$$

i.e. after integration we have

$$T - T_1 = -\frac{\varphi}{2\pi\lambda l}\ln\left(\frac{r}{r_1}\right) \qquad [2.29]$$

If we take $r = r_2$, then $T = T_2$. Generally, we take:

$$R_t = \frac{1}{2\pi\lambda l}\ln\left(\frac{r_2}{r_1}\right) \qquad [2.30]$$

R_t is known as the thermal resistance of matter. This expression allows us to describe relation [2.29] as follows:

$$T_2 - T_1 = R_t\phi. \qquad [2.31]$$

This is a similar relation to Ohm's law $U = Ri$ and, in the same way that we measure total electric resistances, if the object is made of several successive layers their thermal resistances will add up. This can be shown by:

$$\Delta T = (\Sigma R_i)\phi \qquad [2.32]$$

For example, in the case of a cylinder made from three layers of rays, r_1 (internal rays of the first layer) to r_4, with conductivity from λ_1 to λ_3 we have:

$$T_1 - T_4 = \frac{\varphi}{2\pi l}\left(\frac{1}{\lambda_1}\ln\left(\frac{r_2}{r_1}\right) + \frac{1}{\lambda_2}\ln\left(\frac{r_3}{r_2}\right) + \frac{1}{\lambda_3}\ln\left(\frac{r_4}{r_3}\right)\right) \quad [2.33]$$

If the heat transfer regime is not permanent but transitory, quantities of internal heat dQ_i that circulate per time unit dt, $dQ_i = \phi\, dt$ must be taken into account. The variation of internal heat-energy in a basic volume is given as:

$$dQ_i = C_p dm \frac{\partial T}{\partial t} dt \quad [2.34]$$

where dm is the mass of the basic volume, and C_p the specific heat capacity of the material (J kg^{-1} °C^{-1}). The calculation of this quantity can be generalized to the case of a three-dimensional space and it leads to the equation of heat propagation or the following Fourier equation:

$$\frac{\partial T}{\partial t} = \frac{\lambda}{C_p \rho}\left(\frac{\partial^2 T}{\partial x^2} + \frac{\partial^2 T}{\partial y^2} + \frac{\partial^2 T}{\partial z^2}\right) + \frac{s}{C_p \rho} \quad [2.35]$$

In this equation, s represents the power of internal heat sources and ρ the material's density. The quantity $\frac{\lambda}{C_p \rho}$ is the thermal diffusivity, which is not to be confused with thermal effusivity, which is: $\left(\frac{\lambda}{C_p \rho}\right)^{1/2}$.

2.1.3.2. *Convective exchanges*

Convective exchanges are generally made between a wall at temperature T_p and an ambient environment at temperature T_a. Two types of convection can be distinguished:

– *natural convection*, where exchange fluid is naturally moved by the variation of its molecular weight when it heats or cools, when contact is made with a surface; and

– *forced convection*, where the exchange fluid is moved artificially.

No matter what the convection type, the exchange occurs in the intermediary zone between the surface and the undisturbed fluid, which is known as thermal boundary layer. In this layer, the exchange takes place partially by conduction at the surround of the surface, and then by mixing. The heat flow exchanged by convection can be calculated by relation [2.36], where T_a is measured at a point sufficiently far from T_s and h is a coefficient of superficial exchange by convection.

$$\phi = h\,S\,(T_s - T_a) \qquad [2.36]$$

It depends on the form of the surface, the nature and the fluid speed. It is expressed in $\text{Wm}^{-2}\,°\text{C}^{-1}$. The calculation of h follows from the resolution of heat propagation equations in viscous fluids, thermal conductors. There are three equations: the mass conservation equation, the movement equation and the conservation of energy equation. As these relations are not analytically determinable, h can be estimated from calculation of three numbers whose evaluation remains complex, and must be adapted to treatable cases.

These three numbers are:

– The Nusselt number, Nu (equation [2.37]), results in the ratio between heat flux exchanged by convection and artificial heat flux that would be exchanged by conduction.

$$Nu = \frac{hd}{\lambda_{fluid}} \qquad [2.37]$$

This ratio causes interference of h, the thermal conductivity of fluid λ_{fluid} and a dimensional characteristic of the wall d. Practical formulae have been developed to evaluate this, so that the value of h can then be extracted. These formulae cause interference from two more numbers, Pr and Re, which are Prandlt numbers and Reynolds numbers. By way of an example, in the case of laminar flow, for $0.6 < Pr < 15$, the average Nu is given by:

$$Nu = 0.66 Pr^{1/3} Re^{1/2} \qquad [2.38]$$

In the case of turbulent flow, for $0.5 < Pr < 50$, the average Nu is expressed by:

$$Nu = 0.036 Pr^{1/3} Re^{4/5} \qquad [2.39]$$

Other formulae of this type are proposed in specialized literature. They adapt more specifically to particular cases that are encountered.

– The Prandlt's number, Pr, expresses the aptitude of heat to move in the fluid:

$$Pr = \text{kinematic viscosity/thermal diffusivity} = \frac{\mu C_p}{\lambda fluid} \quad [2.40]$$

where μ is the dynamic viscosity of the fluid that is expressed in Poiseuille (kg $m^{-1} s^{-1}$) and C_p, its specific heat capacity.

– The Reynolds number, *Re*, characterizes fluid flow.

$$\text{Re} = \frac{\rho v d}{\mu} \quad [2.41]$$

where v is the flow speed, ρ mass density and d one likely dimension of the surface. According to the form of the wall, the calculation of *Re* allows definition of the flow type. For example, in the case of water circulating on a flat and infinite surface in zero incidence, if *Re* is smaller than 3.10^5, the flow is said to be *laminar*. If *Re* is larger than 10^6, the flow is said to be *turbulent*.

When convection is natural, a fourth number interferes in the determination of *Nu*. This number allows expressing the aptitude of the fluid to move when heated dispite its viscous forces. The Grashof number, *Gr*, is equivalent to the ratio:

$$Gr = \frac{g\beta(Ts - T_a)d^3}{\nu^2} \quad [2.42]$$

where ν is the kinematic viscosity of the fluid (in m/s²) and β its constant pressure dilation coefficient.

In the case of free convection, *h* has the following orders of magnitude:

– for gas (air): 3–20 W m^{-2} °C^{-1};

– for water: 100–700 W m^{-2} °C^{-1};

– for boiling water: 1,000–20,000 W m^{-2} °C^{-1}.

The flow of a current *I* into a resistive probe with resistance *R* results in a heat exchange where power *P* can be calculated by the relation known as the *Joule effect*: $P = R\, i^2$. This phenomenon is called *self-heating*. For temperature probes, it can introduce measurement errors from several 0.1 mK to several hundredths of 1°C if it is not corrected. Self-heating, in an environment at rest, essentially results in natural convection. Its correction must take into account the characteristics of the environment: viscosity, thermal conductivity, flow speed (if there is a relative speed between the probe and environment), etc. This correction is generally made when probes are calibrated. It remains valid at the time of usage if calibration is carried out in conditions similar to those of usage.

2.1.4. *Response time of temperature sensors*

As temperature measurements are measurements of exchanges of quantities of heat, they depend on time. A sensor whose initial temperature is T_c, put into an environment where the temperature T_m is different to the former, is not instantly able to indicate a value representative of T_m. Furthermore, the response time of a temperature sensor can be defined as the necessary duration for the temperature shown, after having been subjected to a sudden variation (step), to no longer differ in its final value by more than 1% of this variation, conventionally fixed.

If radiation phenomena are neglected, the heat flow exchanged between the probe and the environment is shown in relation [2.36], which rules over convection. This heat quantity is produced (or absorbed) by conduction in the thermometer, and it can be evaluated using equation [2.43]. It is proportional to its mass m, its heat capacity C_p (expressed in J kg^{-1} °C^{-1}) and to the difference between T_c and T_m. If T_a is replaced by T_m and T_s by T_c in [2.36], the equality of heat flows exchanged leads to the formulation of a differential equation in which the solution takes the following form:

$$T - T_m = (T_c - T_m)e^{-t/\tau} \qquad [2.43]$$

where τ represents the time constant of the system. It is expressed by:

$$\tau = \frac{mC_p}{hS} \qquad [2.44]$$

where S represents the surface of the sensor and h the coefficient of exchange by convection. If we try and express the evolution of the temperature sensor from its original value, T_c, equation [2.43] takes the following form:

$$T - T_c = (T_m - T_c)(1 - e^{-t/\tau}) \qquad [2.45]$$

The exponentially reversed term that figures in this expression, suggests that thermal equilibrium between the sensor and environment will only be achieved after an infinitely long period of time. It is therefore necessary to specify criteria in order to examine the response time. If we propose $t = \tau$, equation [2.45] gives:

$$\frac{(T-T_c)}{(T_m-T_c)}=1-e^{-1}=0.632 \quad [2.46]$$

τ, therefore, represents the time needed for the temperature sensor to attain 63.2% of its final temperature T_m. τ is known as the response time to 63.2% or the final response. This is a parameter that can be measured by plotting the slope to the origin of the function ($T - T_c$), see Figure 2.14.

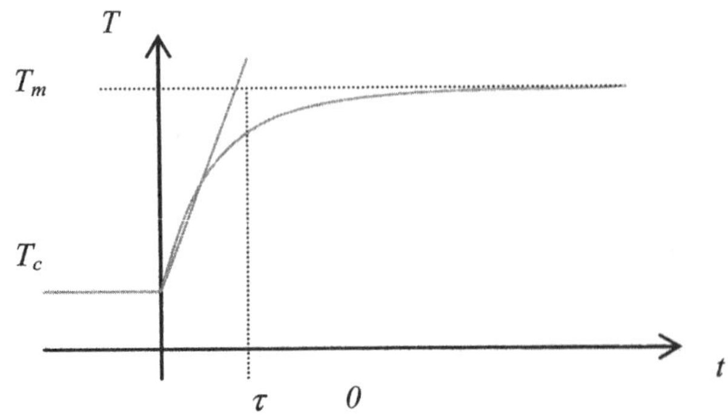

Figure 2.14. *Representation of the temperature shift in the transient state between the sensor, whose initial temperature is T_c, and the environment, where temperature is T_m. τ represents the response time at 63.2% of the final response*

As expression [2.44] introduces variable h, however, the measurement of response time is strictly only valid in conditions of operation that have been determined. As indicated in section 2.1.3.2, h varies in relation to the conditions of measurements across Nusselt, Prandlt, Grashof and Reynolds numbers. Response time therefore differs according to exchange conditions, and the difference in speed of movement that exists between the sensor and environment.

As exchange conditions are often relatively standard, the difference in the speed of movement poses a certain number of problems in oceanography. The problems are particularly significant when temperature is taken in the first few ocean layers with a mobile instrument, such as a CTD suspended by a hoist or placed in a towed support. As a result, several studies where carried out in the 1970s and 1980s in order to study the behavior of different temperature and conductivity sensors at different speeds of movement. As thermistors where known as the "ideal candidates" for *in situ* measurements, several studies were conducted to determine the form of their spectral response at different speeds of movement. Gregg and Meagher [GRE 80] focused on the characteristics of dynamic responses given by thermistors 0.4 mm in diameter and protected by glass beads. Their experimental device allowed them to apply pulses of temperature to thermistors moving at speeds v between 0.08 and 3.80 m/s. The results of these measurements showed that the amplitude of the transfer function obtained in such conditions can be described by a second-order filter (with two poles which time constants are identical) for frequencies below 25 Hz. Above 10 Hz, a first-order filter (with one pole) also provided a good representation. These representations have shown that the variation of the filters' cut-off frequency at –3 dB was proportional to $v^{1/3}$, and that the attenuation of the response to temperature was linked for the most part to the formation of the boundary layer. From this, they deducted that the response time of thermistors, as defined by relation [2.46], is almost equivalent to:

$$\tau \approx 10\, v^{-0.32} \text{ (in ms)} \qquad [2.47]$$

As the variation in the response times stated by various thermistors is to the order of ±10%, however, the authors concluded that it was preferable to take the measurements for each thermistor in order to refine the value of the coefficients in relation [2.47].

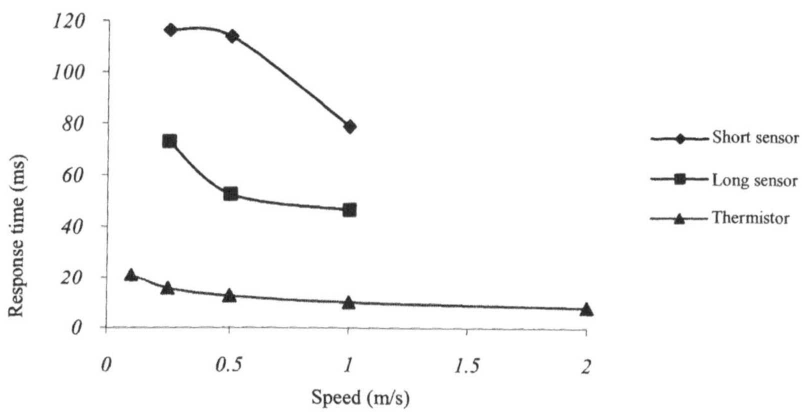

Figure 2.15. *Response time (in s) obtained by measuring the transfer function of a single glass bead thermistor inserted into a "short" sealed tube and a "long" sealed tube [GRE 85] [GRE 80]*

Another study was carried out by Gregg and Hess, this time concerning thermistors inserted into thin sealed steel tubes that were 2.45 cm long with a diameter of 1.1 mm at the level of the sensible part and 3.9 mm long with a diameter of 0.8 mm at the level of the sensible part. These two sensors were tested in the same conditions as before, in order to trace the amplitude of their frequency transfer for three circulation speeds representative of the usage of the sensor on a CTD, lowered by a hoist. Once again, it appeared that the response obtained depended strongly on the movement speed, and therefore on the thickness of the boundary layer. Paradoxically, the longest sensor responded fastest. By supposing equality of the poles of the transfer function when the order is greater than one, the response times depicted in Figure 2.15 are obtained.

Studies that look at the measurement of response time from a temperature step are not always representative of the conditions encountered at the time of measurements in superficial ocean layers. When the sensor is lowered towards the floor or lifted towards the surface, it crosses zones where temperature constantly decreases or increases. This variation is not always linear and differs according to different parts of the world and different seasons (see Figure 2.16). It is therefore difficult to model its repercussions on the response time of sensors. The study examining the effects of linear variation in temperature, however, allows us to understand possible problems. If the sensor moves into an environment where the initial temperature is T_{m0} at time $t = 0$, where it is subjected to speeds s of thermal variation (expressed in °C s^{-1}), the temperature of the environment can be shown by:

$$T_m = T_{m0} + st \qquad [2.48]$$

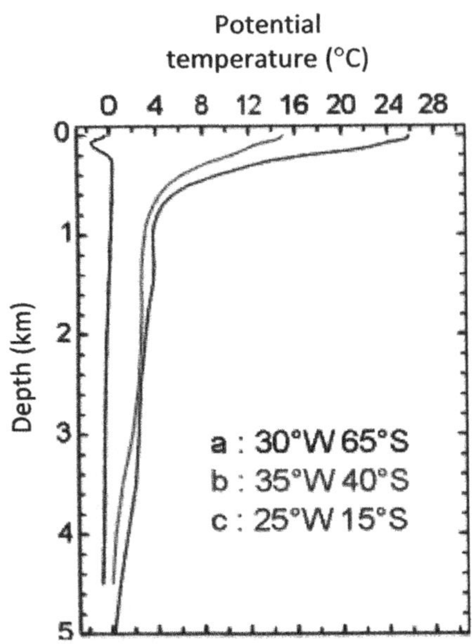

Figure 2.16. *Example of the average yearly temperature profiles in relation to the depth of different regions in the world: a) polar region; b) temperate region; and c) tropical region (Courtesy of © SHOM)*

The equality in exchanged flow leads to a differential equation formed from relations [2.34] and [2.36], in which T_s is replaced by T_c and T_a by expression [2.48]. The solution to this equation is given by:

$$T = (T_c - T_{m0} + \tau s)\, e^{-t/\tau} + st + T_{m0} - \tau s \qquad [2.49]$$

The first exponential term represents the transitional rate during warming of the instrument. The following terms describe the permanent state. The three possible cases and the resulting solutions to this equation are drawn out in Figure 2.17, where the hypothesis is a constant increase in the temperature of the medium. According to the value of T_c and the initial temperature of T_m, which is T_{m0}, it gives: $T_c > T_{m0}$, $T_c = T_{m0}$ or $T_c < T_{m0}$. All curves in the three cases follow an asymptotic direction parallel to the straight *vt*. However, they are separated by a distance τs. The probe therefore introduces an incontrovertible systematic error equal to the product τs. The temperature measured is always inferior to the temperature to be measured if the gradient is positive. In order to reduce this error, either the response time of the sensor must be decreased or, when possible, the relative speed must be decreased. These two solutions can be contradictory, however, as if the relative speed decreases, τ has the tendency to increase, as can be seen in Figure 2.15. A compromise in speed must therefore be made (1 m/s, for example), and an estimation made of the error committed. If S represents the relative difference in speed between the sensor and the medium, expressed in m s^{-1}, s can be presented as in [2.50] and expression [2.51] expresses the error committed.

$$s = S\, dT/dx \qquad [2.50]$$

$$\tau S\, dT/dx \qquad [2.51]$$

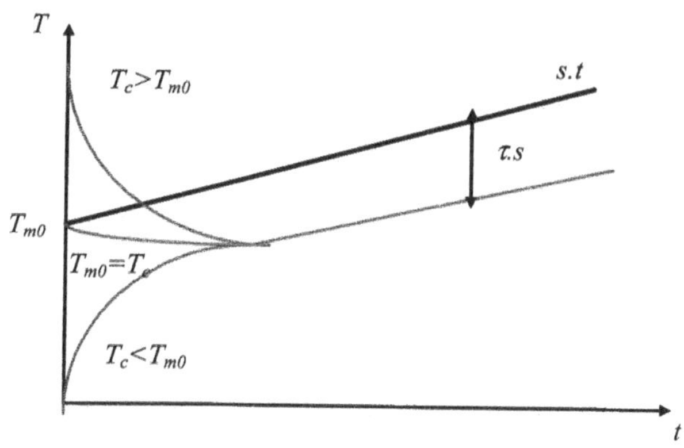

Figure 2.17. *Evolution of the temperature of an instrument immersed at speed v in a medium in which the temperature increase is linear*

If the gradient is negative, the temperature measured will be higher than the true temperature of the medium, but the amplitude of the error will be the same as before. For example, if we take $\tau = 70$ ms for $S = 1$ m/s, when crossing a region in which temperature increases or decreases by 0.25°C/m the instrument will make a systematic error of measure almost equal to $0.07 \times 1.00 \times 0.25 \approx 0.018$°C.

2.1.5. *Viscous heating of temperature sensors*

The principle sources of errors in temperature measurement are the phenomenon of self-heating, as described in section 2.1.3, and the problems with response times, as described in 2.1.4. There is, however, a third source of error, which is linked to the viscous heating of fluid when the temperature sensor passes a certain speed or when fluid circles around the sensor if it is static. The acceleration of flow as it meets the surface of the sensor provokes an increase in temperature related to the fluid's viscosity. It is possible to assess this increase and establish that it is a function of the Prandlt number square root, defined by relation [2.40], and the square of relative speed between fluid and sensor.

The measurements taken by Larson and Pedersen (1996), showed that the warming δ that results is about 0.001°C for a speed of $U = 1$ m/s in laminar flow [LAR 96]. This hypothesis of laminar flow is validated by the fact that for speeds inferior to 10 m/s, the Reynolds Number (equation [2.41]) is less than 20,000 for sea water. They also showed that the relation,

$$\delta t = 1.263\times10^{-4}\, Pr^{0.5} U^2 \qquad [2.52]$$

was able to retrieve the value of this error, in conditions of laminar flow to the nearest 20%, for cylindrical temperature probes. The values obtained with this equation are sensitive to the flux attack angle in relation to the cylinder. This angle of attack can induce variations of 10% of these values. The form of the cylinders' closed extremity (flat or filed end) has an influence evaluated at 5%, and the diameter of the cylinder can also induce a variation of 10%. If no measurement has been taken in turbulent flux, they estimated that flow conditions must reduce the error from factor 2 to 3. The coefficient 1.263×10^{-4} of relation [2.52] would also be replaced by 0.7×10^{-4}.

If the amplitude of the error brought on by viscous heating remains small, it cannot be neglected and needs correcting if we want to follow specifications set out by the WOCE program.

2.2. Determining conductivity

In the field of oceanography, measuring electric conductivity is not an end in itself, but a way of working out the practical salinity of sea water. Instruments to measure temperature that are mounted onto satellites (*Soil Moisture and Ocean Salinity* programs or SMOS by ESA and NASA's Aquarius) that are able to record sea surface salinity, notably by using microwave spectrum (band L), are available. The accuracy of these measurements is limited to 0.2 PSU, however, and is often disrupted by interferences provoked by digital television emitters or cell phones that use very similar frequencies to the Soil Moisture and Ocean Salinity and Aquarius instruments. The calibration of instruments mounted onto satellites is done by data measured *in situ*, notably by thermosalinographs (section 2.1.1.4) or Argo floats (section 3.3.2), and diffused in short times, following the *Global Ocean Surface Salinity Calibration and validation* or GLOSCAL framework programs.

Therefore, *in situ* conductivity measurements remain primordial, and they should be taken at the same time as temperature measurements in order to calculate practical salinity. These measurements are very important as the value of electric conductivity of water depends strongly on temperature. Temperature and conductivity are thus two quantities whose measurements are tightly linked. As these measurements should be taken simultaneously, the notion of sensor response time is of considerable importance and must be the subject of specific treatment, especially when a high level of accuracy is required (section 1.2). In addition, sea water is a complex chemical medium containing life-systems. The chemistry and biology of the medium induce numerous problems in terms of measuring conductivity.

2.2.1. *Principle instruments used*

As for the quantity "temperature", in order to respond to specifications dictated by WOCE HPO (section 1.1), oceanographers must have conductivity instruments usable *in situ* and whose performance are close to the limits that can be reached by the most advanced laboratory instruments. The CTD profiler is among these instruments, either with or without sample bottles, as described in section 2.1.1, and the thermosalinographs (section 2.1.1.4).

Instruments that can be fixed onto cables and then used in anchors to track the salinity of water layers, and Sippican XBT type losable probes that are able to measure conductivity profiles, are also available. These instruments are known as the *Expendable Conductivity Temperature and Depth profiling* system or XCTD (see Figure 2.4). These probes are deployed in the same way as the XBT probes.

They contain a conductivity cell, a thermistor, an electronic conditioner and a biconductor wire bobbin.

The conductivity cell is made up of an alumina composite tube containing four electrodes. This cell is molded in a stiff body of epoxide resin. The electronic conditioner inside the probe converts the variations in resistance of the thermistor and the conductivity cell into variations in frequency, which are transmitted to the boat via the biconductor wire. The conditioner also contains two standard resistances, whose values are periodically sampled throughout the descent.

Each probe is factory calibrated after production, and its calibration coefficients are entered into nonvolatile memory attached to the probe.

The performances of this system are summarized in Table 2.3.

	Range	**Accuracy**	**Resolution**
Temperature	-2.2 to +30°C	±0.03°C	± 0.01°C
Conductivity	20 to 75 mS/cm	±0.03 mS/cm	± 0.01 mS/cm
Depth	0 to 1,000 m	±5 m or 2% of depth	1 m

Table 2.3. *Manufacturer's specifications of the Sippican XCTD-type losable probe (Lockheed Martin)*

Besides instruments used *in situ*, the laboratory salinometer must be cited as the instrument of reference in terms of salinity and conductivity ratio measurements. Its description and mode of function are given in section 1.2.3.3. It is common to find such apparatus on board oceanographic ships that have the use of scientifically-adapted holdings. They are therefore complementary and indispensable for *in situ* instruments of measure as they allow the correction of discrepancies during campaigns at sea.

2.2.2. *Sensors' technologies*

Contrary to temperature measurements, conductivity sensors have no technological classes. Concrete achievements have been made instead that can be classed into two categories: electrode resistive sensors and inductive sensors. This section describes the three most popular "electrode" sensors and the oldest sensors on the market (SBE 4, MK III cell and GCTD cells), of which numerous studies have been carried out in order to model their static and dynamic response. These models allow us to process fitted corrections to the data measured. This section also describes the principles of inductive technology by touching on two achievements made by Falmouth Scientific Instruments, Inc.

Other resistive and inductive conductivity sensors are available on today's market, but the principles of their function are identical to the description given in this section.

2.2.2.1. *The SBE-4 cell*

The SBE-4 cell made by the American company Sea Bird Engineering, Inc. is made from a glass tube that is 190 mm long and has an external diameter of 7 mm. The tube is equipped with three platinum rings placed 50 mm apart and used as electrodes (see Figure 2.18). These rings are covered with black platinum, cast in electro-plating, in order to increase the surface exchange of the electrodes. The glass is covered in epoxide to assure the tightness of wire entry, and is supported by an aluminum plate that improves resistance to mechanical strain. The electric resistance R of the two cylinders of water between the peripheral electrodes, used as floating ground, and the central electrode, is measured with this device. Electric field patterns generated by the electrodes are completely confined by the glass tube. This is therefore a built-in conductivity cell. If l is the distance between the peripheral electrodes and the central electrode, and d is the internal diameter of the glass tube, the resistance R is obtained by the following relation, where χ represents the measured conductivity:

$$R = \frac{1}{\chi}\frac{2l}{\pi d^2} \qquad [2.53]$$

In this relation, the term $2l/\pi d^2$ is the cell constant K. Taking into account the geometry of the SBE-4 cell, $K \approx 2{,}000 \text{ m}^{-1}$.

The electronic interfacing of this sensor is made with a Wien bridge oscillator fixture, similar to that represented in Figure 2.10, where the value of resistance R_2 is given by the ratio [2.53]. Linking relation [2.26] to relation [2.53], it is possible to express the measured conductivity:

$$\chi = 4\pi^2 R_1 C_1 C_2 K f^2 \qquad [2.54]$$

Conductivity is proportional to the square of the frequency, which considerably increases the sensitivity of the process. It does, however, necessitate linearization.

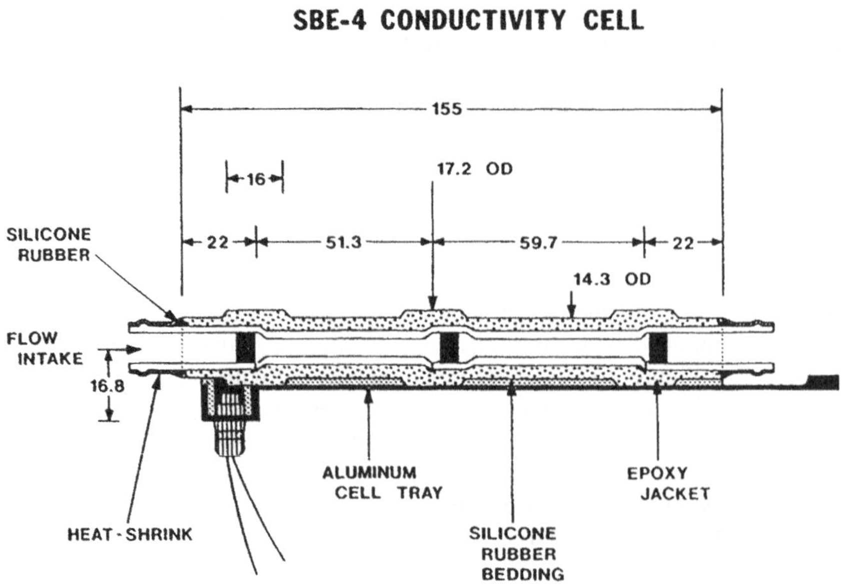

Figure 2.18. *SBE-4 conductivity cell. Glass cylinder covered in epoxide in order to assure tight wire entry, and an aluminum plate that improves resistance to mechanical strain (Courtesy of Sea Bird Electronics, Inc.)*

Like all conductivity cells, the SBE-4 is sensitive to pressure. High pressure exerts compressive stress upon the glass, which contributes to a modification in the value of constant *K*. This effect is taken into account in the calibration equation by a coefficient of compressibility, which is determined by the manufacturer. The measure of conductivity must therefore be coupled with the measure of pressure *p*. However, the gluing of the aluminum plate induces a residual error in salinity that is estimated at 0.001/1,000 dbar, even after correction. The effects of dilation of the cell due to the impact of temperature are also taken into account in this equation. The measure of conductivity must also be coupled with the measure of temperature *t*. The equation of linearization is as follows:

$$\chi = \frac{\left(g + hf^2 + if^3 + jf^4\right)}{\left[10\left(1 + 3.25x10^{-6}t - 9.57x10^{-8}p\right)\right]} \qquad [2.55]$$

where *g*, *h*, *i* and *j* are coefficients determined by calibration.

In fact, the undesirable effect, which is harder to correct and found in all conductivity cells, is that of contamination from all kinds of deposits, such as biological fouling (section 2.2.5), chemical contamination from calcium carbonate, magnesium or hydrocarbon deposits while measurements are taken close to the surface, etc. These non-conducting deposits reduce the internal cell diameter, consequently modifying constant *K*. As the internal cell diameter is 4 mm, it is possible to assess the effect of a deposit of 2 µm on its surface. Formula [2.53] shows that its resistance varies by $1/d^2$. On this basis, the error in salinity measurement will be proportional to the ratio of diameters before and after contamination, squared. For χ = 40 mS/cm, the error ($\delta\chi$) will be $40 \times (1 - 3.998^2/4.000^2) = 0.0399$ mS/cm. For $p = 0$ dbar and $t = 12$°C, $\delta S = 0.039$.

Throughout formula [2.53], at the time of contamination of the SBE-4 cell, the apparent resistance increases, and the conductivity measured becomes lower than it should be. Cell characteristics therefore have a natural tendency to drift, by always measuring conductivity that is lower than the reference.

Sometimes it is possible to observe opposing drifts. This can be due to a loss of the black platinum covering the electrodes due, for example, to the circulation of sediment in the cell, particularly when measurements are taken in coastal zones.

When the cell is brought back to the surface, salt deposits can also be brought up to the level of the electrodes, causing some oxidation and momentarily causing the conductivity measurement values to be too high when the cell is returned to the water. In order to limit these effects, the cells must be filled with distilled water between each usage. Before periods of prolonged storage, they must be cleaned with the non-ionic detergent Triton X-100, then rinsed in distilled water and dried. Despite these precautions, the sensors require regular recalibration to correct the accuracy of the results produced, and attentive tracking of their drift, even during campaigns at sea.

2.2.2.2. *From the MK III cell to the GCTD NBOSI*

The MK III cell, from American manufacturer EG&G Ocean Product, is rectangular in form. It has a length, *l*, of 3 cm, while its interior side *d* is 4 mm. It is made from a ceramic material based on alumina. Alumina is a compound that has the advantage of being a good thermal conductor while also being an electrical insulator. This cell is equipped with four electrodes: two that are able to inject an alternative current of 10 kHz; and two voltage electrodes that collect the variations in the electrical field created by current electrodes (see Figure 2.19).

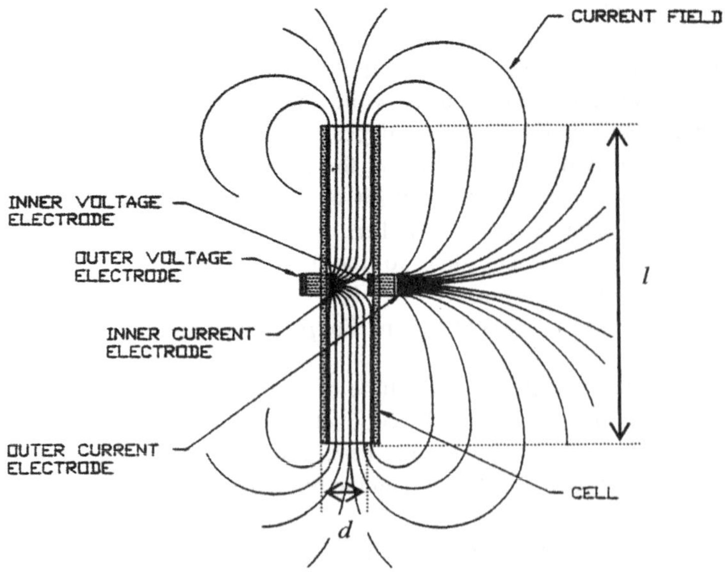

Figure 2.19. *Representation of the MK III C with electric field patterns generated by its current electrodes (Courtesy of Sea Bird Electronics, Inc.)*

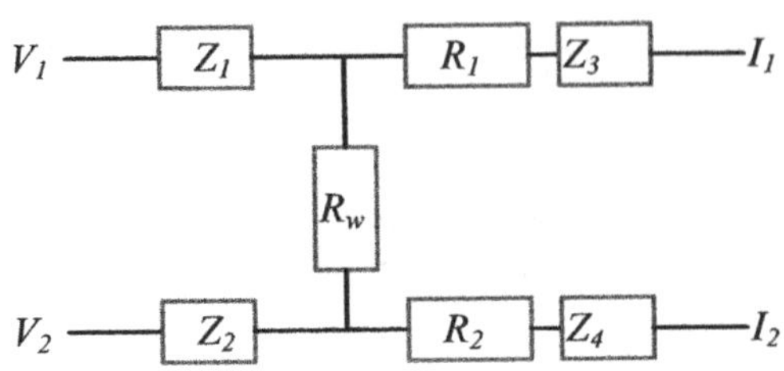

Figure 2.20. *Electrical assembly equivalent to the MK III C cell, according to [GRE 82]*

Current circulating in the current electrodes can be ignored as they are kept at very high impedance, which is advantageous in terms of protection against drifts. Taking into account electrode arrangement, this is an external field conductivity cell. In 1982, Gregg *et al.* proposed an electrical fixture equivalent to this cell and its electronic interface [GRE 82], as shown in Figure 2.20. Z_1 and Z_2 represent voltage electrode impedance and Z_3 and Z_4, current electrode impedance. R_w represents the electric resistance of the volume of water situated between the border of this zone and external electrodes. If we ignore the phenomenon of polarization of electrodes, we can write:

$$R_w = \frac{K}{\chi} = \frac{V_2 - V_1}{I_2 - I_1} \quad [2.56]$$

If χ represents the conductivity of a sample of sea water, and K is the cell constant, formula [2.56] becomes:

$$\chi = K \frac{I_2 - I_1}{V_2 - V_1} \quad [2.57]$$

As the MK III cell is symmetrical in relation to the axis passing through its electrodes, K can be calculated by considering that the two half cells are equivalent to two resistances connected in parallel. Therefore, the expression of K will be:

$$K = \frac{1}{2}\frac{l}{2d^2} \quad [2.58]$$

and have a value of 469 m^{-1}. However, Gregg *et al.* showed that approximately 11% of conductivity measured comes from a volume exterior to the cell [GRE 82].

The electronic fixture of the interface is made from an adjusted current transformer so that currents I_1 and I_2 (see Figure 2.20) are equal at equilibrium. This remains accurate as long as the difference between the electrodes ($Z_2 - Z_1$) remains weak, and the conductivity of water is zero. As soon as the gradients of conductivity form in the cell, disequilibrium is formed between I_1 and I_2 so that there is a difference in voltage of $V_2 - V_1$. The root mean square (RMS) amplitude of this difference in voltage is canceled out by a counter-reaction circuit that injects a bigger current into one of the electrodes. This is shown by

$$V_2 - V_1 = R_w(I_2 - I_1) = V_s \quad [2.59]$$

or

$$\chi = K \frac{(I_2 - I_1)}{V_s} \quad [2.60]$$

Another precision transformer converts this current difference into a difference in voltage, which is then digitalized and treated.

This cell is less mechanically fragile than the SBE-4, and it also has smaller thermal inertia. Its sensitivity is lower, however, as its cell constant is weaker. Furthermore, it is as sensitive as SBE-4 to marine pollution.

The deposits that occur on the interior walls increase the value of K and, as in the SBE-4 cell, lead to measurements of conductivity that are lower than they should be. The same precautionary steps should therefore be taken. Its inventor, Brown, attempted to resolve the problem with such measurements by building a sensor called Mark V, fitted with four cells similar to that of the MK III C. The electric linkup of this sensor was preset so that the volume of water in which conductivity was measured was outside the four cells, and so it was considerably larger than in the MK III C cell, in relation to cell surfaces that could be polluted. The results did not become the subject of any publications. It must be noted that the exterior electric field of the cells induces certain sensitivity to surrounding objects. Certain precautions must be taken, notably during calibration, in order to avoid disruption of the electrical field, and the response given must be modified.

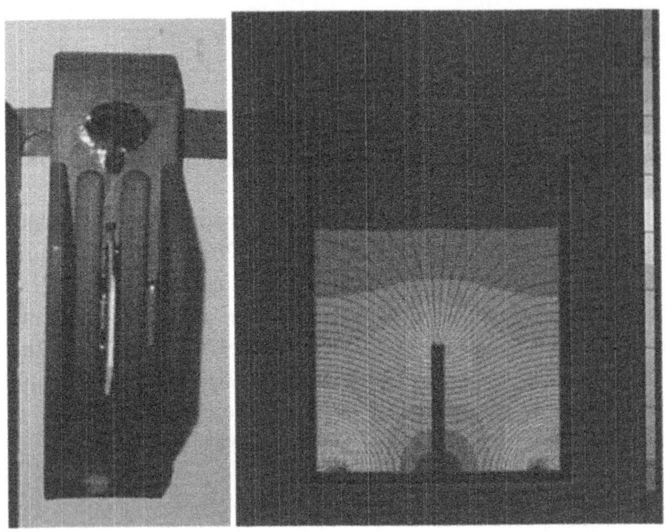

Figure 2.21. *Left: A GCTD sensor viewed from above. Right: Diagram of the confinement of field patterns generated by the central electrode (Courtesy of NBOSI)*

To compensate for problems presented by external electrical fields, Brown submitted a new patent in 2004 that led to developing a cell called GCTD by the company NBOSI. The concepts developed by Brown that address cells with four electrodes (that are relatively insensitive to changes in electrode impedance), made from light material with a good thermal conductivity (therefore possessing a low thermal mass and a good response time), are well adapted to new applications that came into fruition in the millennium, with autonomous underwater vehicles (AUVs) (see section 4.3). AUVs are autonomous instruments that are permanently on the move over long periods of time. The sensors that they carry need to be very light and energy-efficient, which can cause problems for pump systems. If pumping is not available, the sensors must have a very low intrinsic response time. Concepts explained earlier therefore are very interesting in relation to these applications.

GCTD is an outfit made of an "open" conductivity cell, a thermistor attached to the interior of the cell, a pressure sensor and an electronic interface board. The conductivity cell is made from bulkheads in the form of fins in order to confine the electric field generated by the electrode placed in the center, and also to allow easy flow of water during movement of the platform. The bulkheads and the support are made from a light ceramic material that is a good thermal conductor (alumina).

2.2.2.3. *Inductive cells*

A cross-sectional view of an inductive cell is shown in Figure 2.22. It is shaped like a ceramic ring. This ring consists of two or three electric windings that form a transformer, in which the second coil is simply the liquid that surrounds the ring.

These sensors have the advantage, *a priori*, of being less sensitive to marine pollution as they are devoid of electrodes. The simplest configuration is one that contains a single transformer.

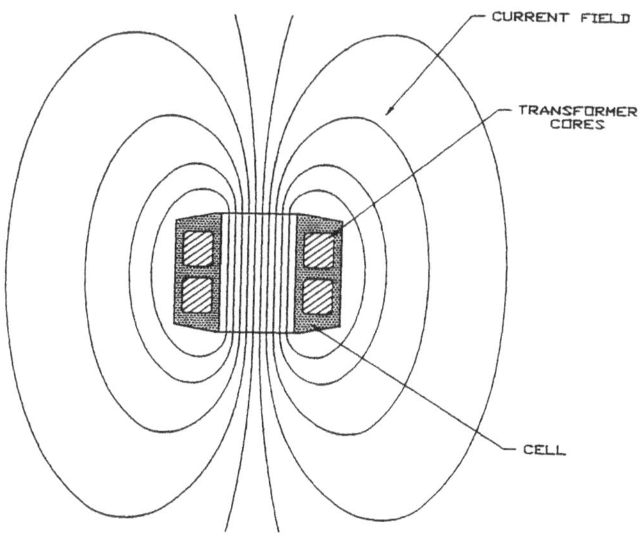

Figure 2.22. *Cross-sectional view of an inductive conductivity sensor with two windings (Courtesy of Sea Bird Electronics, Inc.)*

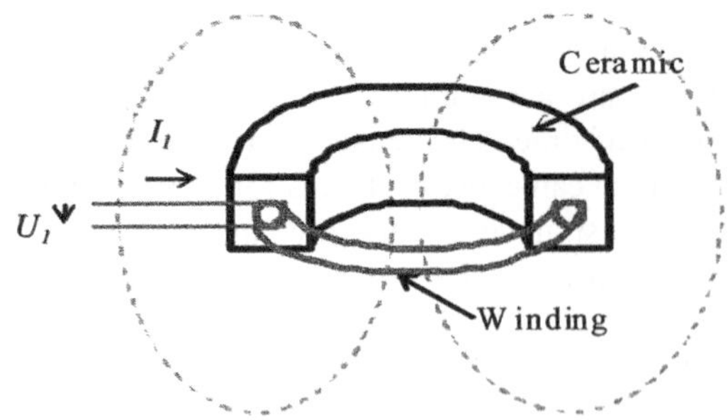

Figure 2.23. *Cross-sectional view of an inductive conductivity sensor with a simple transformer*

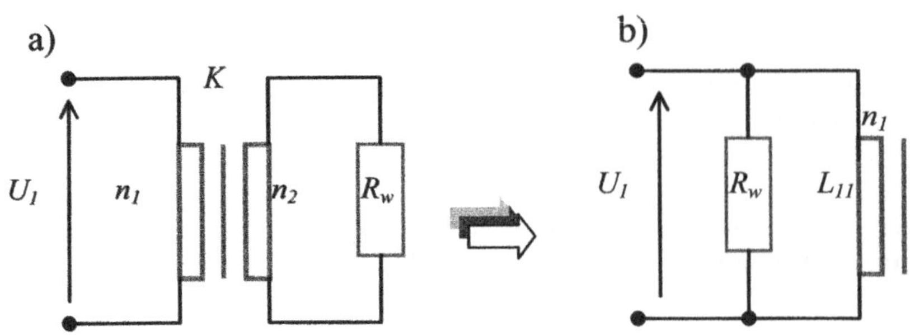

Figure 2.24. *a) Electric circuit equivalent to an inductive cell, with a single winding. As the water only makes one loop, the diagram can be simplified, as in b)*

Figure 2.24 (a) shows the electric circuit equivalent of an inductive cell with a single winding, powered by an alternative voltage U_1, with a frequency, f, of $2\pi/\omega$. The water, in which electric resistance is shown by R_w, only makes one loop, $n_2 = 1$. Circuit a can be simplified by Figure 2.24(b), where it is easy to extract the formula [2.61].

$$I_1 = \left(\frac{1}{n_1^2 R_w} + \frac{1}{j\omega L_{11}}\right) U_1 \qquad [2.61]$$

n_1 represents the number of turns in the first winding, and L_{11} its inductance. This formula contains a first active reel part, which is in phase with voltage U_1, and a second imaginary part, with a quadratic phase shift based on U_1. It appears that the active part of current I_1 is uniquely dependent on R_w. The inductance of the winding and the permeability of the core K do not intervene. Electrical conditioning therefore consists of extraction of the active part I_{1act} of I_1. Several methods can be used in order to do this. If φ is the phase lag between U_1 and I_1, we can use:

$$I_{1act} = I_1 \cos(\varphi) \qquad [2.62]$$

The sensor can also be shunted with a capacity C, so that $\omega C = 1/\omega L_{11}$, to directly obtain:

$$I_1 = \frac{U_1}{n_1^2 R_w} \qquad [2.63]$$

The sensor can also be created so that $\omega L_{11} >> n_1^2 R_w$. As the imaginary part becomes negligible, it leads again to formula [2.63].

This very simple cell configuration does not seem to have been widely used in oceanography, where the "double transformer" configuration is preferred. In this fixture, two transformers are embedded in the ceramic ring (see Figure 2.22). The loop formed by the water is the secondary of the first transformer, which is known as the "drive transformer", and the primary of the second transformer, which is known as the "reader transformer". They are arranged so that the only coupling between the two is the loop formed by the water. Depending on the fixture, either the current or the voltage of the reader transformer is what is measured.

This configuration has an alternative, known as the "double transformer with additional loop". The drive transformer winding in this configuration is connected to the winding of the reader transformer by the supplementary resistance, R_K, which acts as an intermediary forming a supplementary loop. This supplementary loop along with the loop formed by the water creates a "push–pull" fixture (see Figure 2.25). The formula between voltage power U_1 and voltage induced U_4 is therefore very complex. In 1985, Striggow and Dankert formulated [STR 85]:

$$U_4 = \frac{n_4}{n_1} \frac{1}{1 + n_4^2 \dfrac{R_w R_K}{R_w + R_K}\left(\dfrac{1}{R_A} + \dfrac{1}{j\omega L_{44}}\right)} \frac{R_w - R_K}{R_w + R_K} U_1 \qquad [2.64]$$

Different solutions can be used to eliminate the term containing the expression of inductance L_{44}. The solution proposed by Brown consists of using a current realizing condition $R_A = 0$, and choosing a very stable resistance R_K. Taken from equation [2.64], we therefore get:

$$I_4 = \lim_{R_A \to 0} \left(\frac{U_4}{R_A} \right) = \frac{1}{n_1 n_4} \left(\frac{1}{R_w} - \frac{1}{R_K} \right) U_1 \quad [2.65]$$

This expression no longer depends on L_{44}. The term in $1/R_w$ is proportional to the conductivity to be measured and in $1/R_K$ acts as a zero offset. Other studies have shown that it is preferable to replace R_K with a fixed value capacitor. The expression given is therefore rigorously independent of L_{44}.

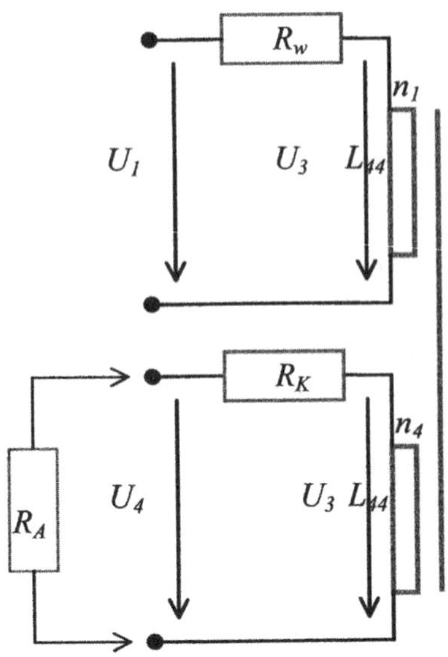

Figure 2.25. *Circuit equivalent to a "double transformer" with additional loop assembly*

"Double transformer" fixtures were made in an attempt to automatically cancel out variations in the magnetic permeability of the core. This permeability varies with temperature and pressure, which makes it possible to vary the value of inductances L_{11} or L_{44}. Electronic compensations that avoid reactive effects, therefore lose their effectiveness. Problems with relaxation and hysteresis are added, so that variations in permeability due to temperature and pressure becoming nonlinear. In order to limit the effects of pressure, sensors made by FSI are equipped with a small deformable bladder that is full of oil. When there is an increase in exterior pressure, the bladder constricts and restores the balance in pressure between the windings and the medium.

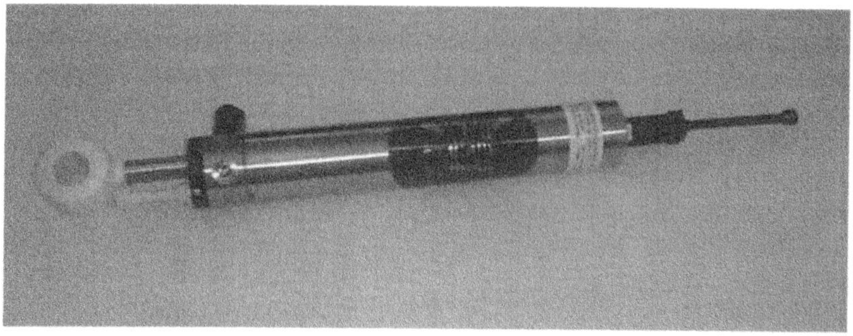

Figure 2.26. *Example of an inductive sensor attached to a CTD probe (Courtesy of FSI, Inc.)*

It has also been shown that chemical and biological deposits affect the resistance R_w that is being measured. Whether they occur on internal or external cell walls, they modify the constant that R_w depends on, resulting in lags and a drift that is difficult to manage once values of conductivity are obtained. The cells can be covered with anti-fouling paint, which is effective but has a limited lifespan and does not prevent deposits on the support parts, where pollution can modify the sensitive volume of the sensor and induce drifts that cannot be managed.

These sensors have a major disadvantage: they are very sensitive to disruptions in their electrical field. These disruptions can be brought on by the simple presence of objects (metallic or non-metallic) in their immediate environment. These objects modify the path of field patterns and can cause major shifts in the values of conductivity measured. This phenomenon is a real nuisance during calibrations. Calibration must be performed in large vats in which there is a minimal distance between the walls and the instrument to be calibrated. It can also make things tight according to the environment they are used in.

To resolve the disadvantages previously cited, FSI recently developed another conductivity sensor named NXIC, which stands for "non-external field inductive conductivity". This sensor is made up of two transformers with double windings mounted in parallel, see Figure 2.27.

To obtain an external field that is zero, from Figure 2.27a, we need to have the condition $I_3 = I_1 - I_2$ with $I_1 = I_2$. As drive transformers T_1 and T_2 are identical (with the same number of turns and same diameter), they are powered by voltages E_1 and E_2, which gives:

$$\frac{R_1}{E_1} = \frac{R_2}{E_2} \qquad [2.66]$$

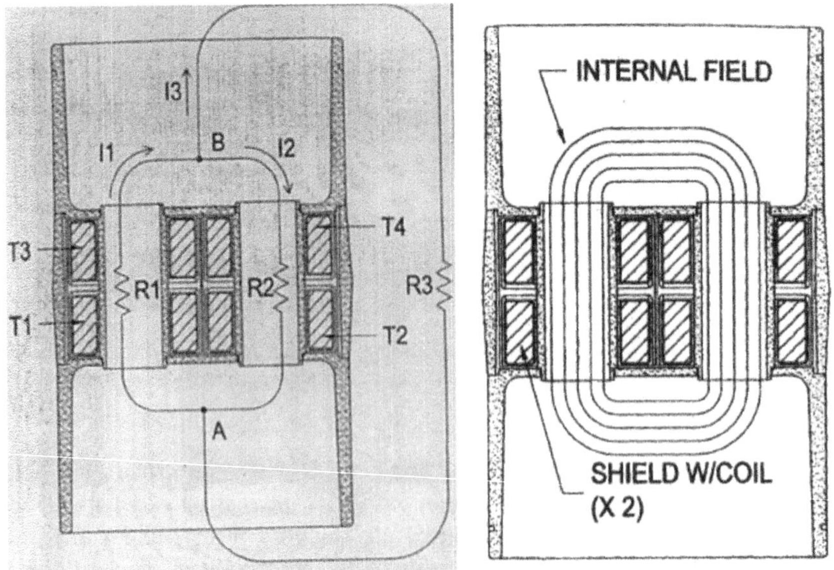

Figure 2.27. *a) Electrical assembly equivalent to the NXIC sensor. b) Representation of the path of field patterns (Courtesy of FSI, Inc.)*

If $E_1 = E_2$, equation [2.66] gives $I_1 = I_2$ and we obtain the image in Figure 2.27b of field pattern paths. The sensor thus made has an electrical field that is completely internal, therefore rendering it insensitive to the outside environment. According to the manufacturer, if the two transformers T_1 and T_2 are not completely identical, it is possible to adjust the values of E_1 and E_2 in order to obtain the equality of formula [2.66]. This leads to the belief that natural chemical deposits are made in a symmetrical way in the two cells, with the risk of modification to their geometry, R_1 and R_2 values, and therefore equality [2.66]. If this is not the case, proximity effects could still appear, causing shifts in the conductivity values measured. The lack of information on this subject prevents the prediction of behavior *in situ*.

2.2.3. *Response time of conductivity sensors*

The problem surrounding the determination of conductivity sensor response time has been the subject of numerous publications, as it is of high importance in the production of salinity calculations without error. These calculations imply that temperature measurements, conductivity and pressure are carried out at the same moment on the same sample of sea water. If the alignment of pressure measurements does not pose any problems, the alignment of responses in the temperature of conductivity and temperature sensors remains delicate, particularly when the sensors cross the first few layers of the ocean, where large temperature gradients and high salinity can be present. The problem is even more complex when we consider that variations in the electrical conductivity of sea water are dominated by temperature to the 80% level (or higher).

If thermal inertia of the sensor means that it does not instantly adapt to the water temperature of the medium, its temperature will have an influence on the water that it crosses, and therefore on the conductivity value measured. As a result, conductivity sensors have a response time that is linked to variations in the salinity of the medium (estimated to be 30 ms for the SBE-4) and a response time linked to variations in temperature. These two effects combine to give a response time in relation to conductivity.

Among the conductivity sensors presented in section 2.2.2, the SBE-4 and MK III C have probably been most frequently studied for the determination of response time and the influence of thermal effects. For these two sensors, the value of response time according to changes in conductivity is largely linked to the ratio L / v, where L represents the length of the sensitive part of the cell, and v the average speed of fluid circulation. This ratio measures the "flushing" of the cell, in other words its aptitude to make water circulate quickly. For the SBE-4, this speed is approximately 0.05 s.

There are, however, two more intervening factors: the thickness of thermal and saline boundary layers and the quantity of heat stored in the cell walls. According to Lueck [LUE 90], as heat diffusivity is approximately 100 times greater than that of salt, the boundary layers in the cells are different. For this reason, even if the cells have no capacity to store or reproduce heat, the response time in relation to temperature and salinity will always be different.

If we are not able to align these parameters, we can attempt to reduce the influence of thermal effects. There are several ways in which we can do this. The average thickness δ_T of the thermal boundary layer of a surface, in laminar velocity, is proportional to its length L. According to Schlichting [SCH 55]:

$$\delta_T = 1.2 P_r^{-1/3} R_e^{-1/2} L \qquad [2.67]$$

where P_r and R_e correspond to Prandlt and Reynold numbers defined by formulae [2.40] and [2.41]. P_r depends uniquely on the physical characteristics of the fluid. For "standard" sea water, $P_r \approx 9.5$. In order to reduce the value of δ_T, we can only optimize the value of L when it is possible to do so. We can, however, increase the value of R_e by increasing the speed of fluid circulation, or even by making circulation turbulent. This is the solution chosen for the Sea Bird SBE-9 probe. Sea water circulated via a pump at a constant speed v of 2.4 m/s in a specially adapted pipe installation (see Figures 2.1 and 2.28). The temperature probe is placed at the entrance in order to create turbulence. Water is then sucked in by a Sea Bird "TC-duct" (TC stands for temperature conductivity) that increases turbulence. The thermal boundary layers are generally thinner with turbulent flow than with laminar flow for an equivalent value of R_e. The heat exchange between the cell wall and sea water is therefore improved, which in turn reduces the amplitude of errors made in measurements brought on by heat stored in the walls. Furthermore, the pump system has the advantage of being able to render the circulation speed of the fluid in the sensor and the speed of the instrument movement in water independent. The response time of the temperature and conductivity sensors is therefore fixed (see Figure 2.15) and software corrections made to the data are more reliable. Gregg and Hess [GRE 85] tested the impulse response of the SBE-4 cell and its pump system. The module of the transfer function obtained acts like a function *sinc*(kL), with:

$$sinc(kL) = \frac{\sin(\pi kL)}{\pi kL} \qquad [2.68]$$

where L represents the effective length of the cell that can be calculated where Q is the flow rate of the pump expressed in $cm^3\ s^{-1}$:

$$L = \frac{1.51v}{Q} \qquad [2.69]$$

In terms of thermal anomalies generated by the cell, the geometry of the MK III C sensor or the GCTD is more favorable than the geometry of the SBE-4 sensor, as it is shorter, its section is smaller, and it is made from material (aluminum) that is a good thermal conductor. In addition, its walls are a lot thinner, creating lower thermal inertia than that of the SBE-4 cell. However, its fixture onto the CTD (or on an AUV in the case of GCTDs) is made without inserting it into a pump circuit.

Besides this, Lueck [LUE 90] showed that the thermal anomaly generated is proportional to $v^{-1/2}$ for laminar velocity, and to $v^{0.8}$ for turbulent velocity. Dependence on the speed at which the instrument moves in water makes sensors harder to correct. Its impulse response has also been studied. The module of the transfer function can be approached by a formula similar to [2.68], where L represents the real length of the cell.

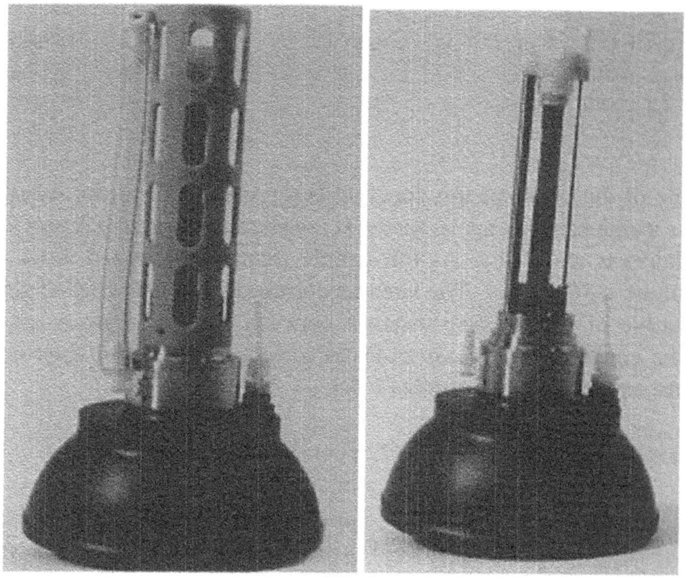

Figure 2.28. *Fixture made by Sea Bird Engineering, Inc. in order to adapt a conductivity cell and a temperature probe to drifting floats. The outfit is called SBE 41. In the photo on the left, the cell is protected by a guard. The exterior TC-duct enables water to circulate through pumping in order to align the response time of the temperature sensor and that of the conductivity cell. The "U" fixture enables conservation of water in the cell while the buoy returns to the surface. This avoids salt and different pollutants from being deposited. The cell drift is therefore better managed (Courtesy of Sea Bird Electronics, Inc.)*

2.2.4. *Aligning the response times of temperature and conductivity sensors and correcting thermal inertia*

No matter how much care is given to the drawing of a conductive cell and its fixture onto a CTD, corrections are always necessary in order to align the responses of conductivity and temperature sensors in space and time. A misalignment of responses results in artifacts affecting calculated salinity values (see Figure 2.29).

Corrections have been determined for the SBE 911 by Lueck and Picklo [lUE 90] from observations made *in situ* in a region of the ocean where variations in salinity and temperature present staircase profiles. These oceanic structures have enabled them to calculate the transfer function H of the CTD conductivity sensor in relation to that of its temperature sensor. H is shown by formula [2.70], where $i^2 = -1$, ω represents the angular frequency in rad/s, and s_T and s_C the cut-off frequencies of the temperature and conductivity sensors:

$$H = \frac{\left(1 + i \, \omega/_{s_T}\right)}{\left(1 + i \, \omega/_{s_C}\right)} \qquad [2.70]$$

The slope of the phase of this function is representative of the delay that exists between the conductivity and temperature sensors. Between 0.2 and 3.0 Hz, the calculated lateness is 0.077 s. As the sample frequency adopted is 64 Hz and the pumping output is 30 $cm^3 \ s^{-1}$. The lateness corresponds to five "water samples" and is representative of the physical distance between the temperature sensor and the middle of the conductivity cell. In order to correct this lateness, they established a numeric filter equation from formula [2.70]:

$$T'(n) = 0.9350 \ T'(n-1) + 0.5872 \ T(n) - 0.5222 \ T(n-1) \qquad [2.71]$$

where $T(n)$ represents the measured temperature of sample n. This filtering only gives correct results if $T'(n)$ is slowed down by five samples, in other words, the following formula should be taken into account when calculating salinity:

$$T''(n) = T'(n-5) \qquad [2.72]$$

This filtering corrects disparities in the short term; these disparities are introduced by the speed of water renewal in the cell ("flushing") and the thickness of boundary layers.

In order to correct the effects of thermal inertia linked to heat stored in cell walls, Lueck and Picklo [LUE 90] determined another formula for filtering. As the effects can last several tenths of a second, they cause disparities in the long term between the temperature measured and conductivity values. This filtration formula was established from the impulse response of the conductivity cell. It allows calculation of the conductivity value $C_T(n)$ corrected from temperature effects:

$$C_T(n) = -b \ C_T(n-1) + \gamma a[T''(n) - T''(n-1)] \qquad [2.73]$$

In formula [2.73], γ represents the sensitivity of conductivity to variations in temperature ($\gamma = \partial C/\partial T_{S,P} \approx 0.1 \ S \ m^{-1} \ °C^{-1}$), $a = 4f_n\alpha\beta^1(1 + 4f_n\beta^1)^{-1}$, $b = 1 - 2a\alpha^{-1}$ and $T''(n)$ is the temperature corrected by formula [2.72]. Here f_n is the Nyquist frequency or the frequency from which sampling leads the aliasing of the spectrum. $f_n = 12$ Hz for an SBE 911, as its sampling frequency is fixed at 24 Hz.

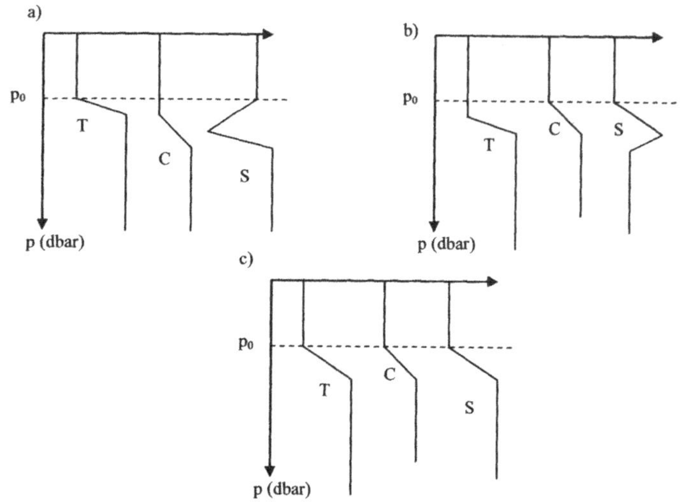

Figure 2.29. *Effects of the misalignment in the response time of temperature (T) and conductivity (C) sensors to the pressure p_0 on salinity S. a) C is late in relation to T. S shows a peak in salinity levels that is too low. b) T is late in relation to C. S shows a peak in salinity levels that is too high. c) T and C are well aligned. S follows variations in the salinity gradient*

Sea Bird Electronics, Inc. recommends using the values $\alpha = 0.03$ and $\beta = 0.14$, but earlier studies show that other values of α and β allow improvements in corrections to thermal inertia. The most recent study [MEN 09] showed from profiles made in zones with strong thermal gradients, that the values $\alpha = 0.0132$ and $\beta = 0.0829$ allowed reduction of maximal errors in salinity due to temperature gradients, from 0.05 to 0.017. A study of the same type was undertaken in 2007 by Johnson *et al.* [JOH 07] to analyze SBE 41 heads being used to equip drifting floats (see Figure 2.28).

$C_T(n)$ is a correction to be applied to the $C_M(n)$ value of sample n, from which the corrected conductivity value is obtained using:

$$C_c(n) = C_M(n) + C_T(n) \qquad [2.74]$$

Giles and McDougall were interested in corrections that could be made to measures taken by the MK III C sensors that equipped CTD profilers by General Oceanic (Neil Brown Sensor) [GIL 86]. These profilers are fitted with a platinum-based temperature sensor. As the response time of this sensor is relatively large (175 ms on average), the profilers are equipped with an unprotected thermistor that is placed at the entrance of the conductivity sensor. As the sensors are not inserted in a water pumping circuit, their response time can vary in relation to the speed of the profiler movement. Even if the boat is equipped with a winch with a regulated unwinding speed, the speed of movement of the profiler can vary considerably due to the rolling movements of the boat. When this speed is low, the response time of the conductivity sensor is much less important than that of the thermistor, while when this speed is higher (> 0.6 m/s), the conductivity sensor responds faster than the thermistor.

Giles and McDougall [GIL 86] have developed a recursive filter algorithm that takes into account the response time of each sensor in relation to the speed of movement. Conductivity values are delayed down or delayed up in relation to temperature values, following this speed. The filtering equation that they developed is as follows:

$$C_c(t) = e^{\left(\frac{-\Delta t}{\tau}\right)} C_c(t-\Delta t) + \frac{\left[1-e^{-\frac{\Delta t}{\tau}}\right]}{\left[1-e^{-\frac{\Delta t}{\tau_c}}\right]} \left\{ C_m(t) - e^{\left(\frac{-\Delta t}{\tau}\right)} C_m(t-\Delta t) \right\} \quad [2.75]$$

where:

- $C_c(t)$ represents the corrected value of conductivity;
- $C_m(t)$ represents the measured value of conductivity;
- τ the response time of the temperature sensor;
- τ_c the response time of the conductivity sensor; and
- Δt is a multiple of the sampling frequency of the profiler.

The thermistor used having a low stability over time, another formula was developed to recalibrate the temperature measured by this sensor with temperature measured by the platinum probe. Once again, the response time of the platinum probe τ_p depends on the speed of movement ($110 < \tau_p < 300$ ms) and the recursive filter adopted takes into account the variations in response times.

2.2.5. *Biofouling and protection of instruments*

Biofouling is the result of the development of microscopic life forms, such as microalgae or bacteria, at the surface of all materials immersed in marine environments. This development follows the adsorption of organic and inorganic macromolecules (proteins, lipids, salt and silicon) by the material. This is followed by a phase called immobilization, where bacteria attach themselves, and then a consolidation phase where exopolymers are produced. The next phase is that of colonization, where the biofilm is formed. As a result, the pollution of surfaces causes the adhesion of multiple and diverse organisms, debris or sediment, before finally causing the attachment of larger organisms such as shells and barnacles. Such developments are more frequent and occur in a shorter period of time in coastal areas and on surface instruments, as microorganisms need nutrients, oxygen and light to develop. These microorganisms also develop faster in warm as opposed to cold water. The surface condition of materials used and the presence of edges or cavities are also of importance. The rougher the surface, the more likely biofouling is to occur. For example, platinum electrodes are covered in an oxide, called black platinum, which is used to increase the active surface. As this substance is particularly rough, it does not favor protection against biofouling.

There are several different methods of protection – active or passive, volumetric and surface protection – but each one has a limited action. The most well-known is antifouling paint. Antifouling paint is a passive volumetric solution. The paint is a coating that contains several components from a base of copper oxide or chemical biocides (pesticides, algicides and bactericides) that can be applied to certain immersed surfaces. Generally, it is not possible to apply the paint to the active surfaces of sensors without disrupting their function, but such paint can be used to cover inductive conductivity sensors. The fact that sensors can be protected in this way is a great advantage, but the degradation of this coating over time leads to a progressive or sudden drift in values measured due to modification of the cell constant of the sensor, and this is followed by the inevitable adhesion of microorganisms as the coating loses its effectiveness. Furthermore, the supports of external field sensors must be taken into account as protection of the sensor alone is not sufficient. The supports can be protected from biofouling by using paint made from fluorinated polymers or silicon.

To protect volumes of work, pellets are used that contain a substance called tributyltin (TBT). This is an active protection that is effective when water can be circulated so that it comes into contact with the pellets and then with the sensor. TBT also has a short lifespan, and its use is not recommended by environmental protection agencies. This has resulted in a ban in the use of TBT to protect against biofouling since 2003. The ions produced can also slightly modify the conductivity values measured.

TBT can occasionally be replaced by copper elements that are advantageous, as this component has biocide properties due to the release of Cu^{2+} ions that act with enzymes in cell membranes, preventing division. External sensor parts that are completely covered in copper to protect them from fouling are also available. Rolls of scotch tape covered in copper, which allows protection rings to be fitted around equipment, can also be found on the market. These systems are proven to protect optic sensors, but carry a risk of disruption to measurements made by conductivity sensors. Glass surfaces can be protected with deposits of SnO_2, a conducting material that produces chloride by electrolysis.

As there is no single technique that is perfectively effective in the fight against biofouling, other, more active and original methods have been developed, such as chlorination by electrolysis or the mechanical and automatic cleaning of sensors. The electrolysis of sea water produces the release of oxygen in the form of O_2, but also hypochloric acid (HC1O) and hypochlorite (ClO^-):

$$2H_2O \rightarrow O_2 + 4H^+ + 4e^-$$

$$Cl^- + H_2O \rightarrow HClO + H^+ + 2e^-$$

$$Cl^- + H_2O \rightarrow ClO^- + 2H^+ + 2e^-$$

The third reaction is dominant when the potential applied is greater than 1 V (measured by an electrode saturated in Calomel). The release of H^+ ions decreases the pH of the sea water, and as the pH drops below 7.4 the second reaction dominates. The HC1O released combines with the bromide ions present in the water to give hypobromite (HBrO) which, in combination with HC1O, is a powerful biocide.

This technique is used on titanium or titanium iridium electrodes that are fixed to the protected surface. An automatism enables periodic application of the necessary potential. In a very aggressive medium, typical cleaning consists of chlorination for 10 min, three times an hour at the rate of 1 mg/l. Once again, its utilization is restricted and even excluded for conductivity sensors.

Different active methods of mechanical cleaning of surfaces have also been analyzed.

The most simple of these consists of the periodic activation of a piston whose extremity is equipped with a silicon ring. The piston penetrates the conductivity cell and removes the deposits that are produced, in the same way as a windscreen wiper, which prevents the development of a biofilm. This technique seems to be effective, but it struggles to remove the components of the biofilm that are tightly bound, and it can only be used in inductive conductivity cells as they present the risk of electrode deterioration when covered in black platinum, for example.

The utilization of ultrasound has also been tested. This method prevents the development of biofilm in the same way as a piston, and can manage to destroy such a biofilm after formation.

Another possibility is the retraction of sensitive elements when it is not necessary to continuously carry out measurements. The elements are therefore protected by a hood that opens automatically before taking measurements.

The drawback of these methods is that mechanical methods generally consume electricity, something that limits the use on autonomous applications that function for long periods of time. For this reason, some of the research into protection against biofouling is currently focusing on nanotechnology, but no solution of this type has been presented as yet.

2.3. Determining pressure

As Chapter 1 indicates, the measurements of pressure directly intervene in the calculation of numerous quantities that allow characterization of the ocean, but these measures can also be used to determine the depth or height of water. As the ocean has a maximum depth of approximately 6,000 m (excluding deep sea trenches), the corresponding pressure is 600 bar (to the nearest 1–2%). In order to respond to the specifications set out by the WOCE, pressure measurements must be taken to the nearest 3 dbar by the most precise instruments available (CTD profilers in this case). This inaccuracy of 3 dbar must be obtained so that the temperature of the sensor can vary from –2°C to +35°C. Numerous pressure sensor technologies are available on the market. Only a small number fit these criteria. This section presents two of the sensors that are often encountered in oceanography.

2.3.1. *Piezoresistive pressure sensors*

Taking into account the formula of pressure definition [1.27], the measure of this quantity comes down to a force F that applies itself to a surface wall S. This wall is the load cell that deforms or moves itself under the action of force F. These deformations are translated into electrical signals that we connect to pressure values by calibration. For piezoresistive sensors, the transformation of electrical signals is carried out with help of strain gauges. When the measuring range is low (<10 bar) the body can be a flush-mounted membrane (see Figure 2.30). For larger measuring ranges, this can be a recessed tube, in which the diameter varies very slightly according to pressure.

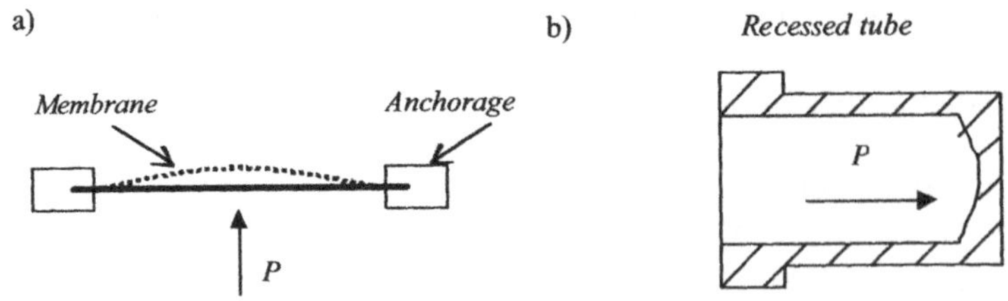

Figure 2.30. *Examples of load cells of pressure sensors: a) an anchored membrane deformed under the effects of pressure P; and b) a recessed tube*

Strain gauges are sensors that translate their own deformations under variations in resistance, which must be that of the structure to which they are attached. They can be formed from a metallic conducting wire or a semiconducting material (see Figure 2.31). They are fixed to an isolative support that can be glued to the structure whose deformation we want to follow.

As the deformations measured in pressure sensors are small, semiconductor gauges are often used to take measurements as they are more sensitive than metallic gauges. Their linearity is not as good, however, and their thermal sensitivity higher, which requires us to compensate for the thermal effects in the sensor.

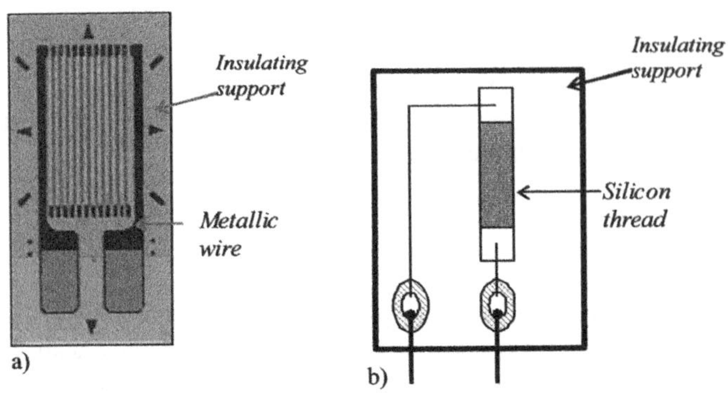

Figure 2.31. *a) Metallic wire resistive strain gauge structure (© VISHAY). b) Semiconductor strain gauge formed from a thread of silicon monocrystal*

The resistance without load of a wire of length L, section s and resistivity ρ is shown by:

$$R_0 = \rho L/s \tag{2.76}$$

Under the influence of strain, the resistance of the wire varies in value ΔR and the following proportional relation can be demonstrated:

$$\frac{\Delta R}{R_0} = k\frac{\Delta L}{L} \qquad [2.77]$$

where k is the gauge factor. Its value is close to two for metallic gauges, while it can be around 100–200 for semiconductor gauges.

Different techniques enable us to compensate gauges for changes in temperature during their making. This compensation is not always possible or sufficient, and it therefore becomes necessary to use electronic conditioning devices that are able to compensate for changes in temperature. In Wheatstone bridge fixtures, (see Figure 2.9), this compensation can be achieved by mounting a resistance or a passive gauge in opposition to the sensor whose value varies in the same proportions as the active gauge, in relation to temperature. For cases where high accuracy is required, the compensation can be made digitally, by directly measuring the temperature of the pressure sensor to correct the gauge factor. When using this mode of function, temperature calibration of the set is necessary.

This sensor technology is used in certain CTD profilers, for example the SBE 25 by Sea Bird Electronics, Inc. that is fitted with the SBE 29. It is a pressure gauge sensor, whose accuracy reaches ±0.1% full scale (f.s). The error in temperature is compensated for by a thermistor, which has a stability of ±0.004% f.s per month. It is sold in different pressure ranges that go from 20 to 7,000 dbar. FSI also fits out its CTD profilers with silicon pressure gauge sensors that are micromachined and have temperature compensation. This technology can reach accuracies of ±0.01% f.s for a stability of ±0.002% f.s per month

2.3.2. *Piezoelectric pressure sensors*

In the case of piezoelectric sensors, the body is a blade or cylinder cut from quartz crystal, PZT ($PbZrTiO_3$) ceramic or a PVF_2 ($((CH_2CF_2)_n)$) polymer. Crystal quartz is different from other materials because of its high stability. It consists of silica (SiO_2) in crystal form. It has a hexagonal section with prismatic extremities. It is an anisotropic structure, as it has three electrical axes that pass through the middle of the sides, and an optical axis that passes through the top of the prisms.

These crystals are known as piezoelectric, because if they are perpendicularly cut at the axis said to be "electrical" and subjected to mechanical strain they develop dielectric polarization. If a blade is made from the same crystal that is properly cut and polished and has two faces covered with metallic frames, electrical charges can be collected whose quantity Q is proportional to the force F exerted on the glass:

$$Q = K \times F \qquad [2.78]$$

where K is a piezoelectric constant with a value of 2.32×10^{-12} C N^{-1} when a quartz blade is perpendicularly cut at its "electrical" axis. To produce a sensor is just to amplify the electrical signal generated by compression. This amplification can only be carried out with particular assemblies, known as charge amplifiers, which cancels out the capacitive parasistic effects caused by charges generators. Inversely, if a potential difference is applied between electrodes, the crystal deforms and generates a movement force. This force can be used to directly generate micrometric movement (in the case of ceramics used on optic test benches to pilot the positioning of certain elements) or acoustic waves (microphones and hydrophones, see section 2.4.1) when potential is applied to a certain frequency.

In order to improve the metrological characteristics of the sensors, certain manufacturers have made oscillators from them. By inserting quartz crystal electrodes into an oscillating electrical circuit, it is possible to move the frequency of the circuit if the blade is subject to compression. If the sides of the blade are perpendicular to an electrical crystal axis, its cut is said to be at the Curie point, and it has two frequencies of resonance, f_1 and f_2: one that depends on its thickness e ($f_1 = 2.86\times10^6 / e$); and the other on its size l ($f_2 = 2.86\times10^6 / l$).

This is the process used by Paroscientific in its Digiquartz sensor, which is installed on the Sea Bird SBE 9 CTD profiler (see section 2.1.1.1). Pressure is transmitted via bellows to a rigid rod connection that may move on a pivot. Its movement sparks the compression of a quartz blade. The rigid rod is balanced by two mobile weights that compensate for the effects of forces generated by shocks and vibrations. The set is placed in vacuum. As the crystal can be subject to strain linked to thermal dilation, it is necessary to compensate this sensor for the temperature effects in order to improve accuracy. In this way, a temperature sensor is integrated into the fixture in order to correct the values of pressure measured.

The ratio, according to Paros [PAR 72], between frequency f generated by the sensor and pressure p exerted can be described by:

$$p = A\left(1-\frac{f}{f_0}\right) + B\left(1-\frac{f}{f_0}\right)^2 \qquad [2.79]$$

where f_0 is the frequency generated at zero pressure, and A and B are two constants. In fact, f_0, A and B depend on temperature. In order to compensate for the temperature effects, the manufacturer modeled f_0, A and B as polynomials. The period t of signal is what is measured. In order to implement the compensation, equation [2.79] is rewritten as follows:

$$p = C(T)\left(1-\left(\frac{t_0(T)}{t}\right)^2\right)\left(1-D(T)\left(\frac{t_0(T)}{t}\right)^2\right) \quad [2.80]$$

where $C(T)$ and $D(T)$ are two new constants and $t_0(T)$ is the signal period at zero pressure. Their dependence on temperature T is expressed by polynomials [2.81], in which constant values C_1, C_2, C_3, D_1, D_2, t_1, t_2, t_3, t_4 and t_5 are provided in the calibration file of the sensor.

$$\begin{aligned} C(T) &= C_1 + C_2T + C_3T \\ D(T) &= D_1 + D_2T \\ t_0(T) &= t_1 + t_2T + t_3T^2 + t_4T^3 + t_5T^4 \end{aligned} \quad [2.81]$$

This fixture, integrated into the digital electronics of the Sea Bird SBE 9 CTD profiler, is able to reach an accuracy of 0.015% f.s per month, and its maximum drift is 0.0015% f.s per month.

2.3.3. *Errors in pressure sensor measurements*

The sensor technologies described earlier allow the accuracy required by oceanographic measurements to be achieved. When dynamic measurements are taken, however, errors can appear because of:

– *Problems with hysteresis*. As the instrument rapidly increases in pressure, then returns to the surface, the load cell of the sensor does not instantly revert to its initial form. For a depth reference (mid-range) the value indicated by the sensor on descent is not the same as the value indicated on return to the surface. For a Digiquartz sensor, for example, this persistence can reach 0.16 dbar for an increase in 6,800 dbar. This phenomenon is visible during calibration with a dead-weight tester but it is rarely taken into account, and calibration coefficients are usually, an average between the values measured on the increase and decrease of pressure.

– *Problems in temperature lags*. Sensors used being compensated in temperature, this compensation can happen with lags or advances on the true temperature of the sensor as the instrument crosses zones of high thermal gradients, which then leads to errors in the measured and corrected pressure. These errors are proportional to the thermal sensitivity of the sensor and can be of high amplitude (several dbar) if they are not taken into account when the instrument is produced. The method used by Sea Bird, Inc. in order to reduce these errors on SBE 9^{+} probes involves the thermal isolation of the Digiquartz sensor so that the temperature compensation is independant of sudden variations in the temperature of the medium. They therefore guarantee errors to below 0.5 dbar for a thermal shock of 18°C.

2.4. Determining velocity

Velocimeters can be classified as oceanographic instruments, as in order to precisely determine the speed of sound in water it is necessary to calculate the temperature and salinity of the medium, or at least to reference the instrument that measures velocity from a velocity reference determined in this way, as indicated in Chapter 1. However, velocimeters are more often used in hydrography to correct the processing of echo sounders determining the speed of sound propagation, and to improve the accuracy of depth measurements.

2.4.1. *Principles of measurement*

There are two methods that can be used to calculate speed of sound propagation in water:

– using equipment to measure temperature and conductivity as described in sections 2.1.1 and 2.2.1, which are: losable probes (XBT, XCTD), CTD profilers, thermosalinometers, etc., and entering the values of temperature, salinity and pressure obtained in the equations of the speed of sound given in section 1.2.4; and

– a direct measurement of the propagation time of acoustic waves between an emitter and a receptor in which the approximate distance that separates them is known.

Instruments that use the second solution are known as velocimeters. Numerous processes are available that can enable the measurement of sound propagation speed in liquids, or at least variations in propagation speed. Generally the only instruments that enable the measurement of absolute speeds with sufficient accuracy are aboratory instruments. Velocity measurements are carried *in situ* out with the help of techniques that have high sensitivity but need to be calibrated to be exact. Their general principle is described in Figure 2.32. It is based on the implementation of a hydrophone.

A *hydrophone* is the marine equivalent to a microphone. Sound is a wave of mechanical vibrations. The effects of the wave cause particles from the medium to go back-and-forth with low amplitude around a middle position. That is how, step-by-step, the wave propagates. Its propagation speed or velocity depends on the density of the medium [1.1].

Like a microphone, a hydrophone is a reciprocal or reverse device as it can be used to generate or capture the waveforms of vibration. However, the number of acoustic transduction techniques that function in water is lower than the number that function in air. This limitation is in part caused by static pressure exerted by the medium, which excludes the usage of certain principles, and in part by the frequency domains used. Piezoelectric technology is one of the only principles that can overcome these constraints (section 2.3.2). Despite this, in order to reduce effects of pressure, the transducer is often placed on the membrane of an oil-filled container so that equal pressure in both parts of the element can be retained.

When quartz is used to manufacture piezoelectric pressure sensors, barium titanate, zirconate titanate or PVF2 type polymers are often chosen to make hydrophones. They enable the production of transducers in various forms and dimensions. Furthermore, they have very weak electrical impedance compared to that of quartz, which allows them to be powered by lower voltages for the generation of an equivalent acoustic power.

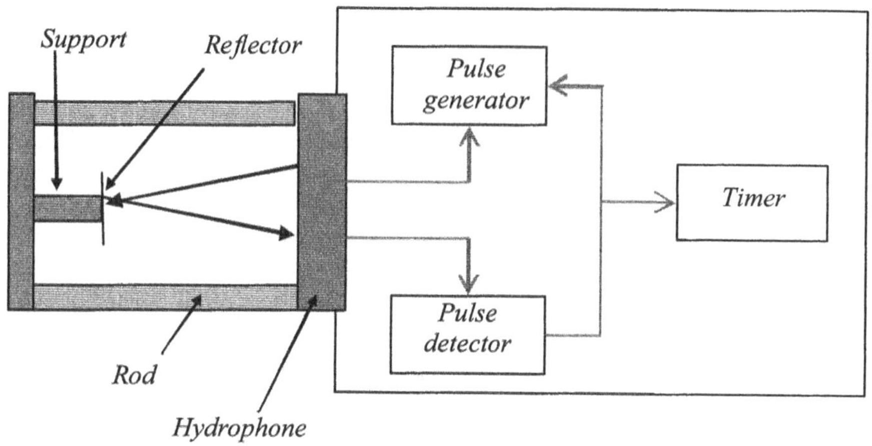

Figure 2.32 *Operating principle of an acoustic velocimeter*

In order to produce a velocimeter, a hydrophone is used as an emitter–receptor. It emits an impulse or sequence of impulses that reflect onto a reflector placed at a fixed distance. The reflected impulses are "listened to" by the hydrophone, as an electronic system enables the transformation to receptor. The measurement globally consists of counting the time *t* that is taken for the wave to go to and return from the reflector. The distance covered by the wave is known as the acoustic path c_a. The speed of sound c is the ratio between the two:

$$c = c_a / t \tag{2.82}$$

In Figure 2.32, the distance that separates the hydrophone from the reflector must be constant, regardless of the temperature and pressure of the medium, if a reliable measurement is desired. In order to ensure this, it is imperative that:

– materials that are not sensitive to dilation are used to produce the rods and reflector support, and that dilation of the reflector support compensates for the rods (their direction of dilation being opposite); or

– this dilation be compensated for by taking temperature and pressure measurements.

The material most often used in the production of the rods is Invar 36, as it has a low thermal dilation coefficient, α. The mirror support can therefore be produced in titanium as, for example, its dilation coefficient is six times higher than that of Invar. Invar is a metallic alloy (section 3.2.2.4) containing 36% nickel. Its name originates from the word "invariable", in reference to its weak coefficient value, α. By choosing an adequate relation between the rod and support lengths, it is possible to compensate for thermal effects.

Another solution has recently been developed. It consists of using a composite material made from carbon fibers (70%) and epoxy resin. The coefficient of thermal dilatation of this material is very low ($\alpha \approx 0.1\times10^{-6}$ °C^{-1}) on the oceanographic temperature scale. It is therefore used to produce rods that directly link the body of the velocimeter to the reflector. The rods are attached to the metallic parts (body and reflector) through a gluing technique.

The distance c_a is often near to 10 cm. In order to obtain accuracy to the nearest 0.1 m/s or lower on c, by deriving equation [2.82] we can show that the correction or compensation of dilatation dl to the nearest $0.1 \times 0.1/1{,}500 = 6.7$ µm must be carried out. Here, α is defined by relation [2.83], where l is the distance between the hydrophone and the reflector. It is often expressed in the form of polynomial relation [2.84], however, which can be used to calculate and compensate for the variation in distance dl produced by the variation in temperature dT, if we choose a solution consisting of temperature measurements.

$$\alpha = \frac{1}{l}\frac{dl}{dT} \tag{2.83}$$

$$\alpha = a + bT + cT^2 \tag{2.84}$$

In the same way, in order to obtain a reliable instrument, the measurement of time must be very stable. Apart from stabilized oscillators, different techniques can be used to generate and count acoustic pulses. If the pulse emitted is made up of a carrying sinusoid of pure frequency f_0, wavelength λ and duration $\pm L$ to the origin point x and at time $t = 0$, it can be written that:

$$u(x,0) = \begin{cases} E\cos\left(k_0 x + \theta\right), & \text{for} \quad -L \leq x \leq +L \\ 0, & \quad\quad\quad |x| > L \end{cases} \qquad [2.85]$$

with:

$$k_0 = 2\pi/\lambda_0 = 2\,\pi f_0/c_p \qquad [2.86]$$

where:

- k_0 is known as wave number;
- λ_0 is the length of the wave emitted;
- c_p is the speed of propagation of the sinusoid phase; and
- θ is an initial phase shift.

It can be shown that when this impulse crosses a dispersive medium, it is subject to a phase shift $\omega(k_0)$xt around the frequency f_0. If L can be sufficiently high in front of λ_0, at time t we have:

$$u(x,t) = \begin{cases} E\cos\left(k_0 x - \omega\left(k_0\right)t + \theta\right), & \text{for} \quad -L \leq x \leq +L \\ 0, & \quad\quad\quad |x| > L \end{cases} \qquad [2.87]$$

Equation [2.87] enables us to state that for a continuous monochromatic source, while crossing the medium, each point of the resulting wave propagates to a phase speed independent of x and t, even when the medium is dispersive. To put it another way, the phase speed of the wave only depends on the dispersive properties of the medium:

$$c_p = \frac{\omega\left(k_0\right)}{k_0} \qquad [2.88]$$

By differentiating relation [2.88], it is also possible to show that around wave number k_0, the group speed of the impulse is equal to phase speed c_p. Consequently, by measuring the propagation time of a single impulse, measuring the time of flight or by measuring phase shift, the speed of sound propagation in water can be determined.

If we look at phase ϕ, it is possible to calculate the difference that exists in the acoustic path travelled $\Delta x = x_2 - x_1$ between two coordinated points x_1 and x_2, by using equation [2.87]. If we know that the phase shift is equal to 2π, if $\Delta x = \lambda_0$, for the quantity $x_2 - x_1$ we have:

$$\varphi = 2\pi \frac{\Delta x}{\lambda} \qquad [2.89]$$

The difference in total phase that will be measured by the velocimeter takes into account a certain amount of other phase shifts that are introduced by the reflector, the detection assembly, the hydrophone, etc. These phase shifts can be regrouped under the supplementary term $\gamma(f)$. If the base of the hydrophone is taken as a reference, the difference in path travelled by the impulse will be $\Delta x = 2l$. Therefore the entire phase shift has a value of:

$$\Phi = 4\pi \frac{l}{\lambda} + \gamma(f) \qquad [2.90]$$

If the wave frequency is constant, the term $\gamma(f)$ is constant, constituting a simple shift. Relation [2.90] combined with relation [2.86], enables us to determine the value of velocity to a nearest constant, which can be quantified by calibration:

$$c_p = \frac{4\pi l f_0}{\Phi - \gamma(f)} \qquad [2.91]$$

If it is possible to change the distance l of quantity Δl, relation [2.90] can be used to obtain absolute measurements of velocity, as we then have $\Delta\Phi = \Delta\phi$. By measuring the slope $m = \lambda/4\pi$ of the straight line and expressing the value of Δl [2.92] from equation [2.89], the value of c_p can be obtained as shown in equation [2.93].

$$\Delta l = \frac{\lambda}{4\pi} \Delta\Phi \qquad [2.92]$$

$$c_p = 4\,\pi\, m\, f_0 \qquad [2.93]$$

Numerous velocimeters function according to slightly different principles that were developed in the early 1960s and known as the "sing-around" principle. They emit a short duration impulse that corresponds to a frequency f. A monitoring window is then positioned to detect the echo of the impulse. As soon as the echo is detected, another frequency f_i is emitted and then detected in turn. The cycle is repeated n times. The sequences of n impulses form an oscillator, whose frequency is proportional to the velocity c being measured. Relation [2.82] shows that:

$$c = \frac{A}{\left(\frac{1}{nf_i}\right) - B} \qquad [2.94]$$

where A is a constant representative of the acoustic path c_a, and B represents the delay in time taken into account by the electronics. A calibration using relations enabling the calculation of reference velocities, c_{ref} (see section 1.2.4), is necessary to precisely determine the value of coefficients A and B. They can be estimated by least-squares regression established from:

$$\left[\frac{1}{(nf_i)}\right] = \left(\frac{A}{c_{ref}}\right) + B \qquad [2.95]$$

Equation [2.94] enables us to take into account the effects of temperature on the length of the acoustic path, if necessary. Coefficient A can be replaced by product $l\alpha$ where α is obtained from relation [2.84] and coefficient B is also replaced by a polynomial relation of the same type. A coefficient k_p, of sensitivity to the pressure effects, can also be added. Equation [2.94] therefore becomes:

$$c = \frac{l(a + bT + cT^2)}{\left(\frac{1}{nf_i}\right) - \left(d + eT + fT^2\right)} + k_p p \qquad [2.96]$$

If there are a sufficient number of calibration points, the coefficients of this equation can also be determined by using the least-square regression method from equation [2.95], where A and B are replaced by their respective polynomials. As this calibration is carried out at practically zero pressure, the term in $k_p p$ can be ignored. Coefficient k_p is determined by measurements in a high-pressure chamber.

It is difficult to obtain measurement uncertainties below 0.25 m/s with this technique, however, as the sending of impulses results in an interference field produced by multiple reflections of the incident wave. These reflections reduce the quality of detection. Furthermore, the interference field that is created is difficult to compensate for as it varies with the density value of the medium.

Analog techniques enable detection of the arrival of impulses by the thresholding and triggering of a timer. Unfortunately, the stability of the detection threshold of impulses is limited by noise superimposed on the signal.

As a single impulse is sent to take the measurement, we have a "measurement of the time of flight". In the case of analog fixtures, this time of flight can be measured by the load of a condenser that is triggered at the moment when the pulse is emitted. This condenser is discharged at the same time as detection of the reflected pulse. The measurement uncertainty of this technique is again limited by the stability of the detection threshold and the signal-to-noise ratio. Furthermore, the response from this type of velocimeter heavily depends on its temperature and is nonlinear.

To decrease these levels of uncertainty, most recent velocimenters use digital processing of the signal that involves calculation of the function of intercorrrelation of impulses emitted with impulses detected. This operation rest's on the programming of the following typical relation:

$$\Gamma_{xy}(\tau) = \sum_{n=-\infty}^{+\infty} \Gamma_n \delta(\tau - nh) \quad [2.97]$$

with:

$$\Gamma_n = E[x_k y_{k-n}] = E[x(t)y(t-nh)] \quad [2.98]$$

where τ represents the delay between signals $x(t)$ and $y(t)$, and δ is the Dirac sampling function. Relation [2.98] shows that the correlation function sampled can be obtained using an analog process to correlate signals $x(t)$ and $y(t)$, or by calculating the correlation function of discrete signals x_k and y_k, in which case to obtain good resolution on the velocity measurements the signal must be sampled and analyzed with great speed. As the signal $y(t)$ is lagged by a quantity nh due to the path travelled by the impulse that is detected, function Γ_{xy} will present a peak at time τ corresponding to the time taken by the impulse to go and return.

Calculation of function Γ_{xy} also enables a considerable increase in the detection of the signal-to-noise ratio. This enabled Valeport Limited to achieve a noise level below ±0.002 m/s (rms) on its SVS velocimeters and a temporal resolution on the correlation function to the nearest 10^{-11} s, which corresponds to a velocity of 0.5 mm/s. These performances enable a decrease in the level of uncertainty of the measurement below that of the principal relations of velocity calculation (that is to say, around ±0.01 m/s to ±0.03 m/s).

2.4.2. *Instruments used at sea*

Instruments used at sea to evaluate velocity can be classified into three categories:

– losable probes;

– hull velocimeters; and

– sound velocity profilers.

Some losable probes are made to measure velocity profiles in the same way that losable probes are made to measure temperature and conductivity profiles (see Figure 2.4). They are manufactured by Sippican, and called XSV, which is short for *eXpendable Sound Velocimeter*. They use the "*sing-around*" technique mentioned in section 2.4.1. Pulses are emitted at a frequency of 6.5 MHz. They are reflected on a copper reflector placed 5.2 cm away. The frequency of the oscillator formed by the pulse sequences varies between 27 and 30 kHz. This signal is transmitted by a lead connected to a computer onboard the boat that counts frequency. Theses probes produce accuracies of ±0.25 m/s in the 1,405–1,560 m/s range. The resolution of the measurement is 60 cm vertically and 0.04 m/s regards velocity.

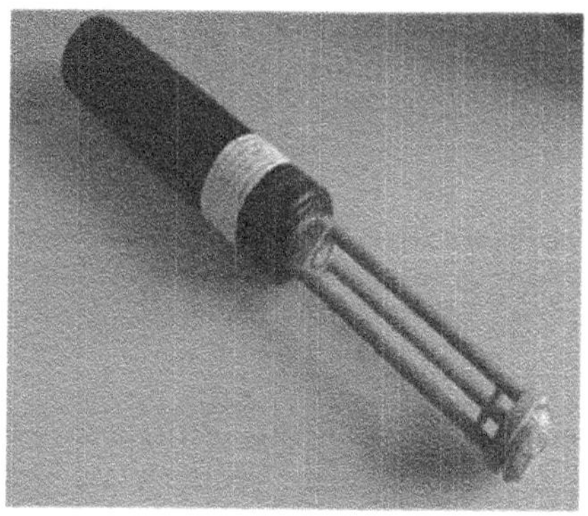

Figure 2.33. *Hull velocimeter SV Smart Sensor by Applied Microsystemd Ltd (Courtesy of © SHOM)*

Hull velocimeters are designed to be installed onto load-bearing structures, such as underwater vehicles, surface ships, etc. They enable punctual velocity measurements to be taken in order to enable beam forming in multibeam echo sounders, for example. They require an exterior power source and transmit data that have already been digitized. An example of this is the SV Smart Sensor by Applied Microsystems, Ltd, Canada (see Figure 2.33). The measurement technique of this instrument is the measure of the "time of flight" of 1 MHz pulses. It is based on direct application of relation [2.82]. In this sensor, the acoustic path does not require any corrections in temperature nor pressure measurements, as the metals used to produce the rods and reflector support are Invar 36 and stainless steel T-316, in which the ratio of dilation coefficients enable a stability of ±5.5 nm/°C of the acoustic path, which results in a negligible error of ±0.0014 m/s in velocity, in the temperature range –2°C to +32°C. The oscillator used is also stabilized in temperature in this range. When calibration is carried out in distilled water, and velocity references are calculated using the Del Grosso relation, the manufacturer forecasts a measurement uncertainty of less than 0.06 m/s (r.m.s) for a resolution of 0.015 m/s.

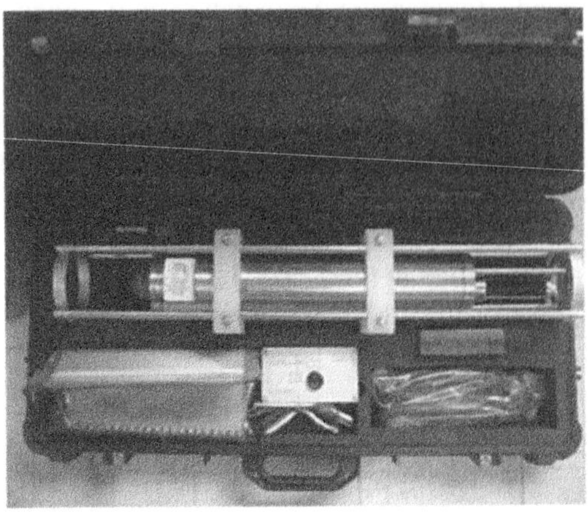

Figure 2.34. *The Valeport MIDAS SVP model sound velocity profiler in its transport case, with interface box used to recharge its batteries and extract memorized data (Courtesy of © SHOM)*

Operational autonomy is the principal difference between sound velocity profilers and hull velocimeters. Velocity profilers are powered by rechargeable batteries and have the ability to store measured data. They are either used in mooring or to measure velocity profiles. An example of this is the MIDAS SVP sound velocity profiler (see Figure 2.34), which is manufactured and commercialized by Valeport Limited. This instrument is entirely digitalized, and measurements are taken by intercorrelation. The measuring head is made from composite material rods with a *t* dilation coefficient close to 0.1×10^{-6} °C^{-1}. It is also equipped with a temperature sensor that can be calibrated to an uncertainty of ±0.01°C, and a piezoresistive pressure sensor, in which the precision shown is ±0.04% of the range of measurement. The stability of its internal clock (20 ppm) is able to guarantee a measurement accuracy of ±0.03 m/s (linearity errors excluded) in the 1450–1550 m/s range and a drift to the nearest ±0.0075 m/s in the first year (see section 2.6.2 for more information about the stability of oscillators).

2.5. Determining current

Circulation of oceanic water mass is due to several kinds of actions: surface wind, density variations at depth, gravitational forces of the moon and rotation of the earth that generate a Coriolis force, solar energy and gravity distributions. Study must therefore be carried out on multiple (from thousands of kilometers to at least 1 m) and temporary spatial scales (from one second to several decades).

Current is also a three- or four-dimensional vector quantity that is linked to two notions: the speed of movement and the direction of movement. Its connection to physical metrology is made by the determination of reference speeds and by the measure of the angular movements of fluid in relation to the equipment and of the equipment in relation to a reference direction given by the magnetic north.

Methods used to make speed and movement direction measurements have come a long way in recent years. Data satellites are now able to locate surface current fields of 100 km^2, but by an indirect method that is based on measurements of water heights, surface wind and temperature combined within physical models. Two general principles of measure are used *in situ*; Eulerian methods, where instruments are dependent of fixed moorings; and Langrangian methods, where instruments are surface or subsurface buoys that we follow (by satellite) the trajectory which is the water mass one's in which they are "anchored". In this section, only Eulerian measurements will be explored (see Chapter 3 for Langrangian methods).

Evolution of Eulerian methods was especially marked by the utilization of acoustics enabling a move from one level of measurements to vertical profile measurements of current speed. Now, certain types of current meters can be installed on ships and used on displacement, which enables a rapid receipt of field speeds. This section reviews several techniques used by Eulerian current meters in different commercial products.

2.5.1. *Rotor current meters*

The measurement principle of these instruments rests on wind speed sensors and the weather vanes currently used in meteorology. The instrument is fixed on a mooring, perpendicularily to the axis of water mass movement, in order to measure horizontal speeds.

Three types of rotor exist. They can be in the form of a propeller with a large pitch, fixed to the front of the body of the instrument containing an electronic timer and recorder. They can be in the form of a "Savonius" rotor, named after its Finnish inventor [SAV 20]. Savonius rotors are made from concave blades (see Figure 2.35) made from light plastic. They are welded to an axis that turns thanks to a pivot-bearing system that is designed to reduce friction. This fixture makes it insensitive to vertical stamping flow. Savonius rotors have a low starting inertia, which is useful when measuring low amplitude currents, but they have a higher constant slowdown speed, which tends to artificially increase the average speeds being measured when the speed of the medium fluctuates. In order to reduce this effect, it is sometimes preferable to use a right blade rotor or *panemone*. This type of rotor's starting threshold is close to that of the Savonius Rotor, and average speeds measured are, *a priori*, closer to average currents speeds.

A counting system, generally made from magnets fixed to the rotor, and a reed switch (ILS or relay reed) fixed in the sealed body of the instrument, enables determination of the number of turns made by the rotor during a certain period of time t. When the reed switch opens and closes, it generates pulses in which the number n is proportional to the rotation speed. In more recent instruments, the rotor's axis is linked to an optical decoder that directly digitizes information from the rotor, which can then be transformed into rotation frequency f.

The ratio n/t is proportional to current speed V. Between n/t and V, there is a linear relation in the 0.01–2.50 m/s range (range of marine currents). It is, however, limited for low speeds by a starting threshold B that is in relation to the friction of the rotation axis and inertia of the rotor. This threshold can provoke nonlinearity in recorded data, particularly when these data are taken in great depths, as the speeds measured are low. The value of B is often in the region of 2.5 cm/s but it can change over time in relation to accretions that can form on the rotor.

The speed and frequency of rotation are linked by:

$$V = Af + B \qquad [2.99]$$

where A is a constant that represents relation n/t, and is also inversely proportional to rotor sensitivity. It is expressed in cm s^{-1} Hz^{-1}.

A and B are determined by calibration of the current meter. This calibration can be carried out by dropping the instrument at a known speed into a vein of static water, or by making water circulate in a vein where the apparatus is in a fixed position. This calibration consists, in the first instance, in counting the number of turns N made by the rotor on a length of L of the basin (usually by optical means) and over a time T. We can therefore calculate the value of A:

A = L / NT.

The value of B corresponds to the zero ordinate of the calibration straight line. Calibration does not take into account the effect of vertical movements, which are supposed to be negligible. *In situ*, their influence can slightly modify the value of A and cause bias in the measure of V.

On rotor current meters, the direction of the current is determined by a fin that is similar to a wind vane. This fin is linked to a compass that enables us to reference its directions to the magnetic north. Internal software authorizes the averaging of the number of turns made by the rotor and associates this value with an average direction. The number of turns and direction are both calculated for a programmable time. This type of current meter is called a *Vector Averaging Current Meter* (VACM) as it provides the average value of East–West and North–South components that then enable the calculation of the vector $\vec{V}$ of average current. The sum of the North–South components is obtained by:

$$[V_1\cos(\alpha_1) + V_2\cos(\alpha_1) + ... + V_n\cos(\alpha_n)] = \sum_i V_i \cos(\alpha_i) \qquad [2.100]$$

and that of East–West by:

$$[V_1\sin(\alpha_1) + V_2\sin(\alpha_1) + ... + V_n\sin(\alpha_n)] = \sum_i V_i \sin(\alpha_i) \qquad [2.101]$$

In these relations, the V_i are obtained from equation [2.99], which takes the calibration of the current meter into account. The module and resulting mean phase of the vector are then calculated and used:

$$\left\|\vec{V}\right\| = \sqrt{\left[\sum_i V_i \cos(\alpha_i)\right]^2 + \left[\sum_i V_i \sin(\alpha_i)\right]^2} \qquad [2.102]$$

$$\phi_V = Arc\tan\left[\frac{\sum_i V_i \sin(\alpha_i)}{\sum_i V_i \cos(\alpha_i)}\right] \qquad [2.103]$$

Figure 2.35. *A Nortek Aquadopp acoustic current meter and a MORS MC 3X0 rotor (Courtesy of Oceano Technologies)*

RCM 7 or 8 by Aanderaa and the MC 3×0 by MORS (Oceano Technologies) are examples of rotor current meters. In order to complete measurements that allow characterization of the medium, these instruments are optionally equipped with temperature and pressure sensors and sometimes conductivity sensors. Their characteristics are given in Table 2.4.

Quantity	**Technology type**	**Range**	**Accuracy**	**Resolution**
Speed	Savonius rotor	2.5–250 cm/s	2% of speed or 2 cm/s	Threshold at 2.5 cm/s
Angle	Wind vane	0–358°	0.25% of the f.s.	0.09° Starting threshold at 4 cm/s
Direction	Fluxgate compass	±30°	± 2.8 °	0.09 °
Temperature	Copper probe	-5 to + 35°C	0.1°	0.25% of the f.s.
Pressure	Piezoresistive	0–100 bar or 0–600 bar	1% of the f.s.	0.25% of the f.s.
Conductivity	Electro-magnetic	0–70 mS/cm	1% of the f.s.	0.25% of the f.s.

Table 2.4. *Manufacturer specifications of a MORS MC 3×0 current meter (f.s. = full scale)*

2.5.2. *Doppler effect current meters*

Progress in underwater acoustics has enabled the production of current meters without moving mechanical pieces. These are less sensitive to the shocks and vibrations that they can be subjected to when installed on mooring lines. They are also less sensitive to the effects of chemical or biological marine deposits.

The ocean is an environment that is generally rich in suspended matter. This matter can be of an organic, phytoplankton or sedimentary origin. It is generally considered that these particles do not move themselves but are moved by the marine currents. As acoustic waves propagate, they behave like a set of diffusers, as much to say, on contact the wave reflects itself in multiple directions. However, a large part of its energy is resent in the direction of the emitter. If the emitter is briefly transformed into a receptor, the acoustic wave becomes an adequate means of following the movement of particles and therefore water mass. The estimated speed of this movement is measured by the Doppler effect, produced on the wave frequency. This effect can only be measured if:

– the medium is loaded with particles (if the water is too pure, no reflection will occur); and

– particle movement has a radial component, that is to say, in the axis emitter–wave receptor.

2.5.2.1. *Doppler effect and its measurement*

The time T taken for an acoustic wave to reflect onto a fixed target placed at a distance d is equal to the ratio:

$$T = d / c,$$

where c represents the propagation speed of the signal.

If the target is now animated by a radial speed v, the time T taken by the signal to find it will be equal to:

$$T' = d / (c + v).$$

The relation between T and T' allows us to write:

$$T' = \frac{c}{c+v} T = \frac{T}{\left(1 + \frac{v}{c}\right)} \qquad [2.104]$$

If the acoustic signal is sinusoidal, its period and wave length λ are proportional to *T*. We have in fact:

$$T = k\lambda / c \text{ and } T' = k\lambda' / c$$

where λ' is the wavelength being detected. This is the apparent wavelength. λ increases or decreases in value $2vT$ according to the direction in which the particles move. It is necessary to multiply the quantity vT by two as the wave leaves and returns. We therefore have:

$$\lambda' = \lambda \pm 2vT \qquad [2.105]$$

In fact, if the particles move towards the source, λ' becomes smaller. If the particles move away, it increases. If f' is the frequency of the reflected wave, f that of the emitted wave and c the speed of sound in water, the general relation $\lambda = c / f$ [2.86] combined with relation [2.105] gives:

$$\frac{c}{f'} = \frac{(c \pm 2v)}{f} \qquad [2.106]$$

As particle speed is so that $v^2 << c^2$, we can write:

$$f' = \left(1 \pm \frac{2v}{c}\right) f = (1 \pm \delta) f \qquad [2.107]$$

by asserting that $\delta = 2v/c$. Here, δ is the *Doppler shift*. If we are able to measure f' and know f perfectly, it is easy to extract the current speed v from this relation.

As f and f' are measured by the same oscillator, drift is not taken into account during measurements, which confers high accuracy. In current meters, the value c is often set to 1,500 m/s, which is a source of error. We can estimate that 1% of error in velocity will also cause an error of 1% in speed. To improve the accuracy of the value v, it is necessary to determine the corrected value of c, which depends on the temperature and salinity of the medium (a 1% error in velocity corresponds to a change in temperature of 5°C). In order to do this, acoustic current meters are sometimes equipped with temperature, pressure and even conductivity sensors as their rotor counterparts. The Medwin relation [1.62] is often used to correct this.

If the acoustic beam crosses layers of water with varying velocities, it is subjected to successive refractions. In using relation [2.107] and the Snell-Descartes law, it is possible to show that the Doppler frequency corresponding to the shift, δ, is not subjected to variation. The Doppler shift is independent of mediums crossed by the wave.

If we take the case of any signal $s(t)$ at a target level placed at a distance d from which it reflects itself, it becomes:

$$s_c(t) = s\left[t - \left(\frac{d \pm vt}{c}\right)\right] = s\left[\left(1 \pm \frac{\delta}{2}\right)t - \frac{d}{c}\right] \quad [2.108]$$

By asserting that $\tau = d/c$ and in assuming a negative Doppler shift, in order to simplify the writing of this equation, at the receptor it is shown that:

$$s_r(t, \delta) = s[(1 - \delta)t - \tau] \quad [2.109]$$

where τ corresponds to the delay of the signal, which is caused by the acoustic path that it followed.

There are several methods that can be used to detect and measure $s_r(t, \delta)$ but it can be shown that the optimal detection of $s_r(t, \delta)$ is made if its combined value $s_r^*(t, \delta)$ is multiplied by $s(t)$ and if the integral of the product is calculated in relation to time. Function $F(\tau, \delta)$ is thus obtained:

$$F(\tau, \delta) = \int_{-\infty}^{+\infty} s^*\left[(1-\delta)t - \tau\right]s(t)\,dt \quad [2.110]$$

$F(\tau, \delta)$ is known as the *ambiguity function.* In the case where $\delta = 0$, it corresponds to the autocorrelation function of the signal $s(t)$. If $\delta \neq 0$, it represents the correlation that exists between $s(t)$ and its duplicate shifted in time and frequency. The calculation of this function enables both the detection and measurement of the signal $s_r(t, \delta)$. If $s(t)$ is a sinusoidal function of pure frequency f_0, $s_r(t, \delta)$ can be writen in the complex form:

$$s_r(t, \delta) = ae^{-2i\pi f_0\left[(1-\delta)t - \tau\right]} \quad [2.111]$$

$$= as(t - \tau)e^{2i\pi f \delta t}$$

where a corresponds to the reduction in the amplitude of the reflected sinusoid. The decomposition of relation [2.111] cannot be used if the frequency shift δf_0 is weak in front of f_0, in other words, if the workings are carried out in narrowband. If signals amplitudes are normalized, $F(\tau, \delta)$ can be written:

$$F(\tau,\delta) = \int_{-\infty}^{+\infty} s(t)s^*(t-\tau)e^{-2i\pi\delta f_0 t}dt \quad [2.112]$$

It is useful to calculate the amplitude of this function at its origin. This information is obtained by calculating the square of the module of $F(\tau,\delta)$ or: $|F(\tau,\delta)|^2$. The value at the origin of the ambiguity function is representative of the signal's energy, and this value corresponds to the maximum of this function. Therefore, all that is required is detection of this maximum and deduction of its corresponding values (τ_0, δ_0) in order to measure these variables. The signal detected by the receptor is never pure. It is embedded in noise $b(t)$, caused by the multiple reflections that it is subjected to along the path it has travelled and by electronic detection. In reality, the receptor detects a signal $r(t)$ as:

$$r(t) = s_r(t,\delta) + b(t) \quad [2.113]$$

Therefore, it is necessary to filter $r(t)$ in order to cancel out noise $b(t)$. This operation can be carried out with an "adapted" filter. The use of this filter enables us to obtain a signal on output that is the ambiguity function. The noise remains a source of error in measurement, however, as the maximum detection can sometimes correspond to noise level, particularly when the energy of the signal $s_r(t,\delta)$ is weak.

2.5.2.2. *The principles of the Doppler current meter operation*

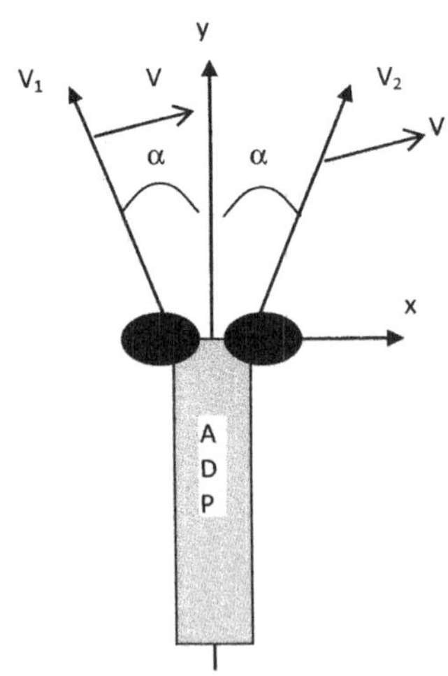

Figure 2.36. *Representation of a Doppler current meter in a reference frame xOy, with beam axes 1 and 2 indicating the direction of speeds measured, V_1 and V_2, in relation to the speed V of the current*

Doppler effect measurements are often made by acoustic pulses with a sinusoidal carrier. The Aquadopp current meter, shown in Figure 2.35, illustrates this principle. It is made from three large surface emitter–receptors oriented according to the perpendicular axis for both of the horizontal emitters and at 45° concerning the emitter providing the vertical speed. The set of horizontal emitters is called Janus, after the Roman god, who had two opposing faces. The large diameter of the emitters in relation to the wavelength of the signal emitted allows us to limit the apeture angle or the beams diffraction and to avoid their overlay, which would lead to crosstalk in the signal being analyzed.

The Doppler effect is a directive phenomenon. Currents perpendicular to the axis of propagation have no influence on the frequency shift that is obtained and the data provided by the three beams are independent from each other. They can therefore be combined in a system of orthogonal coordinates. To simplify the problem, if currents are simply referenced in relation to two directions, x and y, and the two separate beams are both at angle α in relation to the axis $0y$, which corresponds to the central axis of the current meter, the vector speed to be measured will be writen:

$$\vec{V} = \left[V_x, V_y\right] = V_x \vec{r_x} + V_y \vec{r_y} \qquad [2.114]$$

where $\vec{r_x}$ and $\vec{r_y}$ are the unit vectors of axes x and y (see Figure 2.36). Speeds measured according to the beam axes 1 and 2 will be:

$$\begin{aligned} V_1 &= V_x \sin(\alpha) + V_y \cos(\alpha) \\ V_2 &= -V_x \sin(\alpha) + V_y \cos(\alpha) \end{aligned} \qquad [2.115]$$

From relation [2.114] and from the measurement of V_1 and V_2, it is possible to extract the values of V_x and V_y:

$$V_x = \frac{V_1 - V_2}{2\sin(\alpha)} \quad \text{and} \quad V_y = \frac{V_1 + V_2}{2\cos(\alpha)} \qquad [2.116]$$

This two-dimensional system is applied to current meters used to determine horizontal currents in canals or rivers. At sea, it is necessary to take measurements in three dimensions. For a three-beamed system where all beams are inclined to an angle α in relation to axis $0y$, the following type of transfer matrix is therefore obtained:

$$\begin{bmatrix} V_x \\ V_y \\ V_z \end{bmatrix} = \begin{bmatrix} \dfrac{2}{3\sin(\alpha)} & \dfrac{-1}{3\sin(\alpha)} & \dfrac{-1}{3\sin(\alpha)} \\ 0 & \dfrac{-1}{\sqrt{2}\sin(\alpha)} & \dfrac{1}{\sqrt{2}\sin(\alpha)} \\ \dfrac{1}{3\cos(\alpha)} & \dfrac{1}{3\cos(\alpha)} & \dfrac{1}{3\cos(\alpha)} \end{bmatrix} \begin{bmatrix} V_1 \\ V_2 \\ V_3 \end{bmatrix} \quad [2.117]$$

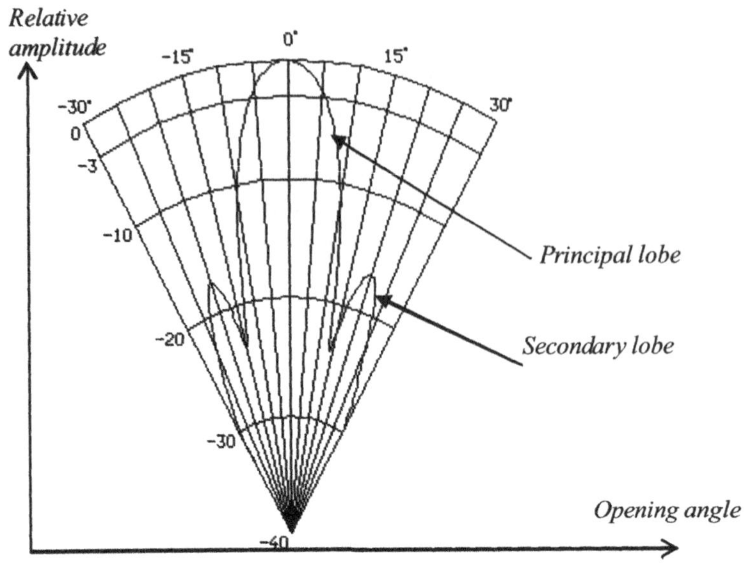

Figure 2.37. *Example of a transducer emission–reception diagram (Courtesy of Reson A/S)*

However, the acoustic transducer emission diagrams often show lobes of secondary emission that also send back a signal that is subject to the Doppler effect (see Figure 2.37). If the amplitude signal backscattered by the principal lobe is weak, i.e. if the medium has few particles in this direction, the secondary lobes can be detected and will induce errors in the measurement, particularly if the current meter is near to the surface, because signal amplitude reflected by secondary lobes is therefore considerable. The thickness of the water canal where measurements cannot be exploited is proportional to the cos of the angle of emission of the beam, which represents 6% of the distance separating the transducer from the surface, in the case when the angle of emission is 20° and 13% in the case when it is 30°. If the current meter is near to the sea floor and is heading toward the bottom, the phenomenon will be the same. It is therefore important to correctly position the instrument in relation to the range of the beams in order to avoid errors in measurement. To solve these problems, Nortek AS has proposed current meters with different beam orientations that can be chosen when placing (made to order) according to the future use of the instrument: seabed mooring, mounting onto a ROV (remotely operated vehicle – see section 4.3), onto a buoy, near to the surface or bottom of the sea, in a canal, etc.

Each transducer of the current meter emits a pulse with frequency of $f_0 = 2$ MHz and a duration of t_i which will determine the depth of the insonified area. Gradually, as the wave travels in the medium, it is reflected. The returning waves are compared to those emitted after a certain period of time t_l corresponding to twice the distance of reflection. This comparison enables a determination of phase shift ϕ of the signal when subjected to the Doppler effect. The ratio ϕ/t_l is proportional to the frequency. It corresponds to the δf_0 [2.107]. We have:

$$\frac{\varphi}{t_l} = \delta f_0 \tag{2.118}$$

From the definition of Doppler shift, δ [2.107], we can form a relation to determine speed:

$$v = \frac{\varphi c}{4\pi f_0 t_l} \tag{2.119}$$

The ratio between t_i and t_l is very important. In order to obtain a good level of correlation between the outbound and inbound waves, i.e. a sure signal detection, t_i must be shorter than t_l. Furthermore, it can be shown that the value of the standard deviation σ_δ on the determination of δf_0, with one single pulse, is inversely proportional to the length of the pulse. In other words: the longer the pulse, the weaker the uncertainty.

$$\sigma_\delta \propto 1/t_i \tag{2.120}$$

It must also be remembered that the longer the pulse, the larger the insonified region and the lower the spatial resolution of the system.

On the Aquadrop current meter, the measurement zone is at 35 cm, and it is 70 cm long, but the user is able change the length in relation to the decrease in distance caused by the medium in which the current meter is immersed. However, in order to properly reduce the standard deviation of the measurements, the speed value provided by the instrument is averaged over several pulses. The standard deviation σ_v of mean values of δ obtained after n pulses is statistically:

$$\sigma_v = \frac{\sigma_\delta}{\sqrt{n}} \qquad [2.121]$$

Other modes of measurement can be used to measure the Doppler effect. In this way, the 3D-ACM current meter made by FSI uses a locked phase loop to determine the phase shift induced by the Doppler effect. This apparatus is made from four emitter–receptors fixed on four arms that are 13 cm long and are placed at a vertical distance of 10.5 cm from each other (see Figure 2.38).

Only three axes are needed to find the value of components v_X, v_Y and v_Z of the current, but a fourth is needed in order to determine the influence of silage provoked by the central support. The transducers emit a sinusoidal signal with a frequency of f_0 = 1 MHz for a period of 2 ms. A measurement technique called "by reciprocity" is used to take an absolute speed measurement that is independent of the instrument. The signal is emitted in one direction in order to determine total phase shift θ_{ab} introduced by current, transducers *a* and *b* and the electronics. Then, the same measurement is made in the opposite direction; the emitter becomes the receiver and θ_{ba} is obtained. The current speed in the axis under investigation is proportional to the difference between these two phase measurements.

$$\theta_{ab} - \theta_{ba} = \frac{2\pi f_0 d}{c - v} - \frac{2\pi f_0 d}{c + v} \qquad [2.122]$$

d being the distance between transducers *a* and *b*. Considering that $c >> v$, the value of *v* is easily obtained from:

$$v = c^2 \frac{[\theta_{ab} - \theta_{ba}]}{4\pi f_0 d} \qquad [2.123]$$

In order to keep the absolute character of the angle measurement, the phase detector is equipped with a device released by a micro-computer that enables us to generate calibrated shifts at 0, 90, 180 and 270°. Signals E_e and E_r emitted and received by the transducers can be shown in relation [2.124] where K_e and K_r have constant values.

$$E_e = K_e sin(2\pi f_0 t + \theta_e)$$

$$E_r = K_r sin(2\pi f_0 t + \theta_r) \qquad [2.124]$$

These signals are sent in a multiplier that gives on output:

$$E_m = \frac{K_e K_r}{2}\left[\cos(\theta_e - \theta_r) - \cos(4\pi f_0 t + \theta_r)\right] \tag{2.125}$$

E_m then enters into a low pass filter that enables filtering of the $2f_0$ component and of obtaining a continuous signal E_{dc}, which contain an offset E_o introduced by the detector:

$$E_{dc} = K\cos(\theta_e - \theta_r) + E_o \tag{2.126}$$

with $K = K_e K_r/2$. If the shift applied to the delivery is $\theta_r = 0°$, E_{dc} becomes:

$$E_0 = K\cos(\theta_e) + E_o \tag{2.127}$$

For the other shifts (90, 180 and 270°), it is shown that:

$$\begin{aligned} E_{90} &= -K\sin(\theta_e) + E_o \\ E_{180} &= -K\cos(\theta_e) + E_o \\ E_{270} &= K\sin(\theta_e) + E_o \end{aligned} \tag{2.128}$$

By carrying out the operations, $E_0 - E_{180}$ and $E_{270} - E_{90}$, a calibrated measurement of θ_e can be obtained:

$$\theta_e = \arctan\left[\frac{E_{270} - E_{90}}{E_0 - E_{180}}\right] \tag{2.129}$$

This measurement is independent of the gain and shift voltages of the electronics, assuring an absence of drift at phase measurement. All of theses operations are sequenced by a micro-computer that enables their repetition in the four channels.

The performances obtained with these two types of current meters are similar. Table 2.5 shows their characteristics relating to the measurement of speed. These characteristics are similar to those shown in Table 2.4 for the rotor current meter MC3X0. The substantial advantage of acoustic current meters is the fact that they have no mechanical pieces that move, enabling them to take measurements of weak currents (with a limit fixed by their measurement resolution) while rotor current meters are limited by their starting threshold, whose value can increase over time.

Figure 2.38. *The 3D-ACM current meter produced by Falmouth Scientific, Inc. in a handling frame. The transducers are fixed by four arms that are 13 cm, each spaced vertically 10.5 cm from each other (Courtesy of FSI)*

	Range	**Accuracy**	**Repeatability**
Aquadopp	± 5 m/s	1% of the measured value ± 0.5 cm/s	0.5–1.0 cm/s 1.5 cm/s at 1 Hz
3D-ACM	0–3 m/s	2% of the measured value	Undisclosed

Table 2.5. *Manufacturer's specifications for Aquadopp and 3D-ACM current meters relating to speed measurement. It must be noted that accuracy can depend on the inclination of the instrument at the time of measurement and on the compensation that is carried out*

2.5.3. *Electromagnetic current meters*

This principle of measurement has not succeeded in replacing the mechanical and acoustic principles, described in sections 2.5.1 and 2.5.2, in oceanography. This is mainly due to reasons surrounding accuracy and the uncertainty of measurement. It is, however, widely used in industries that measure flow, as its intrinsic qualities and theory of operation are well known. Measurement of the speed at which water mass moves using this device is based on Faraday's principle, which states:

> "As magnetic field $\vec{B}$ passes perpendicularly through a conductive liquid at speed V inside an insulating pipe of diameter D, voltage U is induced in the liquid."

If the magnetic field applied is alternating sinusoidal with a pulsation ω, and the current speed is perpendicular to B, this principle is translated by the relation:

$$U = -B\sin(\omega t)DV \qquad [2.130]$$

Electromagnetic current meters are therefore made from a winding, which induces magnetic field $\vec{B}$ in the medium, and two electrodes separated by a distance D. These electrodes are in contact with the medium and allow voltage U to be gathered. As optimal conditions of measurement are met, U varies in proportion to the amplitude of vector speed, which is perpendicular to the magnetic field generated. The optimal conditions of measurement are multiple. First, it is imperative that no object disturbs the path of the magnetic fields generated. Precautions must therefore be taken as the current meters are used in mooring lines. Second, it is imperative to protect the instruments against the animal or vegetal concretions that dirty most instruments immersed in surface layers during long periods of time. These concretions can disturb field lines. Third, it is important that the inclination of the mooring line in relation to current is limited.

Next, if we consider that the distribution of voltage and current in a homogeneous fluid does not theoretically depend on the conductivity of the fluid, it is important to hypothesize that the medium of measurement is electrically homogeneous. The first disruptor of homogeneity to be found is the presence of electrodes themselves. They must be considered punctual, compared to the field tube. If this is not the case, their surface state or the variation of their surface state can create variations in homogeneity and lead conductivity to have an influence on the measurement. In the same way, impurities and diverse deposits can create dissymmetry in voltage received by electrodes, in which impedance is therefore modified. This dissymmetry can cause the degradation of the common mode rejection ratio of amplifiers used to measure U, and therefore lead to errors in the voltage measurement.

As precautions are taken to create favorable conditions of measurement, electromagnetic current meters have advantages over their mechanical and acoustic competitors. First, they do not have mobile parts, and measurement is taken statically. There is therefore no problem with the starting threshold that is found in rotor current meters, or it is reduced to its digital resolution. Measurements of weak currents are possible, and their sensitivity to the shocks and vibrations of mooring lines is lower. As electrodes are at the surface of a current meter casing, the problems with the modification of flow that can be found in certain acoustic current meters are reduced. In the same way, these meters are less sensitive to bubble formation than their acoustic homologues.

Essential characteristics:

Range: 0–350 cm/s as standard
Accuracy: 2% of the reading ±1 cm/s
Resolution: of 0.03–0.43 cm/s according to the range and frequency of sampling.
Starting threshold: identical to the resolution
Direction: given to the nearest ±2° for angles of less than 5°, by a "flux-gate" compass

Figure 2.39. *Example of the S4 electromagnetic current meter produced by InterOcean Systems, inc. It is a sphere that contains four electrodes (black dots) and a frame that enables its suspension from a mooring line (Courtesy of InterOcean Systems, Inc)*

2.5.4. *Doppler effect profilers*

2.5.4.1. *Principles of measurement*

Progress made in terms of underwater acoustic signal processing have enabled the production of instruments that are able to measure current, which is no longer on one level but on vertical profiles. Theses instruments are known as *Acoustic Doppler Current Profilers* (ADCP) or ADP, depending on the manufacturer. Current profilers measure the speed and direction of water mass movement in various parts of the water column.

In order to do this, acoustic energy is transmitted to the medium in the form of pulses or "pings", with help from four transducers, with an inclination of 20° or 30° in relation to the vertical, and in which beams are narrow. Echoes received are continuously sampled. It can be seen that their frequency is outphased compared to the initial pulse. This shift is proportional, to the second order, to the speed of particles met in the path of the beam (the Doppler effect, see section 2.5.2.1).

During the data processing phase, echoes are separated into temporal sections corresponding to spatial cells (see Figure 2.40). If pulses are emitted at time t_0, the signal received between times t_i and t_{i+1} (after the signal that propagates with speed c leaves and returns) will correspond to the section situated at mean distance d_i so that:

$$d_i = \left(\frac{t_{i+1} - t_i}{4}\right) c \qquad [2.131]$$

In this way, the water column can be "cut" into equal sections (usually up to 128) of an adjustable length from 5 cm to several meters, according to the frequency of the instrument. In each cell in the profile, the profiler calculates the mean vector speed by applying a weighting function centered in the middle of the cell. This calculation of speed can be averaged over several pulses in order to reduce the uncertainty in the measurement.

After sending a "ping", it is not possible to immediately set the transducer to record, taking into account its impulse response. This time corresponds to an area found between the transducer and the first cell, which is known as the "white area", where measurements are not possible. This limits the minimum depth at which the instrument can be used (this problem has been solved in the Aquadopp Z-cell current meters produced by Nortek AS, where horizontal transducers have been added to the tip of the instrument in order to measure speed at this level). In the same way, if the impulse reflects on the bottom of the sea, there is an area that is estimated as equal to 15% of the distance between the bottom and the profiler where measurements cannot be made as the intensity of the reflected pulse masks the lowest echo intensities from areas close to the bottom. This phenomenon can also be observed if the profiler is placed on the bottom, pointing towards the surface. The reflections on the surface can mask echoes from layers that are close by.

As speed and direction of the movement of water inside each section are obtained at regular intervals, we are able to follow the evolution of a water column's profile, whose maximum depth can reach several thousands of meters. The user has the possibility of programming the instrument to configure it in relation to the medium being studied.

As in the case of acoustic current meters, in order to determine three components of speed in a cell, the profiler is equipped with at least three transducers. A fourth can be used to track errors (see Figure 2.40). As the transducers insonify different areas, the measurement is based on the hypothesis that the current is homogeneous within the same horizontal layers. This implies that the profiler cannot be used under an angle close to the horizontal due to the risk that a large error will occur during measurement. Devices can, however, be configurated for this particular usage.

Profilers are generally used for bottom tracking or surface tracking. As they track the surface, they are often fixed to an immobile mooring resting on the seafloor. As they track the bottom, however, they can be fixed in several ways:

– to a boat, where they are called hull profilers, as they are integrated and positioned in the hull of the boat; or

– to a cage that is lowered with the help of a winch in order to profile water columns whose depth is greater than the range of the instrument. This technique is called LAD or LADCP, which stands for *Lowered Acoustic Doppler Current Profiler*.

In both of these cases, rolling, pitching and, if necessary, heading must be corrected when measuring speed. To carry out the corrections, profilers are equipped with compasses (section 2.5.5) and tilt sensors.

Problems presented by secondary lobes close to the bottom or surface, and mentioned in section 2.5.2, are even more important to take into account in the case of profilers. The thickness of the section of water where measurements cannot be taken is proportional to the cos of the beam's angle of emission. Thus, for immersion of a transducer at 100 m in which the angle is 20°, the last six cannot be measured.

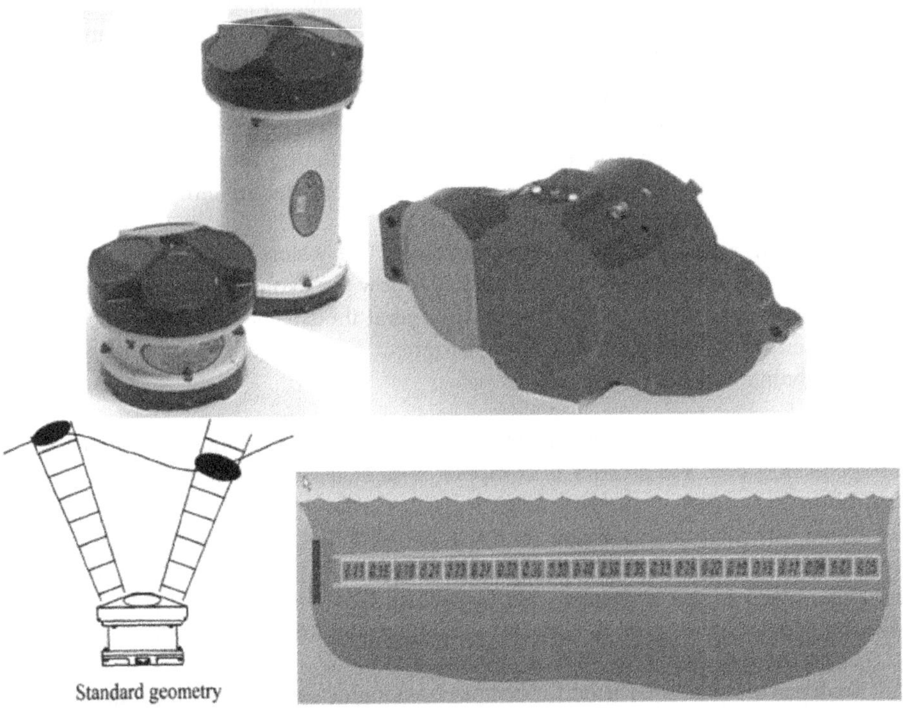

Figure 2.40. *Examples of two types of ADCPs and representation of the geometry of their beams. The one to the left has a classic "Janus" configuration, in order to measure vertical three-dimensional current profiles (speeds according to X, Y and Z at different depths). The one to the right has a special configuration that enables the measurement of two dimensional currents in a horizontal plane (speeds according to axes X and Y). The two devices enable study of the swell by measuring wave-induced orbital speeds (Courtesy of Teledyne R.D. Instruments)*

Profilers are available that cover several emission frequencies. The choice of model will be made in relation to the exploration depth and estimated speed of particles. Depth can vary from one meter to several hundred meters. In oceanography, the maximum measurable speeds are generally below 2 m/s. The maximum radial speed $V_{\max}$ that the profiler is able to measure is in relation to its emission wavelength and to its pulse repetition frequency f_r.

$$\pm V_{\max} = \pm \frac{f_r \lambda}{4} = \pm \frac{\lambda}{4 t_r} \qquad [2.132]$$

V_{max} corresponds to the maximum frequency obtainable while avoiding aliasing phenomena related to the sampling that is being carried out. The frequency, f_r, that corresponds to the inverse of time t_r separating two successive pulses also sets the maximum range r_{max} in which a target position can be detected without ambiguity:

$$r_{\max} = \frac{c}{2 f_r} = \frac{c t_r}{2} \qquad [2.133]$$

Relations [2.132] and [2.133] can be combined allow us to express the ambiguity function range versus speed, as determined by Lhermitte and Serafin [LHE 84]:

$$r_{\max} V_{\max} = \pm \frac{c\lambda}{8} \qquad [2.134]$$

Emission frequency in kHz	Maximum range in m	Diameter of the transducer (mm)
75	400 to 700	280
150	250 to 400	165
300	100 to 250	130
600	50 to 70	100
1,200	25 to 30	55

Table 2.6. *Maximum range of the acoustic transducers of current profilers according to their emission frequency. It must be noted that these values can be considerably reduced if the medium contains a low number of particles*

Relation [2.134] demonstrates the fact that increases in wavelength emission is necessary to improve the value of the ambiguity function. However, a lower emission frequency results in a reduction in the system's bandwith and resolution.

Low-frequency profilers are therefore used when a large range is necessary, i.e. in the deep ocean. On the other hand, high frequency profilers are more frequently used in coastal areas, when high spatial and temporary resolutions are useful. Likewise, the higher the frequency of profilers, the smaller the uncertainty of speed measurements, as their short-term variability is reduced. The reduction in measurement uncertainty is made to the detriment of the range, as high frequencies are significantly better absorbed by the medium (see Table 2.6).

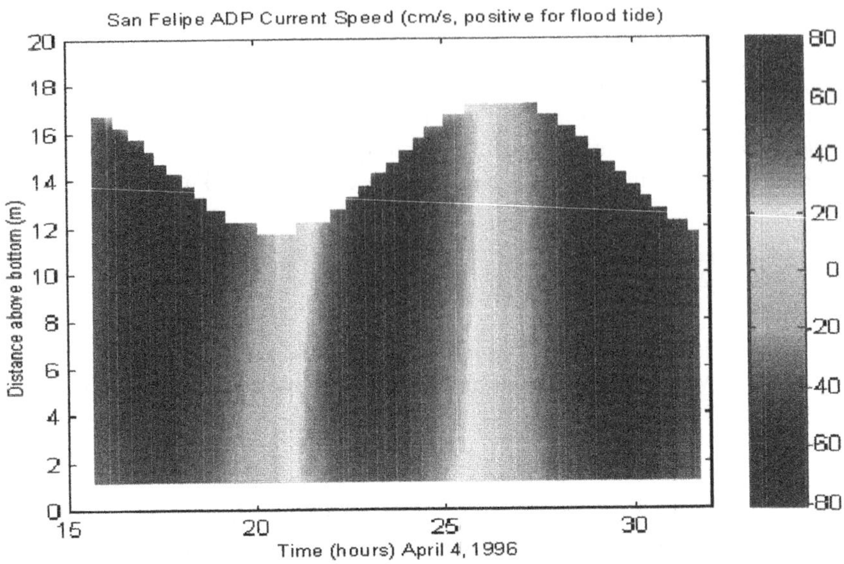

Figure 2.41. *Example of a current speed map that can be obtained with an ADP moored at the bottom. Variation in height observed corresponds to tides (Courtesy of SonTek/YSI, Inc.)*

2.5.4.2. *Operating principles of current profilers*

Conventional profilers, often called *narrowband profilers*, are said to be incoherent as the echoes received from two different pulses are not correlated. These systems are said to work on one pulse. The return time of the echoes is continuously measured, enabling the determination of cell sizes measured along the water column, if the speed of sound propagation c is known. The size of the listening windows of a cell for these systems corresponds to the duration (or length) t_p of the pulse, giving a spatial resolution Δr equal to:

$$\Delta r = ct_p / 2 \qquad [2.135]$$

According to the value of t_p, measured cells can be adjacent or partially overlapped. Therefore, no matter what type of profiler (*narrowband* or other) used, it is necessary to adapt the weighting applied to the measurement in both these cases. When the pulse is short in relation to the cell size, the weighting adopted will be more rectangular; while if the pulse is longer than the cell, there will be a correlation between the cells, requiring more triangular weighting to give more weight to the central values of each cell.

Different tools can be used in order to measure Doppler shift (fast Fourier transform, autocovariance calculations, etc.). The lowest measurement uncertainty obtainable is given by the square root calculation of variance σ_δ of the Doppler noise, in which variation is inversely proportional to that of t_p [2.120]. It can be reduced by multiplying the number of measurements according to relation [2.121], to the detriment of the time required to estimate current speeds in each cell, or by increasing the value of t_p, this time to the detriment of the spatial resolution of the system.

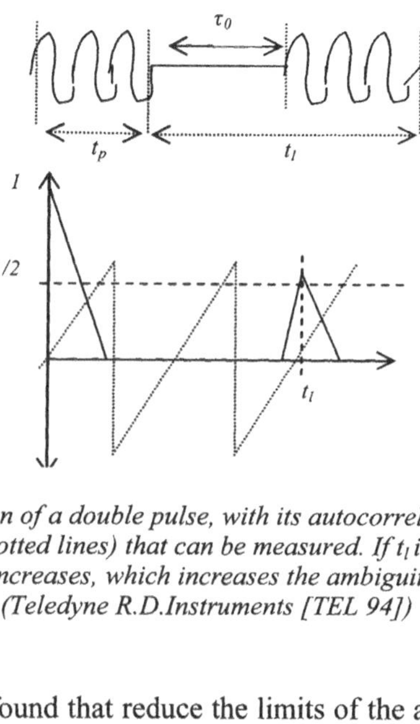

Figure 2.42. *Representation of a double pulse, with its autocorrelation function (solid lines) and variations in phase (dotted lines) that can be measured. If t_l increases, the frequency of the phase function also increases, which increases the ambiguity of speed measurement (Teledyne R.D.Instruments [TEL 94])*

Solutions have been found that reduce the limits of the ambiguity problem range versus speed. They consist of working on a series of two coherent pulses rather than on one pulse. Doppler shift measurement can therefore be made in the domain of frequencies, by detecting the position of the principal peak in the listening window, or in the temporal domain by calculating an estimate of the first moment of the signal power spectrum from the covariance of the two pulses. This technique is called "*pulse-pair*".

Experience acquired in the processing of radar and sonar data has enabled orientation of current profiler data processing towards the estimation method of the signal autocovariance, which consists of calculating the autocorrelation function of two successive pulses, which are making each other coherent (see Figure 2.42). This technique, termed "*pulse-to-pulse coherent*" or "*pulse-coherent*", has the advantage of being simple to program and its results are not biased by the white noise superimposed on the signal. According to Lhermitte and Sarafian [LHE 84], it is able to reduce ambiguity limits from a factor of 100 to 500.

If $C(t)$ is the signal received after the emission of two successive pulses, and $C^*(t)$ is its combined complex expression, the autocovariance of the signal is calculated as follows:

$$R(\tau) = C(t)\, C^*(t+\tau) \qquad [2.136]$$

As $C(t)$ is a complex signal, $R(\tau)$ is also a complex signal, in the form:

$$R(\tau) = R_e(\tau)\, e^{-i2\pi f_0 \tau} \qquad [2.137]$$

In expression [2.137], $R_e(\tau)$ is the amplitude of function $R(\tau)$. It is a real quantity described by a pair functions whose form is given:

$$R_e(\tau) = 2\pi \int S(f - f_0)\cos(2\pi f \tau)\, df \qquad [2.138]$$

and that fill the condition:

$$\int S(f - f_0)\sin(2\pi f \tau)\, df = 0 \qquad [2.139]$$

In expressions [2.138] and [2.139], $S(f)$ represents the standardized Doppler signal spectrum. As reflections are produced by a large number of particles, it is considered that $S(f)$ obeys to a Gaussian distribution. This assumption is questionable, however, as Doppler frequencies and autocovariance function evolve over time, signifying that the samples are not all independent and that the process is not stationary.

$R_e(\tau)$ contains data on the signal spectral width, i.e. its variance. In order to extract them, we need to calculate the amplitude of the autocorrelation function of $S(f)$, which leads us to express the relation:

$$\rho_e(\tau) = R_e(\tau) \,/\, R_e(0) \qquad [2.140]$$

If we assume that the measurements follow a Gaussian distribution, $\rho_e(\tau)$ can be writen:

$$\rho_e(\tau) = \int e^{-4\pi^2 \frac{f^2}{2\sigma^2}} \cos(2\pi f \tau) df = e^{-2(\pi\tau\sigma)^2} \qquad [2.141]$$

Expression [2.141] in this case enables an assessment of the value of the signal spectral variance σ^2, which contains data on the dispersion of measured speeds. This dispersion is due to the random positioning of particles in the medium, introducing fluctuations in the amplitude and phase of the echoes detected, and also producing Doppler noise. For "pulse-pair" signals, the limits $\pm\infty$ of the integral must be replaced by $\pm 1/2\tau_0$ that correspond to the Nyquist limits. Beyond these limits, the signal is subject to aliasing.

In order to calculate σ^2, the value of $\rho_e(\tau)$ needs to be established with the relation [2.140]. For a given value τ_0 of the delay between two successive pulses (Figure 2.42), relation [2.141] gives:

$$\sigma^2 = -\frac{\ln \rho_e(\tau_0)}{2\pi^2 \tau_0^2} \qquad [2.142]$$

This variance is inversely proportional to the spatial resolution of measurements fixed by the value of τ_0.

In order to completely characterize the Doppler shift, we need to estimate the mean Doppler frequency $\overline{f}$ corresponding to the mean speed that we are trying to measure. This calculation can be done as follows:

$$\overline{f} = \int S(f)\, f df \qquad [2.143]$$

However, this information is most often extracted from the phase ϕ of the autocorrelation function that corresponds to instantaneous Doppler shift, and is proportional to the instantaneous speed, as shown in relation [2.119], which is still valid in the case of "pulse-coherent" systems. The time t_l that appears in the expression corresponds to the time τ_0 that separates the two successive pulses. Autocorrelation is estimated from a sequence of M pulse pairs, rather than two successive pulses: $\hat{R}(\tau) = \sum_{n=1}^{M} R(\tau) \Big/ M$. ϕ is obtained by calculating the tangent of the ratio of the real part of $\hat{R}(\tau)$, to its imaginary part.

It must be noted that relation [2.142] is only valid in the case of a Gaussian distribution. In practice, other types of distributions can be obtained. However, this expression, associated with relation [2.119], allows us to show that the standard deviation of measurements of speed can be reduced by increasing the value of t_l (or τ_0) independently of the emission frequency. Increasing the time t_l consequently increases the periodicity of phase variation that can be measured between the pulse emitted and its echoes. The range of speeds measurements is therefore reduced. What is gained in the reduction of uncertainty is lost in measuring range. As phase measurements are limited to the range $\pm\pi$, this interval corresponds to maximum speeds that can be measured by the instrument without ambiguity, i.e. without passing to the following or preceding signal cycle. The maximum speed measurable is given by relation [2.119], where ϕ would be replaced by $\pm\pi$.

The range is equally limited as a result. This limitation can be assessed by relation [2.133], where t_r would once again be replaced by τ_0. Compared to a narrowband system, the limitation of range can be estimated at around 20%.

Particles' movements by advection in the volume of measurements, turbulence, the phase distortions of sent waves and the electronic noise of the receiver circuit can all provoke decorrelation of pulses and lead to measurement noise. Several complex techniques have been designed to reduce this noise, but there are ambiguities in range–speed and noise–resolution that limit the "*pulse-coherent*" technique.

Despite everything, with the "*pulse-coherent*" principle, the duration of pulses becomes a parameter that is independent of t_l and it is therefore possible to use short duration pulses in order to increase the vertical resolution of the determination of speed. Likewise, in a measurement cell corresponding to a section of water, profiler designers can work with several double pulses that are independent of each other in order to reduce the standard deviation of measurements according to relation [2.121], that also allow the calculation of $\hat{R}(\tau)$. However, with short duration pulses that transport little energy, the range of the instrument can again be affected. To limit this problem, "pulse-coherent" profilers use sequences of coded pulses. For sinusoidal signal carrier frequency f_c, it is possible to modulate the phase for a duration T with help from a pseudo-random binary sequence $b(t)$ of N bits, which has good autocorrelation properties. A change of one bit is generated by applying a jump in phase of π to the carrier. The signal therefore takes the following form:

$$s(t) = sin(2\pi f_c t(2b(t) - 1)) \quad [2.144]$$

This type of modulation is known as BPSK, which stands for *binary phase-shift keyed.* Its Doppler resolution expressed at -3 dB and in m/s is given by:

$$\delta_r = \frac{c}{f_c T} \quad [2.145]$$

It is possible to use *Barker* coding, which consists of dephasing each pulse by 0° or 180° in a pseudo-random way (see Figure 2.43). Pulses can be emitted in the form of coded sequences to give the signal enough energy, and then avoid limiting the range of the instrument. They do this while conserving the independent characters of elementary pulses and the possibility of using them separately to amend data as described above, in order to extract speed information. This technique does, however, have limits that are linked to the turbulence of the medium and beam divergence, which can cause a decorrelation of pulses. Interference between pulses can also occur, reducing the resolution capacity of the system.

"Broadband" systems have been developed to resolve the problems related to the ambiguity of speed posed by "*pulse-coherent*" profilers, and to reduce vulnerability to reflections in the medium. They are an impressive compromise between incoherent and "*pulse-coherent*" systems. In these devices, autocorrelation is no longer simply calculated by using two successive pulses, but simultaneously uses several returns from pulses. It is therefore possible to reduce the period between pulses in order to increase the of maximum value of speed detectable, and then to reduce ambiguity, but also to work on pulses with a sufficient distance to cover the chosen measuring range, which is more limited by the signal-to-noise ratio than by the sampling frequency and aliasing problems.

Measurement is still based on calculation of the phase of the autocovariance function of the signal, which is the delay chosen for the calculation corresponding to the period τ_0 between pulses. Speed v is proportional to Doppler dephasing, and its value is obtained by relation [2.119]. In this way, the uncertainty of v is largely proportional to the uncertainty of the phase measurement.

The "*broadband*" technique allows short pulses to be sent in order to increase the vertical resolution of the system. For the same cell of measurement it is therefore possible to increase the number of "sensor" pulses, and so multiply the measurements and reduce the standard deviation according relation [2.121]. However, this concept runs up against the problem of transported energy and the measuring range of the instrument, enforcing the use of pseudo-random coding to generate independent pulses forming a "pattern" in which there is sufficient energy to conserve an optimal range (see Figure 2.43).

An example would be the software proposed by Teledyne R.D. Instruments in the operation of their ADCP. The software contains several programming modes that allow the user to optimize the performance of the device, principally in relation to maximum current speeds to be measured and the uncertainties required for the measurements, in order to manage better the problem of ambiguity in the determination of speed. In these modes of programming, different lengths of pulses and different codings are pre-formatted in order to simplify the choices offered by the extraordinary measurement possibilities of broadband treatments. There is a "standard" mode, which is a compromise that can work for all applications. It is a mode that allows very weak uncertainties on high current levels and, inversely, a mode for slow currents. It has a "bottom track" mode for applications where the current meter is directed towards the bottom and fixed under a boat and a "bottom tracking" mode where each profile can be integrated to calculate the flow of a canal from a boat making a crossing. In these last modes, a pulse emitted to the bottom alternates with a pulse emitted in the water column in order to automatically subtract the speed of the boat.

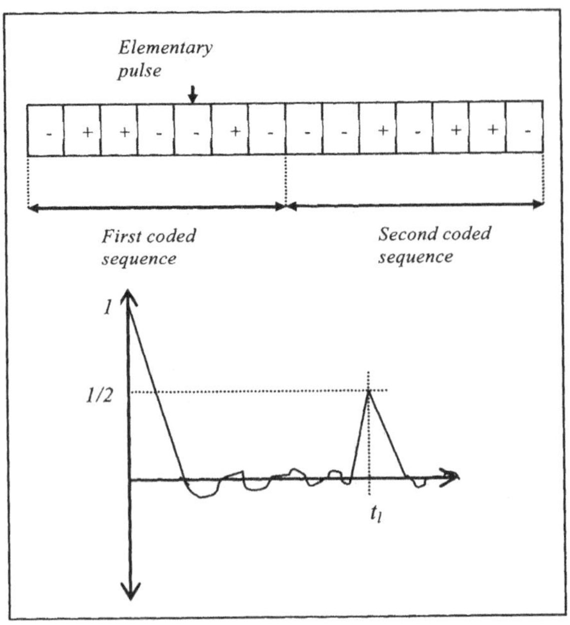

Figure 2.43. *Example of pulse coding and corresponding autocorrelation function (Teledyne R.D. Instruments [TEL 94])*

2.5.4.3. *Range and intensity of current profiler signals*

The range of these instruments largely depends on the intensity of the echoes returned. The intensity *IE* of these echoes clearly depends on the intensity emitted by the source *IS* and that retrodiffused *BS* by particles contained in the medium. In any case it is limited by the absorption coefficient α of sound in the medium being crossed, the divergence of the beam according to propagation mode (resulting in a constant term *TC*) and a term in $20\log_{10}(R)$ where R is the distance of the emitter from the cell being measured. The expression of *IE* comes from the sonar equation, sum of factors in the form:

$$IE = IS + BS + TC - 20\log_{10}(R) - 2\alpha R \qquad [2.146]$$

All terms in formula [2.146] are expressed in decibels (dB). As the decibel is a logarithmic unit, and the last term of formula [2.146] is linear, this signifies that the echo intensity decreases in an exponential way with distance and that the range of detection of instruments is very quickly limited. α is a function of molecular relaxation phenomena provoked by the action of sonar waves on molecules in the medium. It therefore depends on the frequency f emitted. An approximate formula allows the evaluation of its value:

$$\alpha(f) = 20Log_{10}\left(\frac{0.17}{18+t}f^2\right)\Big/1000 \qquad [2.147]$$

Here, f is expressed in kHz, t in °C and $\alpha(f)$ is in dB/km. Likewise, for example, for a frequency of 150 kHz, $0.04 < \alpha(f) < 0.05$ dB/km for $0 < t < 35$°C. In practice, a frequency of 150 kHz emitted with a power of 250 W allows a maximal range of around 400 m.

Taking into account the intervening terms in relation [2.146], it is difficult to extract absolute intensity data from the returned echo. However, studies have been carried out to extract data on the concentrations of plankton and particles in order to estimate the turbidity of the medium (see the definition of this quantity in Chapter 2.10). The relation between the intensity of the echo and concentration of particles is highly dependant on the size of the particles and their distribution in the medium.

2.5.5. *Directional referencing of current measurements*

As direction measurements taken by current meters are relative to the orientation of the instrument in the medium, it is necessary to attach the measurements to a common referencial in order to compare the difference in results obtained from one instrument to another, and to determine the direction of water mass movement in relation to a terrestrial referencial. This referencing is done by an intermediary of the Earth's magnetic field. All current meters are therefore equipped with a compass, which enables determination of their orientation with regards to the magnetic north. Direction measurements are then processed in order for the angular deviation that exists between the instrument and direction of the Earth's magnetic field, then between the direction of this magnetic field and the geographic north or its projection in a system of coordinates to be corrected.

There are numerous physical principles that enable the measurement of the magnetic field. The resulting instruments are called magnetometers. They can be classified according to the type of measurements that they take. They can either be scalar, measuring the total magnetic field, or vectorial, i.e. directional. The most popular technology among vector magnetometers is the "flux-gate" technology (section 2.5.5.5). As only the horizontal component of the field is measured, orientation data must be corrected with or compensated by the angles of inclination of the instrument in relation to the vertical axe.

2.5.5.1. *The Earth's magnetic field*

The principal origin of the Earth's magnetic field is a dipole at the Earth's core. Dipole field $\vec{B}_d$ has a magnetic moment of 8.1×10^{22} A m^2 in amplitude, which is approximately directed according to the North–Souh geographic direction. The amplitude of $\vec{B}_d$ is from 30 µT to the equator and 60 µT to the poles, where it is vertical. Field $\vec{B}$ that is being measured has two other components:

– $\vec{B}_a$, which is the field of global anomalies in which amplitude can reach 10 µT; and

– $\vec{B}_e$, which is a field of local anomalies in which amplitude can sometimes reach that of $\vec{B}_a$.

A fourth component $\vec{B}_t$ added to the fields is representative of temporary variations in which the source can be inside or outside of the Earth. External sources are essentially solar in origin. Internal sources are linked to the Earth's response to external stimulations. Lastly, field $\vec{B}$ also depends on a fifth component, $\vec{B}_x$, which represents anomalies of human origin. They can be static, ($\vec{B}_{xs}$) or variable ($\vec{B}_{xt}$) in time. The measurement of $\vec{B}_{xs}$ allows the detection of ferromagnetic objects, while $\vec{B}_{xt}$ corresponds to frequential fields radiated by electronic or electrical equipment in a general way. In order to illustrate the influence of $\vec{B}_{xs}$, we can determine that a boat weighing 100 tonnes generated a magnetic anomaly of around 500 nT to 30 m. It is no more than 1 nT to 300 m.

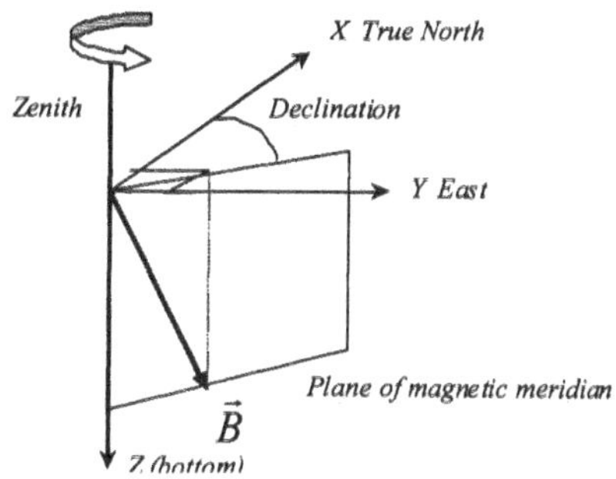

Figure 2.44. *Illustration of reference conventions of the Earth's magnetic field in the northern hemisphere*

Measurement of the horizontal component of $\vec{B}_d + \vec{B}_a$ enables us to reference data in relation to the Earth's magnetic north. The other components, which are detected locally and according to their amplitudes, are sources of errors in measurement.

In order to determine the orientation of a system in relation to the principal direction of the Earth's magnetic field, it is necessary to place three magnetometers perpendicularly. As this system can take any direction, either its movements must be compensated by fixing the magnetometers on gimbals, or they must be linked to a bi-axial *inclination sensor* or "*tilt sensor*". With information on inclination, the magnetic field vector that is being measured can be projected onto a horizontal surface in order to determine the system orientation in relation to magnetic north. The vertical plane that would contain the magnetized needle of a compass is called the "plane of the *magnetic meridian*". The angle shift between the axis pointing North and the axis passing by the magnetic meridian is called the *declination* (see Figure 2.44), and the angle that points the needle horizontally is called the *inclination*. The declination varies between –180° and +180° and is positively counted towards the East. By knowing the magnetic declination of a place, the orientation of the system in relation to true North can be determined. Global maps of magnetic declination are emitted from observations made by specialized satellites, such as Mag Sat, allowing us to reference directional measurements taken at any point on Earth.

Once local declination is known, it is necessary to determine the *convergence* or angle between a geodesic projection system and true North in order to plot the data on a map.

2.5.5.2. *Field measurement and compass calibration*

Current meter software generally has an *autocalibration* function for the compass. To operate it, the instrument must be turned slowly on itself and make at least two full turns. This software gives the maximum error measured during the rotation. It displays measurements in the form of a circle if the calibration of the compass is good and in the form of an ellipse if there is a bias. If the maximum error is greater than 5°, or if an ellipse is displayed, it is necessary to make two or three more turns so that the calibration matrix of the compass is automatically modified. The modifications are supposed to take into account the magnetic anomalies generated by the environment and measured by the compass.

In fact, these pieces of software enable the calculation of a slope and an offset correction according to the horizontal axes x and y. If Φ represents the roll angle, θ the pitch angle and Ψ the orientation according to the axes x and y, and if the magnetometer measures values M_x, M_y and M_z of the field, according to axes x, y and z, the value of horizontal components M_{xH} and M_{yH} can be deduced:

$$\begin{bmatrix} M_{xH} \\ M_{yH} \end{bmatrix} = \begin{bmatrix} \cos\theta & \sin\theta.\sin\varphi & -\cos\varphi.\sin\theta \\ 0 & \cos\varphi & \sin\varphi \end{bmatrix} \begin{bmatrix} M_x \\ M_y \\ M_z \end{bmatrix} \quad [2.148]$$

Angle α that places the current meter in relation to the magnetic north can be calculated from:

$$\alpha = \begin{cases} 90 \\ 270 \\ 180 - a\tan\left(\dfrac{M_{yH}}{M_{xH}}\right)\dfrac{180}{\pi} \\ -a\tan\left(\dfrac{M_{yH}}{M_{xH}}\right)\dfrac{180}{\pi} \\ 360 - a\tan\left(\dfrac{M_{yH}}{M_{xH}}\right)\dfrac{180}{\pi} \end{cases} \text{if} \begin{cases} M_{xH} = 0, M_{yH} < 0 \\ M_{xH} = 0, M_{yH} > 0 \\ M_{xH} < 0 \\ M_{xH} > 0, M_{yH} < 0 \\ M_{xH} > 0, M_{yH} > 0 \end{cases} \quad [2.149]$$

Orientation Ψ is then determined by applying a magnetic declination correction δ to α.

As the Earth's magnetic field is locally disrupted by the presence of metallic mass, continuous currents and magnetic fields generated by electronical components in close proximity to the compass, the values M_{xH} and M_{yH} are biased. The bias can also come from errors in the alignment of the compass and of the instrument. It is therefore necessary to establish a correction matrix from measurements taken during the turning of the set. They enable the determination of offset values O_{xH} and O_{yH}, slope values S_{xH} and $S_{y,H}$, and the corrected values M_{xH}^c and M_{yH}^c of M_{xH} and M_{yH}:

$$M_{xH}^c = S_{xH}\, M_{xH} + O_{xH}$$

$$M_{yH}^c = S_{yH}\, M_{yH} + O_{yH} \qquad [2.150]$$

However, these linear corrections do not enable us to take into account nonlinear effects linked to the presence of disruptive elements in privileged directions, for example. This may be the case when the current meter is fixed in a mooring cage and surrounded by other equipment needed for the assembly, such as acoustic release, battery packs, flash lamps, ballasts, etc. It is therefore necessary to proceed to an absolute calibration. In order to do this, the equipped mooring cage is put on a turning plate. This plate is placed in an area where the magnetic field has been mapped in order to ensure the absence of magnetic anomalies greater than ~20 nT/m in amplitude, knowing that fluctuations of the Earth's magnetic field are ~40 nT/day. If the amplitude of the Earth's magnetic field is 50,000 nT in the area of measurement, this would correspond to a relative uncertainty of 0.08% or to an angle of 0.17°.

It is then necessary to turn the cage by 40°, for example, and measure the exact amplitude of the rotation with a theodolite, in relation to a reference. To create this reference axis, two points are needed for which the coordinates have been accurately determined with the help of a GPS. To convert the "geographic" angle into a magnetic angle, the magnetic declination of the area must be known. Next, we need to calculate the deviation between this reference angle and the angle given by the compass to calibrate, and then repeat the operation in order to cover 360°. A correction polynomial can be calculated from the deviations measured. The uncertainty of this method of calibration has been estimated to be 0.64°, with all corrections made and excluding the uncertainty linked to the instrument being calibrated (see [LEM 07] for more details).

Calibration of the instrument in its configuration of use does not allow the display of errors linked to the area where measurements will be made. The presence of magnetic rocks, wreckages or underwater metallic structures can cause systematic errors that are difficult to evaluate and correct. Furthermore, measurements taken in proximity to the Earth's magnetic poles can be considerably biased due to the lack of sensitivity of the compass and to the fact that the amplitude of the horizontal field components is very small. It is therefore necessary to use another technology that is a lot more expensive in order to adjust current measurements: this is the Sagnac gyro compass technology.

2.5.5.3. *Magnetic compasses with mobile gear*

These sensors are used on the oldest current meters. They are made of two magnetized bars mounted on an axis similar to that of a compass. Placed in a magnetic field, they have the tendency to align themselves in the direction of the field. The rotation couple that produces this alignment depends on the magnetic characteristics of the bars that appear in the form of dipolar magnetic moment $\vec{M}$ and of the amplitude of the magnetic field $\vec{H}$. The module of moment $\vec{C}$ of the rotation couple is written:

$$C = MH\,sin(\theta) \qquad [2.151]$$

$\vec{M}$ is a vector whose direction goes from the South pole to the North pole of the bars. The bars are fixed in a system that allows the determination of their position. The set is called mobile gear.

In the oldest magnetic compasses, this system is an electromagnet that generates a vertical magnetic field whose role is to put a cursor in contact with the track of a fixed potentiometer assembly at the moment of measurement. The set is balanced in order to keep itself horizontal and is bathed in a viscous liquid (oil, glycol, etc.), the role of which is to absorb the movements due to oscillation. The electrical resistance of this potentiometer assembly is calibrated in relation to angle values and in reference to the magnetic north.

In more recent compasses, the potentiometer assembly is replaced by a disk with an optical coding. This coding is a binary coding known as Gray code, or reflected binary classically used to code the angular position of turning systems. It is particular as only one of its bits changes state when the increment is ±1, which is not the case in classic binary coding. The disk is illuminated on one side by electroluminescent diodes and has photodiodes on the other side that enable the light transmitted by each track of the disk, in relation to black or white areas (corresponding to the coding), crossed to be read. The higher the number of tracks, the better the resolution of the coding.

The absence of mechanical action at the time of measure and direct digitalization of the angular position renders the compasses more reliable than potentiometric assemblies.

2.5.5.4. *Hall effect compass*

This compass is made up of a magnetized bar mounted on pivots. The bar can turn and direct itself according to the direction of the magnetic field, like the needle of a compass. Its orientation is measured by four Hall effect sensors positioned around each other at 90° (see Figure 2.45).

The Hall effect is obtained by the passing of current I in a semiconductive material placed in a magnetic field with induction $\vec{B}$ perpendicular to the direction of I. Let us imagine a rectangular plaque made from semiconductive material where electric charges circulate laterally from left to right. The charges deviate towards the top or bottom due to the magnetic field and a voltage V_H then appears between the superior and inferior sides of the component. The amplitude of V_H is given by:

$$V_H = K_H B \frac{I}{e} \quad [2.152]$$

where e represents the electron charge and K_H the Hall constant. As the density of charge carriers of the material is sensitive to temperature, K_H also presents a given thermal sensitivity that is variable according to the material used, but that can reach several per cent by °C.

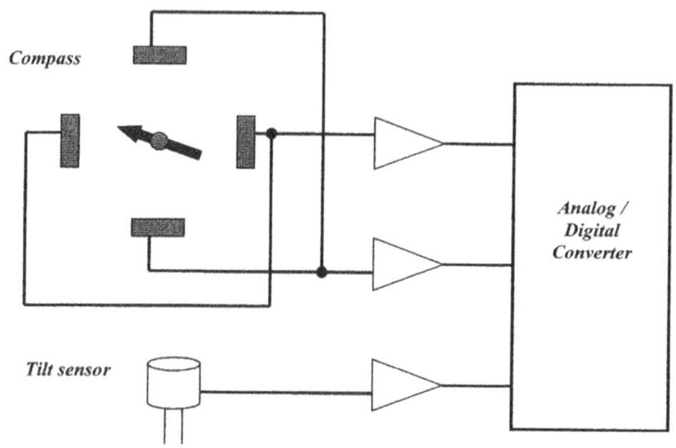

Figure 2.45. *Diagram showing the principle of a Hall effect compass. The orientation of the magnetized rotor is measured by four Hall effect sensors placed at 90° from each other. The measurement of direction is completed by a measurement of inclination of the instrument made with the help of an angle or "tilt" sensor [HOV 96]*

The Hall effect compass therefore presents an equivalent sensitivity to temperature if no precaution is taken to protect it. The RCM9 current meter by Aanderaa Instruments A/S is equipped with this compass technology and has a directional accuracy of ±5° for inclinations between 0 and 15°, and an accuracy of 7.5° for inclinations between 15 and 35°, according to the manufacturer.

2.5.5.5. *Magnetoresistive compasses*

The *magnetoresistive effect* is demonstrated in certain ferromagnetic materials, such as permalloy (an alloy containing 20% Fe and 80% Ni). The resistance of these materials varies from a base resistance R_0 in relation to angle α, which exists between their direction of magnetization and that of the current that runs through them. We therefore have:

$$R = R_0 + \Delta R_0 cos^2(\alpha) \qquad [2.153]$$

The application of a magnetic field of induction $\vec{B}$ perpendicular to the initial magnetization causes rotation of its direction and therefore a variation of α and R according to relation [2.153]. The permalloy is used in the form of a thin film. It is magnetized during production in the same direction as the current that crosses it. Component $\vec{B}$, situated in the film plane and perpendicular to the direction of the current, causes a decrease in the value of R.

In the same way as the Hall effect, the magnetoresistive effect is dependant on temperature. If temperature increases, the value of R decreases, as does the sensitivity to induction. Compasses made from magnetoresistive sensors are therefore sensitive to temperature.

Honeywell commercializes components based on this principle (HMC1001 – 1002). They enable the measurement of magnetic fields according to a perpendicular axis, and therefore to produce compasses. Magnetoresistive resistances are included in two Wheatstone bridges in which differential voltage is amplified and enables measurement resolutions of ~3 nT for a measuring range of $\pm 2\times 10^{-4}$T. These components can present drifts in sensitivity in relation to temperature, relating to about –0.3% of the reading per 1°C and offset drifts of ±0.03% of the reading per 1°C. The error of linearity is between 0.1 and 0.5% of the measuring range.

2.5.5.6. *Flux-gate compasses*

"Flux-gate" technology uses properties from saturable ferromagnetic materials. According to the Ampere theorem, all electrical conductors run through by a current generate a magnetic field and vice versa. If with an electrical conductor, a coil of surface S and n turns is made and placed in a magnetic field $B_{ex}(t)$, this field creates a magnetic flow of $S \times B$ amplitude in the coil and a potential difference $V(t)$ appears at its terminals, according to relation [2.154]:

$$V(t) = -n\frac{d}{dt}(SB) \qquad [2.154]$$

Field B created in the coil is proportional to B_{ex}, as the factor of proportionality is the apparent permeability of the coil μ_a:

$$B = \mu_a \times B_{ex} \qquad [2.155]$$

If a magnetic core of permeability largely greater than 1 is introduced in the coil, the value of B increases, as does the voltage collected. According to relations [2.154] and [2.155], it is shown that:

$$V(t) = -nS\frac{d}{dt}(\mu_a B_{ex}) \qquad [2.156]$$

Likewise, in "flux-gate" detectors the core is made from permalloy, a ferromagnetic material that is easily saturable. In the simplest configuration, this core is a cylindrical bar with an excitation coil and a measuring winding (see Figure 2.46). Most often it has a toric form. The excitation winding is run by an alternating current. This current creates periodic saturation of the magnetic material. When it is saturated, its permeability becomes almost the same as that of air. However, on average it is greater than that of air. If there is a magnetic component in the winding spindle of the measuring coil, the alternance of saturated and non-saturated states creates a variation of magnetic flow in the coil, which is where the name "flux-gate" comes from. If there is no magnetic component in the winding spindle, the variation of flow measured is zero.

The apparent permeability depends on the relative permeability μ_r and on the coefficient of the demagnetizing field D that is linked to the appearance of magnetic poles in the core. Induced voltage therefore takes the following form:

$$V(t) = \frac{-nSB_{ex}(1-D)}{\left[1+D(\mu_r-1)\right]^2}\left(\frac{d\mu_r}{dt}\right) \qquad [2.157]$$

In the presence of a field B_{ex}, the signal removed from the measuring coil has a frequency which is double that of the excitation signal. Different electronic processes can be used to detect the signal, such as synchronous detection or current/voltage converter and demodulator for example. They generally use the base frequency of the oscillator that powers the excitation coil in alternative current as a reference frequency in order to generate on output a voltage that is proportional to the field.

The performances of these sensors are essentially linked to the dependence on the temperature of:

– the electronics used for detection;

– the core composed of ferromagnetic material in which relative permeability can vary;

– the support; and

– the coils, which can dilate.

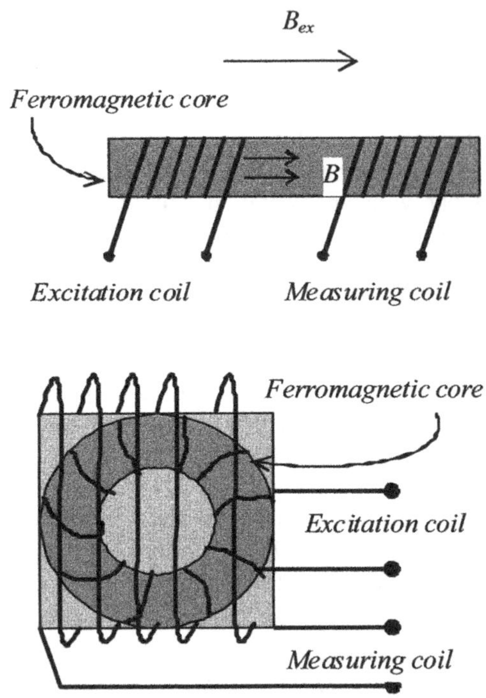

Figure 2.46. *Principle of flux-gate magnetometers with cylindrical and toric cores*

Compensation systems are proposed in the most impressive flux-gate magnetometers. Table 2.7 gives the characteristics of some kinds of flux-gate magnetometers. The accuracy of the conversion measuring voltage–angle in relation to magnetic north is in relation to the resolution of the magnetometer and to its stability with changes in temperature.

Brand	Frequency response	Resolution (≈ noise level)	Temperature Stability	Offset error
Sextant avionique	0–20 Hz	0.2 nT	± 30 ppm/°C	3 nT/°C and ±50 nT to 20°C
Dowty Magnetics	0–200 Hz	3 nT	±100 ppm/°C	<3.5 nT/°C and <10 nT
Nanotesla inc.	0–2 Hz	0.02 nT	±50 ppm/°C	<0.1 nT/°C and <1 nT
Bartington Instruments Ltd	0–1 kHz	2 nT	±36 ppm/°C	0.1 nT/°C and ±2 nT to 20°C

Table 2.7. *Comparison of the principal characteristics of different brands of flux-gate magnetometer*

2.5.6. *Calibration of Doppler effect current meters*

Doppler effect current meters can be used in mooring, in LAD (see section 2.5.4.1) or fixed under the hull of a boat. If they are used in mooring or in LAD, they measure at least three quantities: the speed of currents, their direction in relation to magnetic north and their inclination in relation to the vertical. They therefore need three different calibrations.

Calibration of the inclinometer is relatively easy to carry out. The current meter can be suspended from a support or placed on a plate that can be inclined by 10°, for example, to cover the range of the inclinometer. A comparison of the values of angles read with those provided by the instrument allows us to determine the biases and take into account their corrections.

The directional calibration of electronic compasses is described in section 2.5.5.2. The principles of speed calibration for rotor current meters are explained in section 2.5.1. Its necessity in Doppler effect current meters is arguable, as the Doppler effect is a reference in itself and the resulting phase or frequency measurements are the product of digital processing, and therefore barely sensitive to drift. These measurements do, however, need a time base generated by an oscillator (see section 2.6.2) and this oscillator can eventually lag.

If speed calibration is necessary (to satisfy quality standards for example), it is difficult to carry out calibration in the case of profilers.

Speed calibration of Doppler effect current meters can be done in the same way as it is done for rotor current meters: in a basin equipped with a moving structure on which the equipment can be fixed or in a water "vein" where the speed at which the water moves can be adjusted. There is still, however, the problem relating to the lack of particles in these testing facilities. This results in an absence of Doppler effect or in a considerable increase in the signal-to-noise ratio, which often implies that the standard deviation of speed measurement is outside specifications, particularly for low speeds. Before taking measurements, it is therefore necessary to either charge the test environment with particles or to agitate the particles that should be at the bottom (by blowing, for example).

For Doppler profilers, the depth of the basin or water section generally limits the employment of this technique. Intercomparison in the natural medium is needed. The current meter is put on a mooring line or in a cage placed at the bottom, and its measurements are compared to that of a rotor current meter (for example), previously calibrated in a basin. Although this principle is simple, carrying out and evaluating the related uncertainties of measurement is not. First, it is necessary to properly choose the measurement configuration of the profiler, i.e. its measurement mode (see section 2.5.4.2). This considerably influences the uncertainty relating to speed values that it produces. Second, the reference current meter must be placed in an area where it will not disturb the profiler, while at the same time guaranteeing that it will take measurements in the same area of current. Third, the area of measurement chosen must offer sufficient variations in speed in order to enable good coverage of the measurement range of the current meter. Areas under the influence of tidal currents often have this characteristic. If this is not the case, intercomparison is done around a quasi-constant speed that does not allow the accuracy of the instrument to be judged across the entire range. Finally, this calibration technique needs adequate organization and heavy means of transport (a boat enabling lowering into the water and the recovery of moorings, see section 3.2.3, the reservation of measuring area, etc.) that are expensive and not always easy to set up.

The problem of speed calibration is even more complex in the case of profilers fixed to boat hulls, as to begin with it is necessary to align the profiler well with the axes of the boat. In order to calculate alignment corrections, a technique based on the utilization of least-square regressions was published in 1989 by Joyce [JOY 89]. That same year, Pollard and Read published a method enabling determination of the scale factors and misalignment of a hull profiler from departure and return in the same area [POL 89]. They estimated that uncertainties of less than 0.2° in the alignment correction and 0.3% in the factor scale could be obtained. Other authors, (Alderson and Cunningham [ALD 99]) were also interested in corrections for rolling and pitch. They estimated that over short periods of time, the corrections to be applied following platform movements were small and even negligible (the effect on speed calibration was below 0.005%). However, when it is necessary to take an average of speeds over long periods of time, which is the case for studies of flow transport, an error of 1° in these parameters can lead an error of 10% in the calculation of water volume transported, and if this bias is simultaneous on instantaneous values of these parameters, these errors can be even bigger.

Finally, the speeds measured must be corrected taking into account the speed of the boat. This correction is made by recuperating the GPS positions transmitted by the boat's receiver (see section 2.7.2). As the speed of the boat is not constant, it is necessary to straighten it out with the help of an adaptive polynomial filtering technique [PIE 99]. With this technique, it is possible to reduce noise (rms) linked to the speed of the boat by 16%, as well as reducing the uncertainty of this correction to bring it down to ±0.038 m/s over 20 min.

2.6. Determining time or measuring frequency

As physical oceanographers are interested in phenomena that are perpetually evolving, the dating of data and synchronization of different instruments used to take measurements is important. Data can be recorded by a carrier boat's own central clock, or independently in the instrument if it is autonomous (such as drifting buoys, amongst others). When dating data, synchronization is often carried out to the nearest second; however it is important to have common time references in order to compare records from diverse origins. It is also important that the clocks used for measurement are stable over time, as tests can stretch over several weeks and even months or years. As instruments can be deployed in water at varying depths, the clocks are subjected to thermal strains. Their thermal drift must therefore be managed too.

The application requiring the highest temporal accuracy concerns data space positioning. If instruments are equipped with satellite positioning systems, synchronization and drift management are essential.

2.6.1. *The connection of clocks*

2.6.1.1. *Connection systems in practice*

The different official timescales that are used to reference measurement systems are described in section 1.2.5. In practice, clocks used in various oceanographycal instruments are often adjusted to a particular time, which is that of GPS or TGPS. GPS is a system that was developed by the US Department of Defense (section 2.7.2). The GPS timescale is provided by a master clock situated at the Colorado Springs control center. This clock is synchronized to the IAT (see section 1.2.5), and to one of the American atomic clocks managed by the US Naval Observatory that provides UTC USNO (see section 1.2.5). The UTC ISNO is distinguished on the UTC timescale by the fact that it does not take account of leaps of intercalary seconds that are made once a year on the UTC in order to compensate for the variability of the Earth's rotation in relation to the Sun (this is taken into account in Universal Time, UT_1). TGPS is therefore shifted in relation to the UTC, from 8 to 10 s. However, the GPS signals that comprise said navigation messages enable the user to determine this difference. GPS receivers then have a timescale that can be connected to the UTC. The navigation message equally diffuses coefficients, which enables the user to estimate the drift of the satellite clock. It also contains elements relating to orbit variations, the state of health of the satellite and coefficients of the ionospheric model.

GPS is not the only operational system of global positioning whose time base can be used. The Russian Federation has an equivalent known as *GLONASS,* which stands for *GLObal NAvigation Satellite System*. The GLONASST timescale, is maintained by the National Metrological Institute of Russia, which is situated in Moscow. Contrary to TGPS, the GLONASST is periodically reset according to the UTC. The difference between GLONASST and the UTC is therefore always less than 1 s. However, as dates are processed on Moscow time, there is lag of three hours between the time announced by GLONASS and UTC.

Taking into account the economic impact of these types of systems, Europe will also have a key player in years to come, known as Galileo. Its manufacturers want to produce a tool as accurate as GPS, with better performance with regards to its services. It will be equipped with its own time base that will also be connected to the UTC and IAT. The expected accuracy of this reference will be to within 50 ns for all commercial services.

2.6.1.2. *Errors and uncertainties of the TGPS*

Synchronization between the clock of a receiver and that of a satellite is always marred by a certain number of errors, linked in part to the difference between the geometric distance *r* that separates them (see Figure 2.48) and to the pseudo distance calculated r^{ps}. Pseudo distance is the quantity calculated from the time difference in propagation (of a wave) between the emitter and receiver. The errors that sully this quantity are linked to:

– beam refraction during crossing of the different atmospheric layers, particularly the ionosphere (error Δr_{iono}) and the troposphere (error Δr_{tropo});

– the Earth's rotation during propagation of the signal (error Δr_{rot});

– the deviation in time between the moment of emission and the reference time (error dt_S);

– the deviation of time introduced by elements of the receiver – antenna, clock, circuits, etc. (error dt_r); and

– the relativistic effects (error dt_{rel}) that are explained by the gravitational difference that exists between clocks attached to satellites and clocks attached to land stations.

c representing the speed of light in vacuum, the total error dt is shown as:

$$dt = dt_S + dt_r + dt_{rel} + (r - r^{ps} + \Delta r_{iono} + \Delta r_{tropo} + \Delta r_{rot}) / c \qquad [2.158]$$

In the WGS 84 or *World Geodetic System* of 1984, $c = 2.99792458 \times 10^8$ m s^{-1}. The WGS 84 is the geodetic system in which the coordinates of pursuit stations that form part of the GPS network are referenced. It is a product of the agencies of the US military, is three-dimensional and covers the entire globe.

We must add an error to this due to multiple reflections off terrestrial or marine obstacles before the signal is received. This introduces a multi-path effect and consequently modifies the pseudo distance calculated. This effect can be avoided by adequate positioning of the receiver's antenna. Without adequate positioning, this error can reach approximately 10 m.

Errors Δr_{iono}, Δr_{tropo}, Δr_{rot} and dt_{rel} can be reduced by modeling or calculation. The GPS navigation message is composed of parameters that enable the application of an ionospheric correction model that corrects on average 50–70% of the error induced. Δr_{iono} can then be reduced to a value of between 5 and 20 m. However, with a dual frequency receiver it is possible to measure the amplitude of the error and to practically cancel it out. Several models can be used to correct the Δr_{tropo} error. Most receivers use the Hopfield model, which is recommended in a standardization agreement edited by the US (the STANAG). The remaining Δr_{tropo} error is ~1–3 m. Errors Δr_{rot} and dt_{rel} are taken into account by the global positioning system and by the user. For the GPS, they result in a global error of less than 1 m after modeling.

With the GPS system, *dt* can be determined with an uncertainty of ~10–50 ns when the receiver is fixed, perfectly localized and when observations on one or more satellites are averaged for a few minutes to a few hours. With the GLONASS, this uncertainty is about 1 ms (to 3 σ).

Averaging is necessary as the measurement is noisy. In order to improve this determination, certain receivers enable use of the carrier signal phase. However, this measurement is ambiguous to the nearest few 20 cm and every leap in the cycle leads to an aberrant measurement that often forces us to determine the phase ambiguity. Another solution consists of working on *differentials*, when possible. There are several processes that allow this. Generally, a fixed receiver whose position is known is required for this calculation. This is called a "reference" station that must have the means to transmit corrections. The station measures pseudo distances, calculates corrections and sends them to a mobile user. The mobile receiver applies the corrections to its own measurements of pseudo distances, which allows the calculation of its position. If the fixed receiver and mobile receiver observe the same satellite at the same time, the common terms of error dt_S, Δr_{iono}, Δr_{tropo}, Δr_{rot} and dt_{rel}, cancel each other out by differentiation. Δr_{iono} and Δr_{tropo} are more effectively cancelled out if the mobile receiver is close to the reference receiver. Clock errors due to the satellite are strictly cancelled out. As the system is used to calculate a position, four visible satellites are needed (three for spatial coordinates and one for time) by the fixed station and the mobile receiver, making four differences.

2.6.2. *Time bases of instruments*

Time bases found in oceanographic instruments can be used to date data, position the instrument and measure the frequency emitted by different sensors. As the equipment can be immersed at different depths, it is subjected to thermal strains that, as a result, can lead to clock drifts that add to the natural drifts of the components.

Time bases are often made from piezoelectric oscillators. An oscillator is made up of a periodic signal source or resonator, stabilized by an amplifier. The periodic signal source is an electronic oscillating assembly. Resonators are often produced from cut quartz crystal, but they can also be produced from Langasite or $Li_2B_4O_7$. Theses are piezoelectric materials. Under the effects of strain, their surface is electrically polarized and, inversely, under the effects of electrical polarization they generate a mechanical action (see section 2.3.2). If polarization is alternating sinusoidal, a vibrating phenomenon can be caused that constitutes an oscillator. However, there is a condition that exists in which oscillation can be produced and maintained; the frequency generated must correspond to the resonance frequency of the resonator.

As temperature acts on the structure of materials, it causes a drift in the resonance frequency f of piezoelectric resonators. This drift can be described and compensated for if necessary, by a third-order polynomial given in relation [2.159] where A, B, C and D are constants determined from measurements.

$$\frac{\partial f(T)}{f(T)} = \frac{\left[f(T) - f_{Tref}\right]}{f_{Tref}} = A + BT + CT^2 + DT^3 \qquad [2.159]$$

With the exception of thermal strains, oscillators must also be resistant to mechanical strains related to usage (shocks, vibrations, variation in pressure, etc.). They must be chosen and conditioned for this.

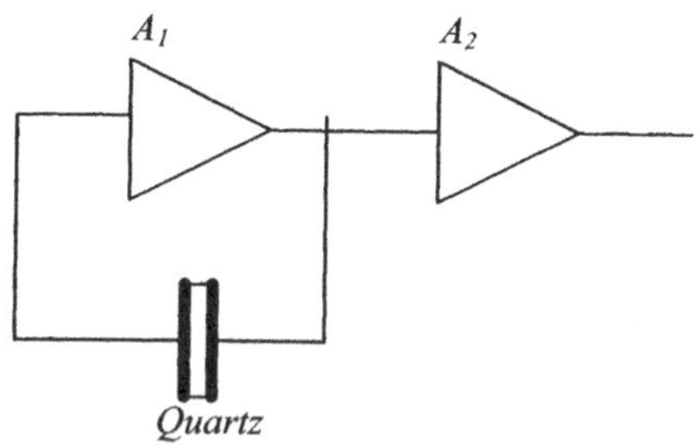

Figure 2.47. *Base structure of an oscillator. A_1 is an electronic assembly that maintains quartz oscillations and A_2 is an assembly that shapes the signal*

In practice, manufacturers of instruments propose different types of oscillators that can be implanted according to stability researched. The most simple and least well performing oscillators in terms of stability are the free oscillators, or XO. They are shown in Figure 2.47. With quartz resonators, any variation in phase ϕ in the oscillation loop results in a variation in frequency f according to the relation:

$$\frac{\Delta f}{f} = \frac{\Delta \varphi}{2Q} \qquad [2.160]$$

where Q represents overvoltage of the oscillator. The higher it is, the weaker the frequency for a given phase deviation. The operation of the oscillator in XO enables us to closely follow the variations in frequency that can be due to variations in temperature.

If we want to manage variations in frequency, voltage-controlled oscillators (VCXO) are needed. They are based on the principle of connecting in series the piezoelectric element with a remote voltage shift element. This element is often a varying capacity diode. The control voltage therefore allows an adjustment of the frequency deviation of the oscillator in a certain range (see Figure 2.48). In general, the voltage-to-frequency relation is nonlinear and these oscillators are manufactured according to linearity tolerances.

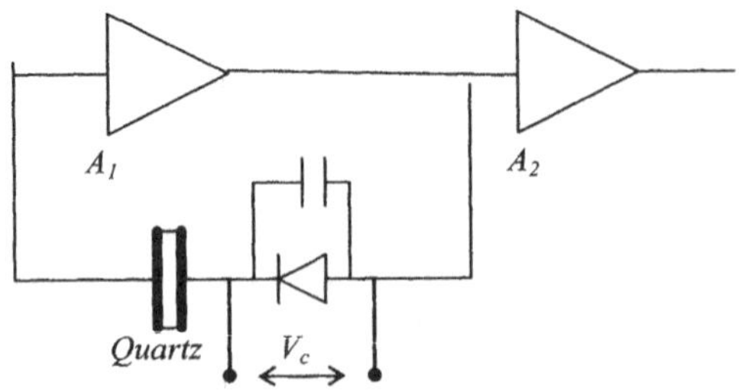

Figure 2.48. *Structure of a control voltage oscillator. A remote voltage shift element is inserted into the assembly*

However, drifts cannot be reduced without using temperature-compensated crystal oscillators (TCXO). The TCXO generates a signal to compensate for the drift in frequency after a variation in temperature. The amplification signal must be modeled by a polynomial of third-order to answer equation [2.159]. It is generated from a temperature sensor that is integrated into the oscillator. The sensor can be a thermistor (CTN), a platinum probe (section 2.1.2) or a silicon element (CTP) inserted into a network of resistances. The output of this network can be linked to a VCXO structure, as shown in Figure 2.48. However, progress in the miniaturization of electronics has enabled the integration of a memory in the oscillator, in which addressing is made from the temperature measured by the sensor. The value from the memory enables us to increment a control voltage. Unfortunately, compensations in temperature drifts produce leaps in frequency function of the sampling interval of the memorized voltage. Other processes are available that enable stabilization in temperature, such as dual resonator TCXO and dual mode TCXO, however they are more expensive.

Finally, when required stability reaches some 10^{-8}, access to oven-controlled crystal oscillators (OCXO) is required. The resonator and a part of its electronics are therefore placed in a temperature-controlled chamber kept at a fixed temperature by a heating–cooling system, a temperature sensor and a control loop. In order to minimize the effects of temperature fluctuations by as much as possible, the chamber is set to the resonator's temperature point of reversal, i.e. a few degrees above the maximum temperature of operation of the OCXO, in order to take self-heating into account. This regulation means that the OCXO is a consumer of energy and its warm up time is elevated. This reduces the number of uses it can be put to, particularly in autonomous instruments. However, OCXOs not only have the advantage of being able to improve the thermal stability of resonators, but they can also slow down their aging, limit phase noise and sensitivity to accelerations. Their size (see Table 2.8) and available space are criteria that also limit their usage as a reduction in size often causes an increase in the amount of current consumed, and a reduction in the stability of frequency.

When management of thermal drift is achieved, one of the essential criteria in the choice of oscillator technologies that remains is its stability in time or ageing. There are multiple origins of this phenomenon, and its consequences can be represented by a simplified equation where A and B are constants:

$$\frac{\Delta f(t)}{f} = A\ln(Bt+1) \qquad [2.161]$$

The logarithmic format of this relation shows that ageing presents a strong slope at the origin, which is progressively reduced over time. It is therefore preferable to buy pre-"aged" components to obtain a temporal drift that is less elevated and more linear. Table 2.8 shows the typical characteristics of annual ageing and thermal stability. Criteria other than ageing also need to be taken into account, such as: stabilization time, linearity of the electronically-controlled frequency and phase noise.

Typical dimensions in mm	Type of oscillator	Typical aging/year	Thermal stability
12 × 12 × 5	XO, VCXO, TCXO	10 to 100 ppm	±25 to ±100 ppm
20 × 20 × 5	XO, VCXO, TCXO	5 to 10 ppm	±1 to ±20 ppm
36 × 27 × 15	TCXO, OCXO	1 to 0.5 ppm	±1 to ±0.5 ppm
40 × 30 × 20	OCXO	$<5\times10^{-9}$	$< 5\times10^{-8}$
78 × 72 × 70	OCXO	$<1\times10^{-9}$	5×10^{-10} to 1×10^{-10}

Table 2.8. *Comparisons of characteristics of the aging and thermal stability of three types of oscillator technologies in relation to the size of their packaging*

In order to illustrate possible usages of these components, an example of the structure of a GPS receiver can be used (Figure 2.51). The receivers are typically composed of a frequency receiver segment, comprising an antenna, a filter and an output circuit designed to extract the pseudo random code from carrier frequencies. This input circuit is synchronized by a TCXO or VCXO reference oscillator, whose quality gives the accuracy of the receiver. The frequency f_{ref} of the reference oscillator is between 13 and 30 MHz, according to its usage. This input circuit sends this information to a GPS processor signal. This process needs a second "real time" clock based on a frequency f_{tr} at 32 kHz. This frequency can be obtained from a second oscillator or from a frequency divider connected to the reference oscillator. The data from the GPS processor are then transmitted to the host processor of the receiver.

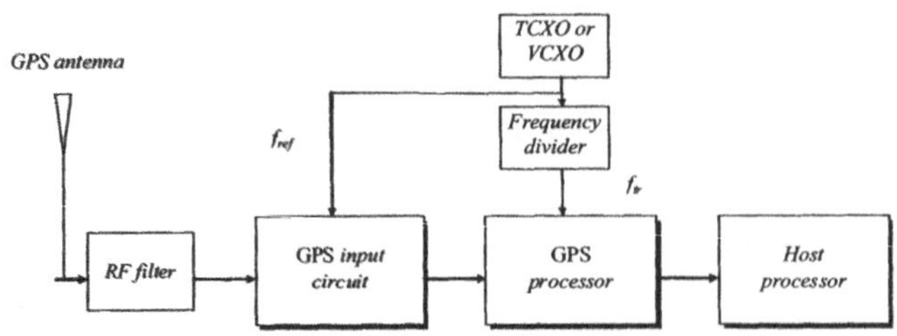

Figure 2.49. *General structure of a GPS receiver (Dallas)*

Although quartz oscillators are still present in most portable and non-portable applications, the development of the first atomic clocks in 2003, comprising a combination of *micro-electro-mechanical systems* (*MEMS*) and *vertical-cavity surface emitting laser* (*VCSEL*) technology, could change this supremacy. Before now, it was not possible to achieve the stability offered by the management of radiation emitted by the hyperfine atomic transition of atoms in mobile receivers, as this type of clock was voluminous and consumed a large amount of energy (see section 1.2.5). In 2003 researchers at the National Institute of Standards and Technology developed a prototype the size of a grain of rice, consuming merely 73 mW, with stability close to 10^{-11}.

Thanks to MEMS technology, the designers managed to create a cavity several microns in length in a silicon substrate, and place a vapor containing alkaline metal atoms (caesium or rubidium) inside before resealing the cavity with small glass lenses. An unconditioned VCSEL is placed under the cavity. It emits a wavelength of 852 nm that enables excitement of the metal vapor. The lenses are covered with a filter to reduce the intensity of the beam laser and a polarizing blade to create circular polarization. The superior lense focuses the beam on a photoelectric detector. In order to obtain a sufficient signal, the atom vapor needs to be heated and kept at a temperature of 85°C. To do this, the lenses have been recovered by a 30 nm *indium tin oxide* layer. This material is transparent to a wavelength of 852 nm and it generates heat, by Joule effect, as it is crossed by a current. To generate a signal on a frequency that allows the excitement of rubidium (6.8 GHz) or caesium (9.2 GHz), the assembly is connected to a VCXO, in which it stabilizes oscillation via a *phase locked loop*.

Symmetricom, Inc. commercialized the first miniature atomic oscillators of this type. The oscillators are presented in the form of a rectangular casing 16 cm^3, weighing 35 g and consuming 115 mW. The frequency they generate has an accuracy of $\pm 5\times 10^{-11}$ and a stability (measured with an Allan variance) above to 5×10^{-12} on 1 hour.

2.7. Determining position and movement

The elements of spatial positioning or geographical referencing are always associated with messages related to physical data measured in oceanography. For measurements taken from a boat, elements are provided with navigation instruments of this carrier. Autonomous instruments that move with the current are equipped with emitters or receivers that enable the trajectory to be followed to be found, and therefore their movement to be determined. This tracking became possible with the development of satellites in the 1970s and 1980s. Besides systems that are specially designed for oceanographic spatial studies (Seasat, Geosat, ERS 1 and 2, Topex Poseidon, etc.), satellite measurements have revolutionized positioning techniques and tracking. As it stands, two principal systems share the market:

– Argos, the only positioning device to collect data; and

– The GPS, whose principal has been explored in section 2.6 regarding dating.

Other systems, such as Iridium, ORBCOMM and OMNISTAR, have appeared in recent years. They are data transmission systems, but combined with a GPS receiver, Iridium, ORBCOMM and OMNISTAR, systems are able to compete against Argos.

2.7.1. *The Argos system*

2.7.1.1. *General architecture*

The Argos system was designed to localize fixed or slow moving platforms (movement speeds inferior to 1 km/day) and to collect physical data measured by such platforms. It was conceived following an agreement between the *National Oceanic and Atmospheric Administration* (NOAA), the *National Aeronautics and Space Administration* (NASA) and the *National Center for Space Studies* (CNES). It has been in operation since 1978, and implies interaction between three subsystems:

– a space segment;

– a ground segment; and

– platforms equipped with *transmitter terminals* (PTTs, Argos systems 1 and 2) or platforms equipped with *platform messaging transceivers* (PMTs, Argos system 3).

Its spatial segment is composed of satellites belonging to the NOAA, which function simultaneously on a low altitude of orbit around the Earth. They are equipped with the *data collection location system (DCLS)*, which was designed to localize and collect emissions from platforms that are found in the visible cone of satellites along their orbits.

The satellites have a circular heliosynchronous orbit, which means that their planned orbit rotates around the axis of the poles at the same speed as the Earth rotates around the Sun. These orbital planes make one rotation per year, allowing periodic crossing at fixed local solar time, above the platforms. The period of their orbit is approximately 102 mins (time needed by each satellite to rotate around the Earth), which enables making 14 crossings per day over the poles (at the equator, this number is reduced to six or seven). The Argos constellation is made up of three older-generation satellites known as Argos-1 (placed at an altitude of 830 ±18 km), and two new-generation satellites known as Argos-2 (placed at an altitude of 870 ±18 km). It also includes three MetOp satellites (launched in 2006, 2010 and 2014 respectively), which belong to the European Organization for the Exploitation of Meteorological Satellites-Eumetsat, plus one NOAA N' satellite launched in 2009, which make up the Argos-3 system.

Contrary to the GPS system, the earth or marine platform is in charge of emitting a signal to enable positioning. The detection cone of the satellite has a diameter of 5,000 km to the Earth's surface. This 5,000 km band passes through the two poles. As the Earth shifts 25° towards the West between two crossings, the coverage of the tracks is 2,200 km at the level of the Equator. This coverage is better if we approach the poles, as the number of daily collections is increased according to the latitude of the platform. However, the duration for which the instrument is visible while a satellite passes is constant, and is approximately 10 mins. This time is sometimes insufficient to collect a complete message, and with Argos systems 1 and 2 several crossings are needed to cover all of the data contained in the message generated by the platform. This problem was resolved by the Argos 3 system, thanks to the development of bidirectional communication. As the satellite passes above a ground receiving station (Wallops Island and Fairbanks in the USA or Lannion in France), it transmits information that has been stored in its onboard recorder. Argos-3 satellites are able to simultaneously receive messages emitted by 1,000 platforms in their receiving fields.

2.7.1.2. *Collection and distribution of data*

The data collection location system (DCLS) localizes platforms by measuring the Doppler effect (section 2.5.2). The frequency detected is slightly higher than that emitted by the transmitters as the satellite approaches them, and is reduced as it moves away. This shift can reach 7.4 kHz for a minimum satellite elevation angle. The nominal frequency of the transmitter is estimated from these frequency measurements. Argos 1 satellites are set to receive a frequency of 401.650 kHz that can be shifted by ±2 kHz in order to take into account the possible clock drift of the emmiters, or a default adjustment of their emission frequency. They have four reception channels of 5 kHz with a sensitivity of –128 dBm, while Argos 2 satellites can receive frequencies shifted from –20 kHz to +30 kHz around 401,650 kHz, and it has eight reception channels of 5 kHz with a reduced sensitivity of –131 dBm. Argos-3 satellites can receive shifted frequencies of ±30 kHz. This broadening of the spectrum has become indispensible because of the considerable increase in the number of emitters and thus the difficulties in reception that this has generated.

Messages emitted by each platform comprise 23 bits of synchronization followed by 1 bit of initialization, 4 bits of message length coding, 14 bits enabling platform identification, 6 bits of code error and 32–256 bits of data (for Argos-2). Recently, the identification of platforms has been made using 28 bits, which has enabled a higher number of identifiers to be used, satisfying a higher number of users. The broadening of identifiers has been made at the expense of the number of bits of data, which are now limited to 248. In Argos-1 and -2, carrier transmitters must be equipped with microcontrollers that are able to form messages of 248 bits, with data produced by their sensors. In Argos-3, the number of bits of data that can be transmitted is between 512 and 4,608 in addition to a "*checksum*", which considerably increases the flexibility of usage.

The transfer speeds in Argos-1 satellites was between 720 and 1,200 bits/s. In Argos-2, it is 2,560 bits/s and in Argos-3 it is 4,800 bits/s.

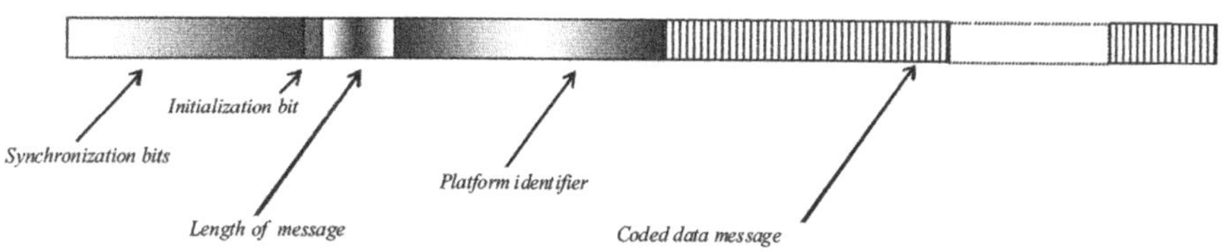

Figure 2.50. *Structure of an Argos coded data message*

Each frequency measurement is dated in relation to the arrival of a data message. Knowing the satellite's orbit and the Doppler shift, it is possible to determine the localization angle. Knowing the platform altitude (0 m for oceanographic instruments), it is possible to determine their sphere of altitude. The intersection of various cones obtained at each measurement with the sphere of altitude, enables their positions to be determined, albeit with ambiguity in relation to the satellite's track. To dispel this ambiguity, more information is needed, such as positions in previous tracking and the possible speeds of movement. A precise position can only be picked up if the satellite receives the transmitter signal at least four times when tracking. This method of localization assumes that satellite positions are known precisely and at all moments in time. The determination is made by orbitography ground stations that are located on geodetic reference sites. The temporal adjustment is made in relation to a dating station equipped with a caesium clock, installed in Toulouse.

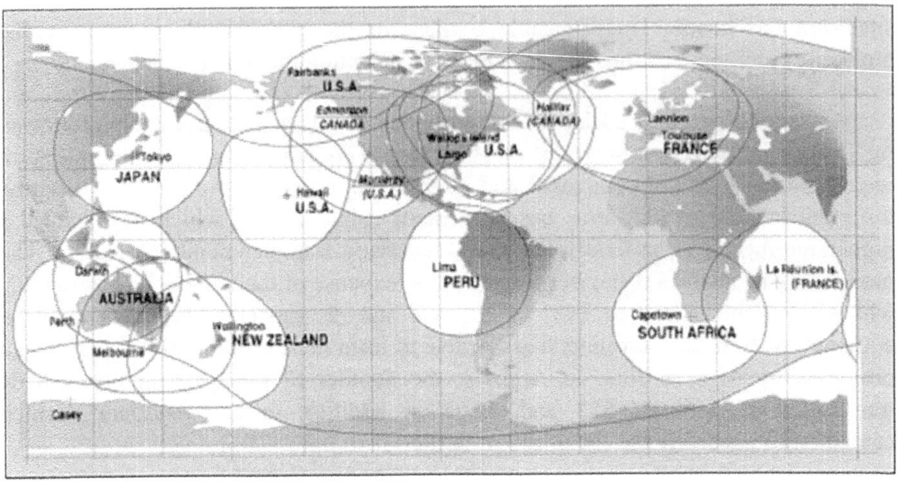

Figure 2.51. *Repartition of all Argos receiver stations on the Earth's surface (Courtesy of CLS ARGOS)*

Distribution of data from the Argos system has been entrusted with "*CLS*", also based in Toulouse, and with its sister company *Service Argos, Inc.*, which is based in Largo, America. The two companies regroup and process the data collected by all receiver stations across the continents (see Figure 2.51). Through the intermediary of existing communication networks, users can also recuperate their data (in approximately 20 mins). *CLS* gives a quality index to messages received. These indexes are a function of the given location, which depends on emission conditions (emission power available, orientation of the antenna, etc.) and the reception conditions. Distribution of positioning errors is assumed to be Gaussian. The quality index given by *CLS* is distributed into three classes of standard uncertainties (equivalent to standard deviations). Class 3 corresponds to a standard uncertainty that is less than or equal to 150 cm, signifying that as distribution is Gaussian, 68% of positions calculated are statistically in the ±150 m range. Class 2 corresponds to a standard uncertainty of less than or equal to 350 m, and class 1 corresponds to a standard uncertainty of less than or equal to 1 km.

In delayed time it is possible to obtain tighter localization gaps (by about 20 m) if the platform is stationary. Data obtained after several crossings are therefore used, frequency is the only quantity that is considered to vary from one crossing to another. Orbit calculation, calculated from three consecutive days of Doppler shift measurements from orbitography platforms, is also used. Interpolation is needed next to obtain the exact satellite ephemeris after crossings over the platform. Ephemerides are elements that enable the variation in relation to the time of the Keplerian orbit of the satellite to be determined. This calculation is also extrapolated in order to obtain the following days' positions and to provide users with data in quasi-real time. Finally, in delayed time it is possible to select measurements from criteria relating to satellite and platform positioning before calculating the position of the platform.

2.7.1.3. *Ground instrumentation and evolution of the system*

Devices exist that are able to receive and decode Argos-1 and -2 messages emitted by platforms on the Earth's surface. They were principally created to test and develop instruments that use Argos transmitters. They can also be integrated into goniometric systems used in oceanography to localize and recover drifting buoys (see Figure 2.52).

Transmitters or PTTs are generally presented in the form of cards to be integrated into instruments (see Figure 2.53). The PTTs produce a message with a short duration (0.36–0.92 s), which modulates a carrier frequency. The message is emitted in a short period of repetition lasting from 45 to 300 s according to the utilization mode of the platform. Drift of the period (or *jitter*) of about 10% of its value is generated so that transmitters that have the same periodicity of emission can be detected without risk of covering their messages.

As previously mentioned, Argos-1 and -2 systems improve the probability that a message can be received by repeating it several times during the satellite crossing. If we estimate that the elementary probability of acquisition is 0.84, and that the message is repeated n times, the probability of acquisition can be estimated with relation:
$P = 1 - (1 - 0.84)^n$. For $n = 3$, for example we obtain $P = 0.996$. Repeating the messages also enables a reduction in the bit errors that alter the system performance. Treatment software enables the elimination of messages that do not conform.

The Argos-3 system does not require repetition. The PMTs are programmed to only transmit when a satellite is in view. The satellite returns a signal of "*acknowledgement*" to the emitter when data have been received without error (verification of "*checksum*"), and the emission stops. The PMTs are equipped with software that forecasts satellite crossings. Likewise, they can calculate the exact moment and duration of the next crossing. Their data are updated by the system operators, via the space to ground link. This system enables the energy independence of the platforms to be increased.

Users can also use the downlink system to pilot their applications. This is possible by sending messages of 128 bits and managed by the, *downlink message management center* (*DMMC*) situated at the CLS center in Toulouse. The user can communicate with the DMMC via a web interface. The DMMC re-transmits messages to ground stations situated in Toulouse, Fairbanks or Svalbard. When messages are successfully received by the PMT, the PMT sends an "*acknowledgement*". The user can remotely activate interrupters and modify sensor configurations or programming.

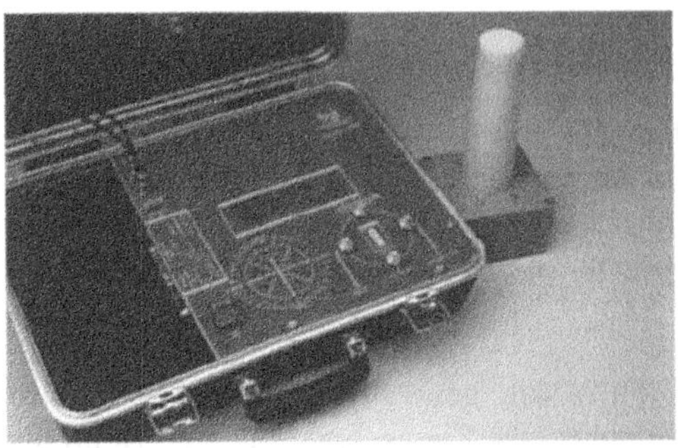

Figure 2.52. *Gonio SERPE-IESM 400P receiver with antenna, enabling localization of drifting buoys (Courtesy of SERPE – IESM). Newer, equivalent systems have been commercialized by Doppler Systems LLC*

The PTT transmitter made by SERPE-IESM is an example of ARGOS-1 or -2 platforms (see Figure 2.53). Associated with an emission antenna and an energy source, it can be used in diverse applications: drifting buoys, platforms following fishing fleets, stratospheric balloons, etc. It consists of an MCX coaxial connector that enables connection of the antenna and a connector for power and the input of data in analog format. It is equipped with a 12-bit analog-to-digital converter. Emission frequency is generated by digital synthesis from a TCXO oscillator (section 2.5.2). It can optionally be equipped with an OCXO, and emits over two frequencies, 401.650 and 401.620 MHz. It also has a channel management system. Its emission power is between 30 dBm and 36 dBm.

The Argos-3 system is compatible with this type of transmitter but it takes all of its power from the PMTs that have been available since 2007. Kenwood and Elta have been chosen to develop these transmitters.

Out of a total of more than 6,300 transmitters currently in existence, only 60% of them are being used to equip oceanographic instruments.

Figure 2.53. *The OEM PTT 07 transmitter made by SERPE-IESM (Courtesy of SERPE IESM)*

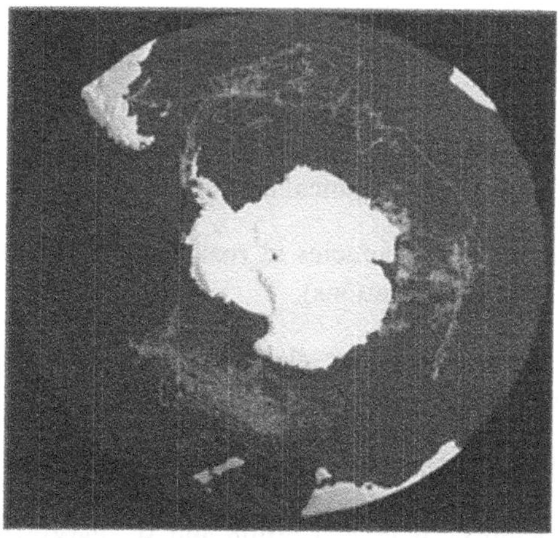

Figure 2.54. *Representation of sea crossings obtained from Argos platforms fixed onto marine mammals. The beacons are also equipped with temperature and conductivity sensors, allowing the collection of a number of important pieces of data on the Antartic Ocean (Guinet C., CEBC-CNRS).*

It is also important to note that the National Space Development Agency of Japan (NASDA) has launched satellites called ADEOS-II that are also equipped with the Argos-3 system. A ground segment was introduced in Japan and has been integrated into the existing CNES/NOAA network.

2.7.2. *The global positioning system*

Two satellite positioning systems are currently in operation: NAVSTARR, which stands for *Navigation System by Timing and Ranging*, is commonly known as GPS, and GLONASS (section 2.6.1.1). A third Chinese system called Beidou is being built to cover the Asia-Pacific region and should be completed in 2020. A fourth European system called Galileo will be in service from 2014 with 18 satellites at 23,600 km from the Earth. It will include 30 satellites by 2020. Today they are grouped under a communal project, the *Global Navigation Satellite System* (*GNSS*), the term otherwise reserved for GLONASS.

Among these three systems, the most well known and most frequently used in oceanography is GPS. It has rendered most other principles used in navigation obsolete, except the *LORAN-C* and its equivalent, the Russian Chayka chain, which continue to be used as backup systems. This means that certain users do not need to be entirely dependant on the GPS and its political and technical management. The GPS's strong points in relation to other positioning systems that were (or still are) functioning, reside in the following facts:

– it enables the determination of latitude, longitude, altitude and time simultaneously;

– it assures world coverage without being particularly badly effected by meteorological conditions or geographical areas;

– it is able to obtain high accuracies in real time (2–10 m), as in delayed time (several mm in fixed-point observations);

– it is free to access and authorize global diffusion of a timescale connected to international references; and

– its receivers can be miniaturized and taken on board vessels or platforms.

Two American scientists, Ivan A. Getting and W. Parkinson, are considered to be the fathers of GPS. Getting came up with the concept when he was manager of Raytheon's research and development department in the 1950s, and Parkinson was its architect in the 1970s, while he was a project manager in the US Department of Defense.

2.7.2.1. *Space, ground and user segments*

In 2010, the GPS constellation was composed of 32 operational satellites. These satellites are classified into four categories according to their launch date and onboard technology:

– category IIA is made up of 11 satellites launched between November 1990 and November 1997;

– category IIR contains 12 satellites launched between July 1997 and November 2004;

– category IIR-M includes eight satellites launched between September 2005 and August 2009; and finally

– category IIF contains one satellite launched in August 2010.

These satellites are in circular orbit at a nominal altitude of 20,183 km, which gives them a rotation period of 11 hours and 56 mins. It is an aging constellation, as today a number have exceeded their nominal lifespan, which is an average of five years. However, the system was established in order to function with 21 satellites, and three active backups, spread over six orbital planes with an inclination of 55° on the equatorial plane. They are equipped with rocket boosters that enable their position and orbital stability to be readjusted. In terms of electronics, each satellite is equipped with transmitter–receivers, two caesium or rubidium clocks (the latter in the newest satellites) and a calculation unit. Several numbering systems allow their identification. Their number is contained in the message provided by satellite to users. By comparison, the GLONASS system is made up of one constellation of 27 satellites launched between December 2004 and December 2010. All of the satellites are equipped with caesium clocks. Only 20 of them have been declared fit for service.

The ground segment is made up of tracking and calculation stations and message download stations. The tracking stations are situated in Hawaii in the Pacific, Colorado Springs, Ascension in the Atlantic, Diego Garcia in the Indian Ocean and Kwajalein in the archipelago of the Marshal Islands. Colorado Springs is in charge of recording data collected by the other stations and controlling satellites' orbits. All of stations record signals emitted on the two GPS frequencies, L_1 and L_2, and they decode theses signals with the help of stabilized cesium clocks. The master station at Colorado Springs is in charge of calculating the ephemerides of the satellites, the correctional parameters of the clocks and the coefficients of the model that enable corrections to ionospheric propagation. These data are retransmitted to the satellites by stations in Ascencion, Diego Garcia and Kwajalein with the help of a link radio on band S over frequencies 1.783 7 GHz and 2.227 5 GHz.

Users have unrestricted and free access to the system. All that is required is possession of an adequate receiver (see Figure 2.51) in order to get:

– *absolute positioning*, which is based on the measurements taken from signals emitted by satellites and identified in a system of coordinates positioned and orientated in relation to the Earth; or

– *a position* said to be *relative* or *differential* in relation to another fixed reference point. This positioning is known as RTK or *real time kinematic*.

The receiver measures a pseudo distance, as the value obtained is affected by an error function of the shift between the receiver and satellite clocks (section 2.6.1.2). The positioning of the receiver, or more precisely its antenna, can be done in real time, or in delayed time if we want to reduce the measurement uncertainty from averaging. In all cases, we need to solve four equations (three for positioning coordinates and one for clock shifts). The receiver must therefore be in the view of four satellites at the same time in order to resolve the system. The tracking capacity of the satellites is an important element when choosing a receiver, but the number of frequencies it can receive, the codes that it can interpret and its capacity to process differential corrections if necessary must also be considered.

2.7.2.2. *Coding*

Each satellite is equipped with two high stability onboard clocks (caesium or rubidium) that provide a fundamental frequency of 10.23 MHz (see Figure 2.55). This frequency is multiplied to generate two carriers known as L_1 and L_2 (as they are in band *L*). L_1, which corresponds to frequency 1.57542 GHz, and L_2, which corresponds to frequency 1.22760 GHz, are then modulated in phase according to a process known as *phase shift keying* (*PSK*). This process is used to transmit a digital signal and consists of the generation of phase shifts of 180° or 90° according to the particular PSK code used (modulation BPSK, DPSK or QPSK). The modulation chosen enables two pseudo random codes to be sent known as *pseudo random noise* (*PRN*) and data messages that indicate the epherimides of the satellite, clock parameters, coefficients of the ionospheric correction model, the health of the transmitter satellite, the deviation time between GPS time and UTC-USNO time and other data required for good operation of the system. Navigation messages are emitted in the two carriers in binary format, at a speed of 50 bit/s. A complete message contains 1,500 bits and is made up from five sub-frames of 300 bits.

L_1 transports two types of PRN code: code C/A (*coarse acquisition*) and code P (*precision*), while L_2 only transports code P. Code C/A is used for the *standard positioning service* or SPS. These services are accessible to all users, while code P is reserved for the American Army and certain authorized users as it defines *precise positioning services*. Access to code P is protected by keys.

All signals transmitted agree with each other as they are a product of the same clock, and the different codes have a phase relation known with carriers L_1 and L_2. They follow a coding known as Gold with maximum length. The period of C/A gold codes is 1 ms, and they are emitted over frequency 1.023 MHz. Until May 2000, the C/A code and message content was voluntary degraded by the American administration in order to limit civilian usage of GPS (see the official US position at http://igscb.jpl.nasa.gov). This is called *selective availability* (SA). SA limited considerably the accuracy of absolute positioning measurements.

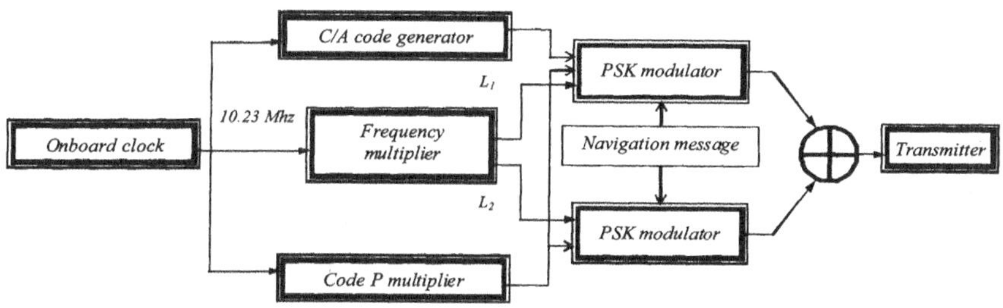

Figure 2.55. *Simplified diagram of the GPS emission signals system*

2.7.2.3. *Measurements of distances*

Two modes of positioning can be put in place with the GPS system: absolute positioning and relative or differential positioning. The first is generally used in autonomous oceanographic instruments, as their position is directly defined by measurements taken with signals from satellites (see Figure 2.55). The second mode requires a reference point equipped with a fixed receiver and a means of radio (or satellite) communication between the reference point and the mobile receiver. Due to the accuracy they are able to reach, and the means to be used, differential techniques are frequently used in geodetics, in hydrography for the positioning of boats and, more recently, in systems that measure sea level.

In absolute positionning, C/A or P codes are used by the receiver so that it can localize itself. It must also extract the navigation message, which is the orbital position of satellites at the time of measurement. This position is expressed as a terrestrial benchmark. The receiver must then measure time and deduct the distance that separates its antenna from the satellite's emission antenna. We say that the receiver takes a measurement of "pseudo distance" (section 2.6.1.2). In fact, the distance is obtained by multiplying the time calculated with the speed of light, as it is defined in the geodetic system WGS 84.

This measurement process is based on the utilization of a loop of servo-control code or *delay lock loop*. Code detection is achieved by correlation between a local code generated by the receiver and that emitted by the satellite. As soon as the receiver "fixes" a satellite, i.e. as soon as correlation between both codes is optimal, it interprets and locally reproduces (with an accuracy authorized by its clock) the C/A or P code sent. By "locking" the repetition–interpretation loop, it can then "track" the satellite that precisely dates the data transition of the C/A or P (time t_e) codes sent. The receiver reads the arrival time of the transition from its clock (time t_r), known as *time of arrival* (TOA) and deducts the propagation time from it $t_r - t_e$. It corresponds to the shift made on the local code that aligns it with the code sent by the satellite.

Due to this mechanism, we can consider that the measurement of pseudo distance is principally effected by the error due to time shift that may occur between the satellite's clock and the receiver's clock. The shift is due to the drift of the receiver's clock, and to the differences in speed and gravity between the satellite and the receiver, causing relativist effects. The system takes these effects into account at synchronization by shifting lightly emitted signals. Crossing of the ionosphere and the troposphere also causes errors. The receivers contain models that enable these errors to be corrected, but this correction cannot be perfect, considering the lack of knowledge of water content in the atmosphere, and ignorance of solar activity. If dt_s is the sum of residual errors in the time difference between emission and reference, and dt_r is the deviation of time introduced by elements of the receiver (antenna, clock, circuits, etc.), the time measured δt will take the following form:

$$\delta t = (t_r - t_e) + (dt_S - dt_r) \qquad [2.162]$$

The pseudo distance r^{ps} is obtained by multiplying each part of relation [2.162] by the speed of light c. The expression $c(t_r - t_e)$ that is obtained therefore represents the geometric distance of the satellite–receiver r, to which the error term Δt is added:

$$r^{ps} = r + c(dt_S - dt_r) = r + c\Delta t \qquad [2.163]$$

The value of r^{ps} is unambiguous to the nearest 300 km. Its uncertainty of measurement is limited to several meters (50 cm to 5 m to 95% confidence) by errors linked to measurements of time and propagation conditions that add noise to its evaluation. To improve this uncertainty, receivers enable the averaging of r^{ps} values over several minutes and, likewise, a reduction of the standard deviation of measurements. Evidently, this is only possible if we work in delayed time or in a slightly delayed time.

More complex techniques are also used in order to improve the uncertainty of measurements. The most sophisticated receivers enable sinusoidal carrier phases L_1 and L_2 to be worked on directly and accessed with milimetric precision. On arrival, the signal phase emitted by the satellite is compared to the phase of a reference signal generated by the receiver. The difference in phase that exists between the two corresponds in wavelength to a surplus distance. If the total number of periods made by the signal from its emission point can be determined, it is possible to directly obtain the transmitter–receiver distance with a high accuracy and a resolution of about a millimeter.

However, as wavelengths L_1 and L_2 are about 20 cm (19 cm for L_1 and 24 cm for L_2), the phase measurements are accurate to the nearest 20 cm. It is possible to widen the field of non-ambiguity by creating beating between these frequencies. This technique is used in bifrequency receivers used in geodetics. It is also possible to combine pseudo distance and phase measurements in order to benefit from the non-ambiguity of pseudo distance measurements and low noise levels in phase measurements. This combination is done by calculating the instantaneous differences between the distance value deduced from the code and that deduced from the phase. Values of differences are then digitally filtered by a simple first-order filter whose time constant can be adapted to apparent code fluctuations. The output of the filter is recombined with the extracted measurement of phase distance to provide a smoothed pseudo distance value. According to the price and the complexity of the GPS receiver, these operations may or may not be carried out.

They are carried out in real time kinematic utilization. The fixed receiver exploits phase shift of L_1 and L_2 signals and transmits corrections to mobile receivers situated in a radius of 30 km. Ambiguities are reduced to less than 10 s when six satellites are in view. Accuracy of positioning is therefore centimetric. It is estimated to be 1% of the wavelength of L_1 and L_2, so 2.0 mm and 2.4 mm. $10^{-6} \times D$ is added, where D is the distance between the fixed station and the mobile receiver, in order to estimate the accuracy of the positioning of this receiver.

2.7.2.4. *Calculating kinematic position*

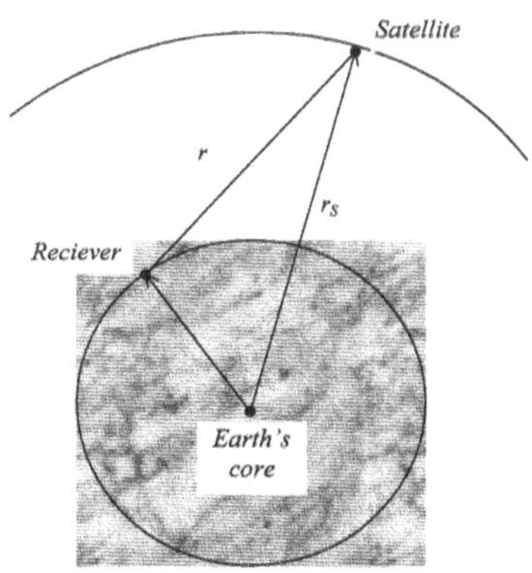

Figure 2.56. *Principle of absolute positioning of a receiver*

This calculation is based on the resolution of a system of equations with four unknowns: X, Y, Z that represent the position of the receiver, and Δt, which is defined by equation [2.164]. Therefore before anything else, the receiver must have tracked at least four satellites in order to give a position value. If more than four satellites are tracked, it will be possible to make a more reliable estimation of the uncertainty of measurement. The number of position equations to be solved therefore depends on the number of satellites tracked. The satellites are considered as immobile at the time of measurement. Their coordinates are calculated in relation to the Earth's core (see Figure 2.56). Each satellite sends its epherimide to the receiver so that it can determine its position.

If X_i^S, Y_i^S and Z_i^S represent coordinates of satellite i taken from a benchmark centred at the Earth's core, if r_i^{ps} represents corresponding pseudo distance, and if n satellites are tracked, it is necessary to resolve n equations of this type:

$$r_i^{ps} = \sqrt{\left(X_i^S - X\right)^2 + \left(Y_i^S - Y\right)^2 + \left(Z_i^S - Z\right)^2 + c\Delta t^2} \qquad [2.164]$$

If $n = 4$, the system can only be resolved by successive iteration. The iterations are carried out by assessing the value of an origin position and that of a true position corresponding to the measurement. The calculation is made from an "innovation" or *re-keying* equation. If T is the renewal period of the calculation, we have: $t = kT$, where k is an integer. Innovation I_k, which in fact corresponds to the difference between the distance measured minus the distance estimated at moment t for a satellite takes the form of equation [2.165], where b corresponds to the measurement noise and $D_{(k-1)}$ to the distance estimated to the previous iteration:

$$I_k = r_i^{ps} - D_{(k-1)} - c\Delta t_{(k-1)} + b \qquad [2.165]$$

$$D_{(k-1)} = \sqrt{\left(X_i^S - X_{(k-1)}\right)^2 + \left(Y_i^S - Y_{(k-1)}\right)^2 + \left(Z_i^S - Z_{(k-1)}\right)^2} \qquad [2.166]$$

The number of iterations, which corresponds to the time needed by the receiver to relock, will therefore depend on the initial value that can be allocated to $D_{(k-1)}$. If this value is close to the value we are looking for, the algorithm will rapidly converge. If the first perimeter of the calculation is high, convergence will be slower.

In fact, values X, Y and Z that we are looking for can be split up as follows:

$$X = X_{(k-1)} + \Delta X_i$$

$$Y = Y_{(k-1)} + \Delta Y_i$$

$$Z = Z_{(k-1)} + \Delta Z_i \qquad [2.167]$$

ΔX_i, ΔY_i and ΔZ_i are therefore new unknowns to be determined. They correspond to coordinates of the difference $r_i^{ps} - D_{(k-1)}$. With this formula, it is possible to calculate the limited development of the function r_i^{ps} in Taylor serials. If this development is limited to the first order, we have:

$$r_i^{ps} = D_{(k-1)}\left(X_{(k-1)},Y_{(k-1)},Z_{(k-1)}\right)+\frac{\partial D_{(k-1)}\left(X_{(k-1)},Y_{(k-1)},Z_{(k-1)}\right)}{\partial X_{(k-1)}}\Delta X+\frac{\partial D_{(k-1)}\left(X_{(k-1)},Y_{(k-1)},Z_{(k-1)}\right)}{\partial Y_{(k-1)}}\Delta Y$$
$$+\frac{\partial D_{(k-1)}\left(X_{(k-1)},Y_{(k-1)},Z_{(k-1)}\right)}{\partial Z_{(k-1)}}\Delta Z \qquad [2.168]$$

Partial derivatives of relation [2.168] can be calculated from equations [2.167], and we have:

$$\frac{\partial D_{(k-1)}\left(X_{(k-1)},Y_{(k-1)},Z_{(k-1)}\right)}{\partial X_{(k-1)}} = -\frac{X_i^s - X_{(k-1)}}{D_{(k-1)}} = a_X$$
$$\frac{\partial D_{(k-1)}\left(X_{(k-1)},Y_{(k-1)},Z_{(k-1)}\right)}{\partial Y_{(k-1)}} = -\frac{Y_i^s - Y_{(k-1)}}{D_{(k-1)}} = a_Y$$
$$\frac{\partial D_{(k-1)}\left(X_{(k-1)},Y_{(k-1)},Z_{(k-1)}\right)}{\partial Z_{(k-1)}} = -\frac{Z_i^s - Z_{(k-1)}}{D_{(k-1)}} = a_Z \qquad [2.169]$$

By combining equations [2.168] and [2.169], the innovation equation [2.165] can in turn be rewritten:

$$I_k = a_X\Delta X + a_Y\Delta Y + a_Z\Delta Z - c\ \Delta t_{(k-1)} + b \qquad [2.170]$$

If four satellites are tracked, we will have four relations of this type to solve, which leads to the writing of a matrix equation:

$$\underline{I} = \underline{\underline{A}}\,\underline{x} + b \qquad [2.171]$$

Noise measured, b, is assumed to be centered Gaussian with zero average and a variance of σ^2_b. It is also occasionally known as UERE or *user equivalent range error*. $\underline{I}$ represents the innovation vector, $\underline{A}$ is the measurement or transfer matrix, and $\underline{x}$ the vector of states to be assessed. They are shown in the following form:

$$\underline{I} = \begin{bmatrix} I_k^1 \\ I_k^2 \\ I_k^3 \\ I_k^4 \end{bmatrix} \quad \underline{A} = \begin{bmatrix} a_X^1 & a_Y^1 & a_Z^1 & -1 \\ a_X^2 & a_Y^2 & a_Z^2 & -1 \\ a_X^3 & a_Y^3 & a_Z^3 & -1 \\ a_X^4 & a_Y^4 & a_Z^4 & -1 \end{bmatrix} \quad \underline{x} = \begin{bmatrix} \Delta X \\ \Delta Y \\ \Delta Z \\ c\Delta t \end{bmatrix} \qquad [2.172]$$

The position difference vector estimated $\hat{\underline{x}}$ is obtained by inversing matrix $\underline{A}$:

$$\hat{\underline{x}} = \underline{A}^{-1} I \qquad [2.173]$$

It presents a noise whose variance–covariance matrix is:

$$\underline{\mathrm{var}}(\hat{\underline{x}}) = \left(\underline{A}^T \underline{A}^{-1}\right) \sigma_b^2 \qquad [2.174]$$

(The symbol ‘T’ represents the transpose of the matrix.)

If $n > 4$, the previous system can be resolved from a least-square regression technique, which consists of minimizing the differences between the measured and estimated points. The application of this technique leads to an estimation of the position difference vector:

$$\hat{\underline{x}} = (\underline{A}^T \underline{A})^{-1} \underline{A}^T \underline{I} \qquad [2.175]$$

However, as measurements are subjected to a noise in which variance can be different from one satellite to the next, it is necessary to allocate a weight w_i to them, such as:

$$w_i = \alpha^2 / \sigma_{bi}^2 \qquad \text{with: } \frac{1}{\alpha^2} = \sum_{i=1}^{n} \frac{1}{\sigma_{bi}^2} \qquad [2.176]$$

Therefore, the relation showing the position difference vector becomes:

$$\hat{\underline{x}} = (\underline{A}^T \underline{wA})^{-1} \underline{A}^T \underline{wI} \qquad [2.177]$$

Its variance–covariance matrix is as follows:

$$\underline{\text{var}}(\hat{\underline{x}}) = \left(\underline{A}^T \,\underline{wA}^{-1}\right) \qquad [2.178]$$

In order to find coordinates of the position being researched, all that is needed is the addition of $\hat{\underline{x}}$ to the calculated vector position of the previous iteration.

2.7.2.5. *Uncertainties in the calculation of position*

From the variance–covariance matrix of the differences in vector position represented in equation [2.174], it is possible to estimate the uncertainties of the measurements taken. This matrix takes the following form:

$$\underline{\text{var}}(\hat{\underline{x}}) = \begin{bmatrix} \sigma_X^2 & C_{XY} & C_{XZ} & C_{X\Delta t}c^2 \\ C_{YX} & \sigma_Y^2 & C_{YZ} & C_{Y\Delta t}c^2 \\ C_{ZX} & C_{ZY} & \sigma_Z^2 & C_{Z\Delta t}c^2 \\ C_{\Delta tX} & C_{\Delta tY} & C_{\Delta tZ} & \sigma_{\Delta t}^2 c^2 \end{bmatrix} \sigma_b^2 \qquad [2.179]$$

where σ_i represents the standard variation of variable i and Cij the covariance of variables i and $j \in \{X, Y, Z, \Delta t\}$. As the variables are independent of each other, the trace calculation of $\underline{\text{var}}(\hat{\underline{x}})$ enables an estimation of the uncertainty of the measurements. It is known as *dilution of precision* (*DOP*). Different (*k*) DOPs are also defined:

$$\textit{geometric dilution of precision}\ (\text{GDOP}) = \sqrt{\sigma_X^2 + \sigma_Y^2 + \sigma_Z^2 + \sigma_{\Delta t}^2 c^2} \qquad [2.180]$$

$$\textit{Position dilution of precision}\ (\text{PDOP}) = \sqrt{\sigma_X^2 + \sigma_Y^2 + \sigma_Z^2} \qquad [2.181]$$

$$\textit{Horizontal dilution of precision}\ (\text{HDOP}) = \sqrt{\sigma_X^2 + \sigma_Y^2} \qquad [2.182]$$

$$\textit{Vertical dilution of precision}\ (\text{VDOP}) = \sigma_Z \qquad [2.183]$$

$$\textit{Time dilution of precision}\ (\text{TDOP}) = \sigma_{\Delta t}\, c \qquad [2.184]$$

In order to determine the uncertainty of geometric positioning, the GDOP must be multiplied by the noise of measurement σ_b or σ_{UERE}. Likewise, in order to determine the uncertainty of horizontal or vertical positioning in time, HDOP, VDOP or TDOP must be multiplied by σ_b.

All of these variables have the dimension of a distance. The GDOP represents an estimation of the uncertainty in the pseudo distance calculation. PDOP, HDOP, VDOP and TDOP enable estimation of the influence of components of this uncertainty. The greatest uncertainty relates to the VDOP, as it is difficult to define the component in z. Several studies have been carried out in an attempt to improve this. The most effective solution consists of integrating an atomic clock (caesium or rubidium) in the GPS receiver. This is known as *clock-aided* GPS. It has been demonstrated that there is a strong correlation between the stability of the receiver's clock and the error in vertical positioning [YAN 11]. A high-stability clock also enables a reduction of the HDOP, but to a lesser extent, and it also increases the reliability of positioning when only three satellites are in view of the receiver.

Furthermore, it is important to know that the component in z is referenced in relation to the reference ellipsoid. The reference *ellipsoid* is a fictive plane surface generated by an ellipse that would rotate around the axis of the poles. It is characterized by an equatorial ray and flattening out. It is a "convenient" base from which a geodetic system can be defined, by relative positions of points situated at the Earth's surface.

Yet, in oceanography it is necessary to reference height in relation to the *geoid*, which corresponds to the mean sea level in the absence of all movement and astral attraction. This mean level represents an equipotential field of gravity, which is combined with mean sea level, reduced by the dynamic effects of wave movements. The geoid is an estimation of the true physical figure of the Earth. As it does not have a uniform density, the surface of the geoid is not flat, but is composed of a series of "hills" and "valleys" that are sharp images of the submarine environment and mass distribution inside the Earth. The "hills" correspond to the excess of mass inside the Earth, and the "valleys" correspond to its deficits. Undulations caused by ocean and Earth tides, and general circulation of the ocean, are added to mean sea level.

The difference between the ellipsoid and the geoid is given by a local calculation model that adds yet another uncertainty to the determination of height. Even if this variable is not fundamental in the exploitation of oceanographic instruments, it is of great importance for the study of variations of water height, and is of much interest to oceanographers.

2.7.2.6. *"Augmentation" systems and GNSS*

The performances of a navigation system can be characterized by its accuracy, availability, continuity and integrity. In order to enforce the last point, which underpins a system that does not deliver any false information, the *Federal Aviation Administration* developed the *wide area augmentation system* (*WAAS*) and the *local area augmentation system (LAAS*). Its aim, at the beginning, was to make the GPS positioning system more reliable in airplanes, in order to securely use it in all stages of the flight. The WAAS and LAAS are now available through the GPS receivers of boats and through several public receivers. The WAAS principally covers North America. Europe has an equivalent system known as the *European Geostationary Navigation Overlay System* (*EGNOS*) and Japan uses the *MTSAT* system, which stands for *Multifunctions Transport Satellite*. These services enter the notion of GNSS.

WAAS, EGNOS and MTSAT make GPS observations from fixed receivers and diffuse differential corrections to registered clients with the help of geostationary satellites of communication systems by organizations; INMARSAT (section 3.1.3), ESA (European Space Agency) or NASA. Increase in the reliability of GPS measurements is only possible if there are a sufficient number of satellites visible at all times. In order for this to be achieved, WAAS and EGNOS have, or will have, the possibility of working with the GLONASS and Galileo systems in addition to GPS.

In order to ensure the integrity of data, they can observe the "health" of the satellites in these systems, detect breakdowns in their clocks and track their drifts. These data are sent to users in the form of a classic GPS signal that delivers supplementary receiver equipment to users. However, transfer speeds are 250 bits/s while GPS transmits messages at 50 bits/s, and the organization of words and coding are different. Therefore, not all receivers are capable of using these services. When they are, the accuracy of positioning is or can be permanent at 1 m or less. As GPS receivers are constantly advancing the level of miniaturization and performance, centimetric accuracy is possible with equipment whose price is continuing to fall.

2.7.2.7. *Positioning systems and data transmission*

Taking into account the market that represents global positioning needs, satellite communication and transmission systems have adapted by offering systems equipped with GPS positioning. Likewise, a high number of oceanographic applications (drifting buoys among others) are equipped with positioning boards and data transmission, such as Iridium, ORBCOMM or OMNISTAR.

– The ORBCOMM constellation is currently composed of 29 satellites positioned at an altitude of 775 km. They offer global coverage day and night. The

Iridium constellation consists of 66 moving active satellites that are positioned at a low altitude of 760 km on a polar orbit. They also offer global coverage and will be replaced in the coming years with 81 new satellites (by Thales Alenia Space technology) that will form Iridium Next. The satellites can communicate with each other and send their data to the ground via the Ka transmission band. The transmission is carried out by modem, according to three modes:

– SBD mode or *short burst data*, which is bidirectional messaging of small text or binary messages (< 2 KBYTE);

– Dial-up mode, which enables direct communication from modem to modem (via an Iridium receiver satellite and a driver installed on a client computer) with a TCP-IP protocol. Data transfer is therefore carried out at 2.4 kbits/s.

– *Router based Unrestricted Digital Internetworking Connectivity Solutions* (RUDICS) mode. With this solution, data is sent directly to the Internet server of a computer.

Other systems have been developed in order to better exploit satellite communications, such as VIZADA, a service that enables the interconnection of different satellite operators (Inmarsat B and C, Iridium, etc.), thanks to the development of five optimally-positioned satellites.

2.8. Determining the height of water

The determination of water height is important information in numerous areas: navigation, river hydraulics, bathymetry, oceanography, etc. Sea level varies principally in relation to the position of stars in relation to the Earth. This is what generates the tidal phenomenon. This phenomenon can be modeled and result in reliable predictions. However, other elements such as climate evolution and its consequences on glacial melting or steric volume, river flows, meteorological events (low atmospheric pressure and strong winds that cause waves) or movements caused by earthquakes or ground movement, can locally and globally modify sea levels. Therefore, it is necessary to complete astronomical predictions with *in situ* and satellite measurements. Such measurements, which are more specific to hydrology, are used in oceanography in order to determine reference levels that are of high importance in studies linked to global climate change (the study of variations in mean sea levels) and also to establish measures against the risk of submersion.

In terms of space, the arrival of satellites equipped with altimetric radars revolutionized measurements, most notably in 1992 with the launch of the NASA-CNES mission known as TOPEX-POSEIDON, and then later Jason-1. Although these measurements are of a high accuracy (the variation in mean level can be observed, *a priori*, to the nearest ±1 mm/year), they remain subject to uncertainties that are difficult to evaluate in the long-term stability of altimeters, in geophysical corrections to be applied to data, and in calculation techniques. In this way, *in situ* measurements continue to remain indispensable and are always references in the subject.

Along the shore, water height can be measured with the help of fixed structures implanted into ports. These are known as *coastal gauges* and *tide stations*. These gauges enable determination of the lowest levels of water, which is very important information in navigation. Offshore, it is necessary to turn to variation in pressure measurements from which variation in water height can be deducted via the use of submerged tide gauges. In this field, new technologies have appeared, such as buoys equipped with GPS receivers, or the exploration of signals from GNSS. These subjects are explored in section 2.9.2.2.

Problems with the referencing of these measurements can sometimes occur. In effect, the mean sea level, contrary to popular belief, is not a plane surface of constant height, but a variable quantity that depends on the local field of gravity among other things (see the definition of geoid in section 2.7.2.5) and the period within which it is calculated. Its value varies according to the position of stars, meteorological conditions, freshwater inflows, etc. Another quantity is therefore necessary in order to reference tide gauges, namely chart datum, as defined by the International Hydrographic Organization. It corresponds locally and theoretically to the lowest extreme astronomic sea level:

> "Chart datum is the reference of sea level below which charted depths on a nautical chart are measured, positively towards the bottom, and above which tidal levels are represented, negatively towards the surface."

This definition was introduced in order to ensure navigational security and indicated sea levels in tide tables are brought in relation to this reference. If we add the values of probes indicated on nautical charts and sea levels, we obtain the theoretical depth at a given moment. In terms of tidal gauges, chart datum is defined by its dimension in relation to leveling benchmarks known as *tidal bench marks*, which are durably installed.

2.8.1. *Tide gauges*

2.8.1.1. *Principles of measurement*

Sea surface is rarely plane and numerous causes other than tides can modify its state (wind, rain, etc.). Tide gauges must therefore be installed in adapted locations, known as tide stations (see Figure 2.57). Tide is a slow phenomenon. Stations are therefore designed to filter or eliminate high-frequency disruptions that could damage the quality of measurements. For this reason, they contain a vertical duct known as a stilling well that communicates with the ocean by one or more horizontal inlets, known as admission tubes, situated below the lowest sea level. In order for the filtering to be efficient, the diameter of the tubes must be low in relation to that of the stilling well. The well and the material needed to take measurements are placed in shelters.

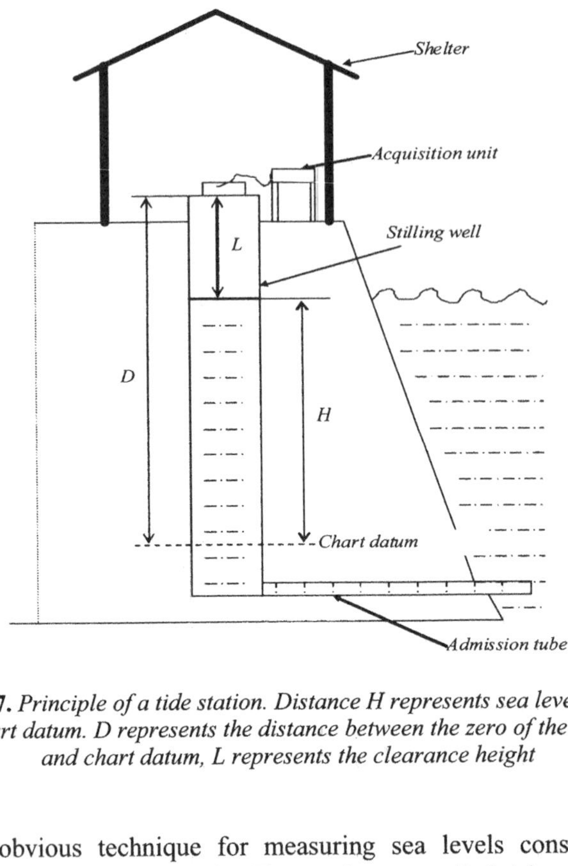

Figure 2.57. *Principle of a tide station. Distance H represents sea level in relation to chart datum. D represents the distance between the zero of the sensor and chart datum, L represents the clearance height*

The most obvious technique for measuring sea levels consists of using a graduated ruler that is fixed correctly. These rulers are called *tide scales*. They are a wooden or plastic surveyor's rod formed by two checkerboards with sides of 10 cm, in red and black alternatively, in order to facilitate the reading. This scale must be placed vertically along a quay or according to its biggest slope if there is one. The location must be protected from swash and waves, which can considerably reduce the quality of the reading. The reading is made at a fixed time by an operator who has a timepiece synchronized to a reference (GPS time, talking clock, national radio, etc.). The uncertainty of measure obtained with these instruments depends principally on this. For calm seas, it is a few centimeters. This mode of measurement, as rudimentary as it is, is still used today essentially in order to connect autonomous tide gauges.

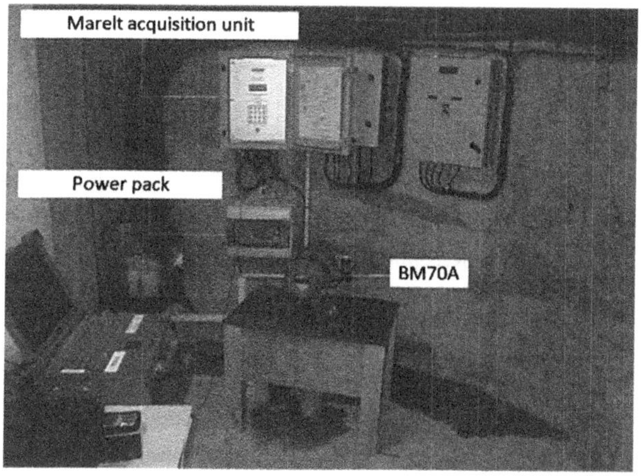

Figure 2.58. *Interior view of a coastal tide gauge shelter (Courtesy of © SHOM)*

Another technique, just as rudimentary and as effective, consists of unraveling from a quay, a tape measure whose end is fixed to an electronic contact. As contact is at sea level, a bulb lights up on the reel of the tape measure, the graduation corresponding to the fixed tidal mark on the quay is then read in order to deduct the height of the water. This instrument is called a *luminous probe*. It is also used to calibrate and join autonomous tide gauges to international chains of length measurements, as luminous probes can be calibrated in reference metrology laboratories specialized in length calibration.

The first autonomous tide gauge was invented by the hydrographic engineer Chazallon, who was also the author of the first tide tables in 1839. It is a *float tide gauge*. The majority of historical sea level measurements were acquired by this type of instrument. Autonomous coastal tide gauges are generally placed in stilling wells in charge of filtering the waves and rough sea (see Figure 2.57). For floats' tide gauges, the float is placed directly into the well. It is suspended by a flexible steel wire that is counter-balanced by a pulley and a counter-weight. The axis of the pulley drives a smaller pulley that ensures vertical movement of the pen that writes variations in sea level directly onto the recording drum. The drum rotates to the rhythm of the clock and either turns once a day or once a week. The paper fixed to the drum provides a tidegraph whose abscissa axis is graduated in hours, and whose ordinate axis is in meters.

The float is semi-immersed as the buoyancy compensates for about half of its weight. Buoyancy works in relation to the density of sea water. If variations in salinity are locally weak, variations in water temperature can cause variations in density that can modify sinkability of the float and generate errors in measurement.

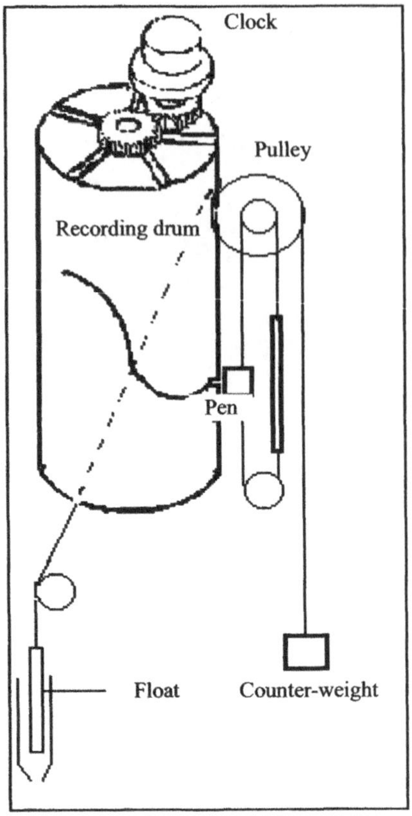

Figure 2.59. *Principle of a floating autonomous tide gauge (Courtesy of © SHOM)*

In this way, this type of installation has been progressively replaced by instruments known as digital tide gauges (see Figure 2.58). Digital tide gauges are made up of an acquisition unit in charge of sampling, dating, storing and transmitting acquired data, but also transducers placed in or above the stilling well (see Figure 2.57). The sampling period and duration of integration of the samples are configurable with the unit's software. This period can be less than 1 s, while integration can be programmed between 1 s and 4 min. Integration aims to filter certain undesirable variation frequencies of sea level. The archiving period can also be programmed. It must be below the weakest of semi-periods of tide signals to be analyzed. Transmission of data is done with the help of a modem board and local telephone network or with help from transmission satellites, such as those described in section 2.7.2.7, when the digital tide gauge is in an isolated area.

It must be noted that such digital tide gauge stations are also equipped with atmospheric pressure sensors. A decrease in atmospheric pressure causes an increase in sea level (as the weight of air becomes lower over the ocean) and inversely, an increase in pressure resulting in a decrease of this level. Variation of 1 hPa (0.75 mmHg) is approximately equivalent to a variation of 1 cm in water, constituting a significant cause of variation. As mean atmospheric pressure is 1,015 hPa (and so 761.41 mmHg), a reduction to 995 hPa will cause an increase in height of around 20 cm. Mean levels used in global studies must therefore be corrected from variations in atmospheric pressure, whose accuracy is in relation to that researched in these studies.

2.8.1.2. *Functioning of digital coastal tide gauges*

In addition to the acquisition unit, the essential element of a digital coastal tide gauge is its transducer. This can be in the form of an immersed pressure sensor. The principle of these measurements is explained in section 2.8.2. It can also be in the form of two contactless measurement technologies: acoustic or radar.

Acoustic transducers emit pulse trains in the frequency range 40–50 kHz. The wave of these pulse trains that is reflected to the sea surface is detected by the transducer during the listening window set up by the acquisition unit. By knowing the velocity C of sound in air, it is possible to calculate the level of the clearance height L of the stilling well from the measurement of the time interval Δt separating the emission of the pulse train from the detection of the reflected train:

$$L = C\,\Delta t / 2 \qquad [2.185]$$

As the transducer is fixed at distance D of the chart datum locally determined from the lowest sea levels, sea level H can easily be deducted (see Figure 2.57):

$$H = D - L \qquad [2.186]$$

However, the velocity of sound in air varies with temperature (around 0.17% per 1°C) and humidity. It is therefore necessary to calculate its value from a simple relation established for sinusoidal waves, while excluding nonlinearities of the propagation equation:

$$C = 331.2\left(1 + 0.97\frac{U}{p} + 1.9\times10^{-3}t\right) \quad [2.187]$$

In this equation, U represents relative air humidity, t temperature in °C and p atmospheric pressure expressed in hPa. In theory, ultrasound measurements of sea levels must be coupled with measurements of temperature, humidity and atmospheric pressure. However, the influence of these parameters on the value of H is not identical for all of them, and a rapid calculation shows that temperature is the quantity that has the greatest influence. It is therefore necessary to find out its variations and even gradients in the stilling well to limit errors in the measurements of H that are caused, and estimated to be 2 cm/°C for a transducer placed at 10 m above sea level.

Level measurement can also be made with another technology: radar. According to its development, this technology uses three different principles of measurement:

– running time measurement of an impulse between an antenna and sea surface;

– the *guided wave radar*: the wave is therefore guided by a cable or a rod that descends to the lowest level measurable and

– a *frequency modulated continuous wave* (*FMCW*, see Figure 2.61).

The modulation adopted is that of a linear frequency ramp. During the emission of this ramp, the sended frequency f_e is instantly compared to the frequency f_r of the reflected signal. The difference Δf between the two is proportional to the time taken between the wave departing and returning and therefore to twice the distance d to be measured:

$$\Delta f = f_r - f_e = a2d + b \quad [2.188]$$

where a and b are constants determined by calibration. The beat frequency constituted by this difference can be treated with different methods. Generally, the FMCW process enables the generation of signals whose bandwidth is higher than that of classic radar impulses, therefore enabling us to obtain a narrower temporal resolution. It also allows easier elimination of parasitic reflections and enables us to work with lower antenna diameters for the same measuring range.

An electromagnetic wave transmitter–receiver is therefore placed at the top of the stilling well in place of the acoustic transducer. The influence of temperature, humidity, pressure and therefore density of air in the stilling well is negligible over the wave frequencies emitted. Several types of antenna are available:

– rod antennas that only allow workings over small distances and have a lower accuracy than the other types; and

– cone, parabolic antennas or *planar* antennas that enable high directivity.

The base of the antenna constitutes the zero mark. A metallic plate enables this wave guide to be fixed and aligned with the uppermost surface of the stilling well. The plate can be pierced to allow a luminous probe to descend in order to intercompare measurements of distance provided by the radar.

The Krohne BM 70 transducer (Figure 2.60) is an example of a radar that can be installed in digital coastal tide gauges. Its emission–receiver antenna is a conic wave guide. Signals emitted by the transducer according to the FMCW technique are situated in the 10 GHz band. The electronics of the Krohne sensor make a Fourier transform whose result gives a measurement of the travel time. Thanks to its treatment, the Fourier transform enables the elimination of parasitic echoes that are produced by multiple reflections, and so errors of measurement.

Figure 2.60. *BM 70 sensor by Krohne, identical to those used in certain digital coastal tide gauges. (Courtesy of Krohne Messtechnik GmbH)*

In order for the process to retain a high accuracy, it is necessary to ensure the perfect linearity of frequency modulation. For this, the oscillator that enables the generation of frequencies is inserted into a phase locked loop. This control technique measures the frequency emitted and automatically adjusts its phase, in turn reducing

98% of the signal's nonlinearity (Figure 2.61) and ensuring good reproducibility of frequency ramps.

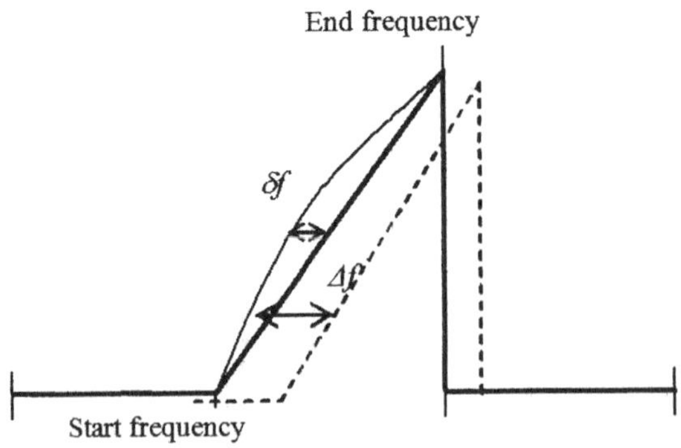

Figure 2.61. *Illustration of the FMCW principle and of a difference δf in linearity. The thin lined ramp represents emission and the dotted line represents reception. For a given frequency, the shift Δf is proportional to the running distance of the electromagnetic wave*

	Measuring range	Uncertainty of measurement
Luminous probe	0–20 m	1–2 cm according to the state of the sea
Tide scale	0–20 m	2–3 cm according to the state of the sea
Float tide gauge	0–15 m	5 cm
Acoustic tide gauge	0–15 m	<2–<3 cm over air draughts of less than 3 m; otherwise 2–4 cm
Tide gauge radar sensor	0–20 m	1–4 cm

Table 2.9. *Comparison of coastal tide gauge techniques. The uncertainties given (at 2σ) are only an indication of possible performances. They can vary according to installations and measurement conditions (Courtesy of © SHOM [LER 04])*

There is a means that can be used to determine systematic errors found in tide measurement installation. Known as the *Van de Casteele* test, it emphasizes the comparison of records made simultaneously by the tide gauge being tested and a standard independent measuring device positioned nearby. Differences in levels measured between the gauge and measuring device must remain constant and equal to zero if the tide gauge and the standard system have the same vertical reference.

The Van de Casteele diagram produced reporting the differences between the standard and the tide gauge in x-axis and the sea levels in ordinates, enables visualization of a certain number of errors. While a perfect tide gauge would give a curve reduced to a straight line parallel to the ordinate axis, an incline shows, for example, a problem with scale factor. A curve whose points are not superimposed between the rise and fall of sea levels shows problems with hysteresis. A curve whose first and last points do not meet up shows problems with shift. These faults can combine to produce a disjointed and inclined curve along, which represents a problem with scale factor in part of the measuring range and problems with hysteresis.

2.8.2. *Tide gauges with pressure sensors*

The principles of determining sea level measurements from one or more pressure sensors have always been and are still applied in coastal tide gauges, but also in instruments immersed at long distances from the coast (see Figure 2.62). The cages are placed on the seabed where the bathymetry and composition are already known. The mooring site is chosen in relation to the current chart of the place (currents must be weak in order to avoid disruption to static pressure measurements), to the local variability of water density (river mouths are avoided) and to the risks of trawling or disappearance of the cage's marking-out.

Pressure p measured by a submerged sensor at depth h is obtained by:

$$p = p_{atm} + \rho g h \qquad [2.189]$$

where p_{atm} represents the pressure exerted by the air column, ρ the water density and g the acceleration of gravity.

The value of h is extracted from relation [2.189]:

$$h = (p - p_{atm}) / \rho g \qquad [2.190]$$

In order to do this calculation, it is necessary to know the local atmospheric pressure, water density and acceleration of gravity. Atmospheric pressure can be recuperated by the intermediary of the onboard meteorology unit or from semaphores or weather services. While atmospheric pressure varies faster in relation to altitude than in relation to distance, it is important that the record is made in a ray less than 20 km from the tide gauge. Variations in atmospheric pressure throughout measurement are more important to determine than their absolute value, as the values of water level calculated are zeroed relative to chart datum coordinates.

Estimating density or its variation over time is more problematic as it is not always constant along the water column below the tide gauge. The value of ρ used in relation [2.190] must therefore be a mean value. It is calculated from equations of state of sea water from 1980 ([1.65] to [1.74] given in section 1.3) or from the TEOS-10 equations. Salinity, which intervenes in this calculation, can be evaluated from measurements taken with losable probes (section 2.1.1). However, it is often cheaper to determine salinity from laboratory measurements of water samples. As the tide gauge remains submerged over a long period of time, these measurements must be renewed periodically in order to estimate the variation in density. It is estimated that a variation of 2% of its value provokes an identical variation in pressure measurements.

There are several types of pressure sensor tide gauges. Instruments that can be submerged away from the coast, on the seabed where depth goes from 10 to 200 m (Figure 2.62), are equipped with high-resolution and good linearity pressure sensors, as the important thing to know is not sea level but variations in the percentages of this level over time, which may be low if the cage is installed at great depths. In order to resolve these problems, paroscientific quartz sensors, which are temperature compensated, are most regularly used. The way in which they function is described in section 2.3.2. They can also be equipped with a conductivity sensor and a temperature sensor in order to follow local variations in the density of the medium. Typical performances of this type of instrument are shown in Table 2.10. As tides are a cyclic phenomenon, and as the pressure measurements that they take are reported to chart datum after processing, their repeatability, their low drift in temperature and over time, the linearity of their measurements regardless of mooring depth and the stability of their clocks are more important characteristics than the accuracy of calibration.

Pressure sensors are also used in coastal tide gauges. A system known as a pneumatic bubbler gauge has been used for years, principally in English-speaking countries, as a replacement for float tide gauges. This type of gauge contains pressurized air in a tube of known volume and length. This tube discharges near the bottom in an open and submerged reservoir. Pressurized air pushes the water column that is in the tube, until it escapes by bubbling from the reservoir through a small opening. The pressurized air that is sent corresponds to the pressure exerted by the water column at the depth of at which the tube opens. Measurement with the help of a differential sensor enables sea levels to be determined using relation [2.190]. The pressure sensor directly measures the difference ($p - p_{atm}$). Once again, it is necessary to evaluate the value of ρ, by measurements or sampling, from the equations of the state of sea water. The gravity g of the area must also be known.

Figure 2.62. *Example of submerged pressure sensor tide gauge (SLS 23 by Mors) installed in a ballasted cage (Courtesy of © SHOM)*

Pressure sensors can also be directly used to measure sea level. They are placed in a protective copper or titanium tube, designed to limit biofouling. These tubes and the sensors are slipped into stilling tubes in order to filter the effects of the waves. If the sensor being used is absolute, it measures the weight of the air and water column, and it is necessary to use a second absolute sensor to measure atmospheric pressure. Both sensors must be synchronized to carry out the operations of relation [2.189]. If the sensor used is differential, the point open to atmosphere must be connected to a pipe discharging at the surface. With submerged tide gauges, sensors used must be well compensated for temperature. If this is not the case, measurements of air and water temperatures to the nearest 0.1°C are necessary in order to correct pressure data. Whatever the pressure sensor, it is subject to drift phenomena over time. As tide stations function continuously over long timescales, it is necessary to dismantle the sensors periodically to calibrate and correct these drifts.

To compensate for drifts and follow the rise (or fall) of sea levels, in 1996 Woodworth *et al.* described a differential system consisting of an absolute A pressure sensor placed at depth in a tube, and a second sensor, B, identical to the first placed at a height determined by leveling and corresponding to the local mean sea surface [WOO 96]. A third sensor, C, measures atmospheric pressure. The pressures difference A – C enables the determination of tidal amplitude, while the difference B – C enables the mean level variations to be tracked when sensor B passes from the submerged state to the free air.

	Pressure	Temperature	Conductivity
Measuring range	13 ranges, form 0–1 m to 0–6,800 m	-5 to +35°C	0–70 mS/cm
Accuracy	0.01 % of full scale	0.02°C	0.01 mS/cm
Resolution	0.2 mm for 1 min of integration and 0.01 mm for 15 min of integration	0.01°C	0.0002 mS/cm
Repeatability	0.005% of full scale	-	-
Hysteresis	0.005% of full scale	-	-
Time base stability	±2 ppm/year and ±2 ppm/°C		
Clock stability	15 seconds/month		
Memory capacity	2 Mbyte of static RAM		

Table 2.10. *Principal characteristics of a SBE 26 submerged tide gauge produced by Sea Bird Electronics, Inc. (Courtesy of Sea Bird Electronics, Inc)*

2.8.3. *Keying and uniting of tide gauges*

Tidal and sea level observations, whether a product from coastal or submerged tide gauges, must be keyed to a well determined reference level and reference time. Therefore they must undergo initial keying relating to height and time. Keying and uniting in time consists of synchronizing observations with a common timescale (section 2.6.1). The internal clocks of digital tide gauges have a drift of about 2 ppm, and so they must be regularly adjusted to retain an accuracy to the second.

Keying and uniting techniques relating to height are more complex. For coastal tide gauges, levelling consists of alignment of the instrumental zero to physical marks whose heights are determined by geometric leveling and attached to a general or local leveling network. These tide marks form an integral part of the observatory. They create the dimension of the chart datum to which zero instrumental is linked. Dimensions of the General Leveling of France (NGF), for example, are given in relation to the geoid. Tidal data in the first instance are then linked to this benchmark. However, it is important that the chart datum is not defined uniquely by its coordinate in the NGF benchmark, as successive leveling operations could produce different results due to the evolution of measuring techniques and the vertical movements of the Earth's crust (the tide itself can provoke vertical ground movements of about 20 cm). Uniting of these marks by leveling must therefore involve referencing to an altimetric system. In France, for example, the legal geodetic system is known as RGF93 (the *French Geodetic Network of 1993*). It is a three-dimensional, geocentric reference system of centimetric accuracy. It is spread across the territory by a *permanent geodetic network* (RGP), which is currently composed of 200 stations equipped in partnership with different bodies (including all stations of the Teria geometer network), which constantly observe the positions of points and unite them to the RGF93. In the case where tide gauges are not in immediate proximity to these points, their coordinates are united by highly accurate GPS observations. It must be remembered, however, that GPS coordinates are expressed in the geodetic system WGS-84. Coordinate conversions are therefore necessary. In the case of tide gauges with pressure sensors moored a long way from the coast, the only way of uniting their observations in relation to a reference level is by observing the concordance of their data during low seas with that of the closest port tide gauges.

2.9. Determining waves and swell characteristics

Determining and predicting the state of the sea surface is of great importance in numerous fields. This state is an essential component of the coastal oceanographic dynamic. It plays a fundamental role in the hydrodynamics and morphodynamics of sediments. There are methods of the prediction of wave amplitude and frequency and the direction of surface currents, but measurement remains essential to calibrate and verify the results of these models. The results are used in navigation security, the determination of coastal erosion, the conception of oil rigs and to determine the drift of surface pollution or atmospheric exchange conditions.

2.9.1. *Factors relating to the origins and modeling of swell*

Sea surface movements can be classified according to scales relating to time and space. When their speed varies slowly, this produces currents. As their speed varies rapidly, they result in the formation of waves. Approximately 30% of these movements are due to tidal currents, but the principal cause is wind action, which results in capillary waves whose forces produce surface tensions. The capillary waves transform into swash with a period of 0.3 s under the effects of gravity. If the wind is reinforced, the swash creates the swell, whose amplitude can reach 10 m over a period of 8–20 s. Waves that are formed propagate freely at the sea surface. However, surface oscillations result at depth in a pressure wave. On the principal axis of the pressure force, water particles move in circular vertical movement. They therefore have an orbital speed that can be measured. This measurement can be taken at depth and enables a deduction of the swell characteristics. The residual movement of orbital speed is called Stokes drift.

Waves that come from offshore and propagate at the surface can be classified according to their wavelength λ and period in:

– short wave: $\lambda < 100$ m and period < 8 s;

– medium wave: $\lambda = 100$–200 m and period 8–10 s; and

– long wave: $\lambda > 200$ m (400 m maximum) and period > 10 s (20 s maximum).

They are transformed as they meet the continental slope, therefore entering into the coastal field. They are subject to other transformations in coastal areas according to the incline of the sea floor and obstacles on the coastal border until their breaking. *Breaking* can be characterized by oceanic surface geometry (slopes). It causes shear currents, and occasionally the elevation of sea level, known as a "surge". Waves, currents and sea level are parameters that are intimately linked.

It is possible to model swell from monochromatic linear wave theory. This analysis is based on the "plan wave" hypothesis, which assumes that we are sufficiently far away from the place of wave formation and that the effects of bathymetry are weak or negligible. In other words, if a corresponds to a level half way between the peak and the hollow, and h to depth, it can be shown that:

$$2a/\lambda << 1 \text{ and } 2a\lambda^2/h^3 << 1 \qquad [2.191]$$

This hypothesis is required in the study of speed and pressure fields. It must also be assumed that the wave is monochromatic, i.e. that it is composed of a principal frequency $f = 2\pi/\omega$, ω being its angular frequency or pulsation. If a wave propagates in direction $0x$ in a medium in which depth h is constant, this equation can be written by a complex function:

$$\varphi(x,z,t) = \frac{aC(z)}{k} e^{i(\omega t - kx)} \qquad [2.192]$$

with

$$C(z) = \omega \frac{\cosh(k(z+h))}{\sinh(kh)} \qquad [2.193]$$

and

$$\omega = \sqrt{gk \tanh(kh)} \qquad [2.194]$$

where g corresponds to the acceleration of gravity and k to the wave vector: $k = 2\pi/\lambda$. This wave propagates with phase speed $C\varphi = \omega/k$. Equations [2.193] and [2.194] show that wave amplitude and frequency are dependent on their wave vector, the depth of the medium h and the observation depth z. Relation [2.194] is also known as the relation of dispersion.

From expression [2.192], it is possible to calculate wave speed and pressure fields. By measuring speed or pressure, the corresponding relations enable an estimation of the amplitude a. The field of speed according to axis $0x$ is obtained by:

$$v_x = \frac{\partial \varphi(x,z,t)}{\partial x} = -iaC(z)e^{i(\omega t - kx)} \qquad [2.195]$$

According to axis $0z$, we have:

$$v_z = \frac{\partial \varphi(x,z,t)}{\partial z} = aS(z)e^{i(\omega t - kx)} \qquad [2.196]$$

with:

$$S(z) = \omega \frac{\sinh(k(z+h))}{\sinh(kh)} \qquad [2.197]$$

The hydrostatic relation enables us to get to the pressure field:

$$p + \rho g z = -\rho \frac{\partial \varphi(x,z,t)}{\partial t} = -ia \frac{\rho \omega C(z)}{k} e^{i(\omega t - kx)} \qquad [2.198]$$

If we measure wave speed according to axis $0x$, at depth z, its amplitude will be:

$$a = \frac{|v_x|}{C(z)} = \frac{|v_z|}{S(z)} = \frac{k}{\rho \omega} \frac{|p|}{C(z)} \qquad [2.199]$$

where $|p|$ is obtained by measurement p_m of the pressure wave:

$$p_m = |p| e^{i\varphi_p} \qquad [2.200]$$

Likewise, measurement of current speed according to $0x$ or $0z$ or that of the pressure at a determined depth, enables us to obtain the waves amplitude values.

In reality, however, sea states are usually considered to be random, and we have to resort to statistical quantities in order to describe them. These quantities can be: average height, heights, periods and significant directions. They are extracted from periodograms by detection of the point at which the crossing of the signal is zero or from spectral analysis of the frequential and temporal domains, based on the Fourier transform. The WMO has edited a book [WMO 76] in which these methods of calculation are standardized.

Usage of the Fourier analysis is based in the linear approximation according to which the real swell is formed by the superposition of pure waves or Airy waves. It enables the determination of their direction θ by the calculation of their *directional energy spectrum*, $E(f, \theta)$, where f corresponds to their frequency. If x_e, x_n and x_h represent coordinates according to the East, North and vertical axes of a float, or of a point on the surface of a moving ocean, the autocorrelation function $A_{ij}(\tau)$ can be calculated (where i and $j = e$, n or h), during period T:

$$A_{ij}(\tau) = \lim_{T \to \infty} \frac{1}{T} \int_{-T/2}^{+T/2} x_i x_j (t+\tau)\, dt \qquad [2.201]$$

Fourier transformation of equation [2.200] gives:

$$C_{ij}(f) = \int_{-\infty}^{+\infty} A_{ij}(\tau) e^{j2\pi f\tau} d\tau = K_{ij}(f) - jQ_{ij}(f) \qquad [2.202]$$

From relation [2.202], the spectral power densities can be calculated: $C_{ee} = S_e(f)$ according to the East axis, $C_{nn} = S_n(f)$ according to the North axis and $C_{hh} = S_h(f)$ according to the vertical axis. As the movement of a surface float is sinusoidal, phase differences between x_e and x_h and between x_n and x_h are 90°. Between x_e and x_n they are 0 or 180°. From equation [2.202], $K_{he} = K_{hn} = Q_{en} = 0$ is deducted, and:

$$C_{he} = -jQ_{he}$$

$$C_{hn} = -jQ_{hn} \qquad [2.203]$$

$$C_{en} = K_{en}$$

The directional energy spectrum that enables the description of wave energy according to wave frequency and direction is obtained by multiplying spectral power density $S(f)$ by its angular distribution function $D(f, \theta)$:

$$E(f, \theta) = S(f)\, D(f, \theta) \qquad [2.204]$$

with:

$$S(f) = \int_0^{2\pi} E(f,\theta)\, d\theta \qquad [2.205]$$

and

$$\int_0^{2\pi} D(f,\theta)\,d\theta = 1 \qquad [2.206]$$

$D(f, \theta)$ can be decomposed in Fourier serials:

$$D(f,\theta) = \frac{1}{2}a_0 + \left(a_1\cos(\theta) + b_1\sin(\theta)\right) + \left(a_2\cos(2\theta) + b_2\sin(2\theta)\right) + \varepsilon \qquad [2.207]$$

The value of coefficients a_1, a_2, b_1 and b_2 can be determined from equations [2.202] and [2.203]. We have:

$$a_1 = \frac{Q_{he}}{S_h(f)}$$

$$a_2 = \frac{S_e(f) - S_n(f)}{S_e(f) + S_n(f)}$$

$$b_1 = \frac{Q_{hn}}{S_h(f)} \qquad [2.208]$$

$$b_2 = \frac{2K_{en}}{S_e(f) + S_n(f)}$$

As the value of $D(f, \theta)$ is known, the mean direction of the swell can be defined as being that for which $D(f, \theta)$ is maximal. We have $\partial D / \partial\theta = 0$ for:

$$\theta = a\tan\left(\frac{b_1}{a_1}\right) = a\tan\left(\frac{Q_{hn}}{Q_{he}}\right) \qquad [2.209]$$

2.9.2. *Instruments used to measure the state of the sea*

The state of the sea can be estimated from space by radar altimeters installed on Missons TOPEX, Jason-1, GFO or ENVISAT satellites (see section 2.8). These satellites enable the determination of significant wave heights in ranges in which the diameter varies according to the state of the sea, from 3 to 13 km, but they cannot determine the direction or wavelength of wave components. Furthermore, these measurements are limited to the direction of the nadir of the instrument, which means that a single altimeter provides a simple trace following the trajectory of the satellite, each trace being very distant from the other because of the Earth's rotation. In order to cover an entire oceanic area, images must be superposed from several satellites. In the case of tsunamis, satellites equipped with altimeters are able to detect and measure wave amplitudes using appropriate image processing, however these methods remain minimally adapted to real-time detection, as they cannot cover all oceans constantly, and the processing times of their data are not viable for the integration of such methods in a warning information network.

Likewise, in this domain, *in situ* measuring instruments remain indispensable. It is possible to directly measure the height and direction of waves by using ultrasound transducers similar to those used for tides (section 2.8.1.2), as these are installed at the bottom of the sea. Resolutions of several centimeters can be obtained for distances limited to around 50 m. This technique is therefore only possible in coastal borders. Offshore, instruments to be used to determine the amplitude and direction of the waves, directly measure these quantities or, according to the theory explained in section 2.9.1, fluid speed according to one or two axes and (or) the pressure at a depth that has been determined. Speed can be measured directly by current meters or deducted from acceleration measurements. Current meters can also be equipped with pressure sensors. Other techniques based on the use of GPS receivers also make an appearance. However, as it is necessary to cover large geographical ranges, high-frequency radar technology is more regularly employed.

2.9.2.1. *Wavemeter buoys*

Wavemeter buoys are basically surface floats connected to a mooring line (see section 3.2.2.2). The line is a cable that is long enough to avoid hindering surface float movements. This cable is also equipped with floats, so that its weight does not modify the amplitude of the vertical movements of the buoy. Between the buoy and the cable, a length of line made from elastomer is adapted, enabling the buoy to avoid interference from the weight of the cable. Anchoring from a fixed position is done with the help of a heavy steel chain that sits on the bottom with ballast.

The buoys are equipped with accelerometers and a magnetic compass (section 2.5.5), which enables the determination of the horizontal direction of movement. They can also contain an antenna and a GPS receiver (section 2.7.2) in order to track their position. Their data are transmitted to the ground by a high-frequency Argos transmitter (section 2.7.1) or another type of communication system, such as GSM or Orbcomm.

Accelerometers are sensors that directly relate the application of a force to the movement or deformation of a load cell. Accelerations related to waves have a relatively weak amplitude and low frequency, so it is necessary to have an accelerometer that can pass zero frequency, or in other words the continuous.

Piezoresistive strain gauge technology is particularly well adapted to such measurements and is widely used, but other inductive and capacitive processes can also be used.

As these sensors must pass the zero frequency, they are sensitive to the influence of temperature, which leads to drifts of their "zero" (error of about 1% of the measuring range for variations of 50°C). In the case of piezoresistve sensors temperature also has an influence on their sensitivity. The coefficient of strain gauges used to track deformations of the load cell is not constant in relation to temperature, but these variations can be compensated for in high-quality sensors.

It is also necessary to take the sensitivity of sensors to lateral accelerations into consideration. This can be about 0.05 g/g, where 1 g represents an acceleration of 9.81 m/s. They can also present linearity errors (≈ ±0.2% of their measuring range) or hysteresis (≈ ±0.3% of their measuring range). Finally, if their bandwidth is insufficient, amortization of measured accelerations can be introduced.

Wavemeter buoys are equipped with triaxial accelerometers. They enable the direct determination of vertical movements, i.e. wave amplitude and horizontal movements, by double integration of the signal. Wave direction is obtained by a correlation calculation between the horizontal and vertical components. The accuracy of measurements is between 0.5% and 10% of the read value for waves in which the period can spread between 1.6 and 30 s.

2.9.2.2. *Buoys equipped with GPS receivers and the utilization of GNSS signals*

Satellites equipped with altimeters (Jason, ENVISAT, etc.) do not often pass over the coast. The necessity to calibrate and validate their data has given way to the need for buoys placed offshore in order to measure sea level and wave height. The buoys have been equipped with GPS receivers (section 2.7.2) to take these height measurements. The GPS has enabled instantaneous measurements to be taken, and thanks to the buoys this property is used for the early detection of tsunamis.

The buoys used are in the form of metallic triangles with floats at each end, with the GPS receiver and the data transmitter situated at the center of the triangle. This light assembly can be dropped from a boat. The buoys can also be more voluminous and can be in the form of a round metallic base attached to an instrumented mast, as the application is used for moorings with a long duration.

In order to estimate sea level, GPS receivers are used in kinematic mode as it is the only method that allows the elimination of all sources of errors (epherimedes, ionosphere, clock drift, multiple reflections, etc.). This assumes that buoys are placed at a distance from reference stations situated on the coast (<20 km), that is compatible with the accuracy required. This accuracy is reduced with increasing distance from the station. Sources of errors have time constants going form 10 s to 1 hour. A high-pass filter with a cut-off frequency of 0.02 Hz therefore enables the elimination of these disruptions, and the usage of GPS in kinematic mode enables the instantaneous height to be determined with an uncertainty of just a few centimeters. The exploitation of these levels in the 0.05–10 Hz spectral band enables the determination of buoy movements and the estimation of wave amplitude. Work is in process to determine sea level by integration of the signal over long periods.

American GPS, Russian GLONASS, Chinese BEIDOU and European Galileo positioning systems form what is known as the GNSS (section 2.7.2). The reflection of the electromagnetic wave emitted by those systems in the band 1.277 to 1.575 GHz is very high at interfaces between air and water, air and ice, and air and snow, which enables their exploitation in passive reflectometry or interferometry systems in order to record multiple data, such as the height of the point of reflection.

Receivers are placed in airplanes, balloons or satellites, and data relating to level are obtained by measuring the propagation time of the reflected signal in comparison to the code of the incident signal. The correlation function of these signals is relative to the slope of the reflecting surface. From the study of this function, it is therefore possible to deduce the surface state of the sea, or in other words wave amplitude, but also the speed and direction of the wind. For several years satellites have been equipped with antennas directed towards the nadir, and studies have been carried out to make this technique more reliable, as accuracy varies from 1–12 centimeters, according to the technique and altitude of the carrier used.

2.9.2.3. *Measurements taken by current meters and pressure sensors*

According to elements discussed in 2.9.1, the measurement of the speed of current according to axes $0x$ or $0z$ (also called components u and v) or the pressure to a determined depth enable the value of wave amplitude to be tracked. Measurements of pressure can be taken with the help of tide gauges (section 2.8.2), if they allow a sufficiently high measurement sampling rate, or with the help of current meters if they are equipped with pressure sensors with a sufficient resolution. Current meters can be the same as presented in sections 2.5.1 to 2.5.3. They are therefore moored to the bottom, depth permitting. However, current profilers have a higher flexibility of usage as they procure speed profiles for the entire water column, and for this reason they can be moored at a deeper level (see section 2.5.4.1).

In fact, wave energy propagates at a determined depth under the surface. It is therefore necessary to choose a convenient immersion depth z for the instrument in order to measure wave energy. Furthermore, high-frequency waves do not penetrate to the same depth as waves with smaller frequencies. For this reason, their measurement depends on frequency as it is not taken at the surface, see Figure 2.63. It is estimated that in order to take a measurement of a wave height with an accuracy of ±5%, it is required that:

$$0.3 < z/\lambda < 0.5 \qquad [2.210]$$

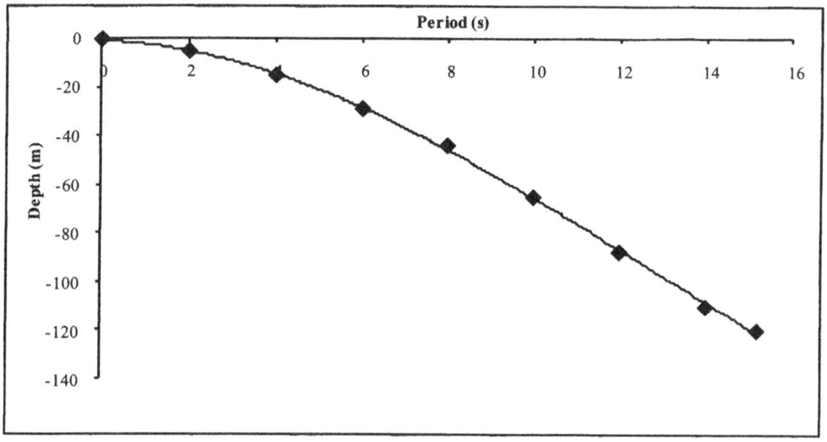

Figure 2.63. *Representation of the maximum depth of penetration of wave energy relative to wave period*

Another source of error is the presence of subsurface currents. These currents modify the path of water particles, which are moved by the wave of pressure generated by waves. The speeds and dynamic pressure measured are higher, which causes an overestimation of wave height. These errors increase with increasing depth of immersion. It is therefore necessary to measure the vertical component, *U*, of subsurface currents and correct dispersion relation [2.194], which becomes:

$$\omega = kU_v \cos\alpha + \sqrt{gk\tanh(kh)} \qquad [2.211]$$

In order to correctly estimate wave height, the characteristics of subsurface currents along the entire water column below the current meter must be determined. This is another advantage of working with profilers.

In relation [2.211], α is the angle between the direction of waves and the current. It can be calculated from relation [2.209], which enables us to determine the principal direction of propagation. However, this evaluation can be misleading in the case where there are several sources of wave flows that propagate in different directions. The result obtained is a mean direction that does not necessarily correspond to that of maximum energy.

To solve this problem, it is necessary to deploy several spatially-independent instruments to create a measurement network. Some ADCPs or ADP's, such as the Workhorse Wave Array by Teledyne R.D. Instruments or the AWAC by Nortek, enable the formation of such a network with a single instrument. They are equipped with a pressure sensor that enables the determination of sea level. Measurement cells (at least one per beam) situated near the surface are therefore automatically used by the equipment's software to estimate wave direction. As the beams have an angle of at least 25° in relation to the vertical, they constitute three independent instruments.

In order to find the angular distribution of wave orientation, the cross-spectrum function $C(f)$ is calculated:

$$C(f)=\int H(f,\theta)D(f,\theta)H^{*}(f,\theta)d\theta \qquad [2.212]$$

$H(f,\theta)$ is the spectral power density measured by each beam, which should correspond to the response obtained with a plane and uniform wave of unit amplitude propagating in direction θ; $H^{*}(f,\theta)$, and it's Hermitian transposed. $D(f,\theta)$ is the angular distribution function to be determined. It contains a multitude of solutions that satisfy equation [2.212]. Its determination cannot be done directly. Access to methods such as the *maximum likelihood method* is required. It enables us to obtain a solution in the following form for each frequency and each direction to be explored:

$$D(\theta)\propto\frac{1}{H(\theta)^{*}C^{-1}H(\theta)} \qquad [2.213]$$

2.9.2.4. *Measurements taken by high-frequency radar*

First used to measure marine and aerial currents along the coast, high-frequency radar measurements are now used to determine the characteristics of waves.

These systems, whose concept dates to the 1950s, are composed of emission and receiver antennas placed on a coastal site. The antennas traditionally emit over the high-frequency radio band corresponding to a range of 3–30 MHz and have wavelengths λ between 10 and 100 m. This band is situated between the bands used for television emissions and radio transmissions. On more recent installations, the range has been extended to 50 MHz in order to increase the resolution of measurements, and to 100 MHz in some cases to avoid disruption due to emissions from other sources.

Since the 1980s, several systems of wave emission and transmission have been tested. In all cases, the electromagnetic energy reflected or retrodiffused by the sea surface is measured. It has been shown that penetration depth d of high-frequency waves in the ocean can be estimated by the relation:

$$d = \sqrt{\frac{\lambda}{8\pi}} \quad [2.214]$$

where the maximum explorable depth is around 2 m.

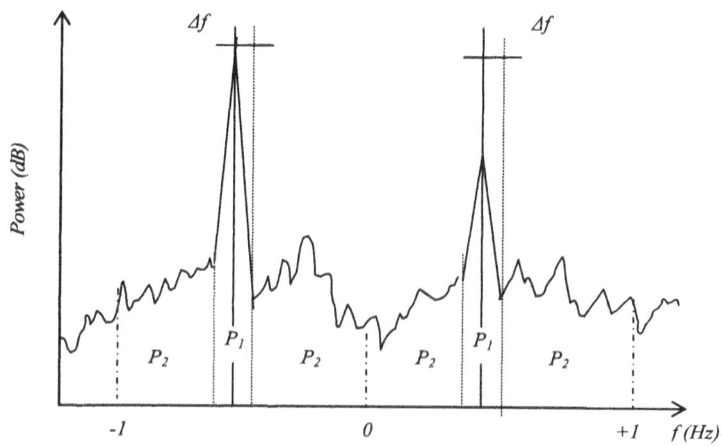

Figure 2.64. *Representation of a signal spectrum retrodiffused with two Bragg peaks. The peak of negative frequency in relation to central frequency is obtained when waves draw away from the receiver, and the other when they draw towards the receiver. The P_1 zones between the dotted lines correspond to first-order spectral bands, and those between the dotted and solid discontinued P_2 lines correspond to second-order bands*

Interpretation of the energy spectra of these waves allows the extraction of sea surface data. A phenomenon is therefore studied: the *Bragg* reflection. Waves (or swell) in which the wavelength corresponds to half of that emitted by the antenna behave as Bragg diffraction gratings, which are identical to those found in optics or crystallography with X-rays. Very narrow maximum intensity is observed in directions where the diffracted waves are in phase accordance. Spectral analysis of the signal retrodiffused by waves allows the observation of two peaks arranged in symmetry with emission frequency (see Figure 2.64). The peak with negative frequency corresponds to waves that draw away from the receiver antenna, and positive frequency corresponds to those that draw to the receiver. D. Crombie first spotted this in the 1955, and was able to provide an explanation [CRO 55]. The place of the maximum of these peaks enables the determination of the Bragg frequency of waves to the nearest Δf, which corresponds to the Doppler shift (see section 2.5.2.1) provoked by the movement of water masses by surface currents.

Bragg waves measured in the high-frequency band correspond to short gravitational surface waves, in which it is assumed they propagate like deep water waves. This assumption enables the exploitation of data contained in the Doppler shift observed, which can be used to determine the value of the current. They move with the phase speed c of deep water waves, whose value is shown in the following relation:

$$c = \sqrt{\frac{g\lambda}{4\pi}} \qquad [2.215]$$

where g represents the acceleration of gravity and λ the wavelength of the energy that is transmitted.

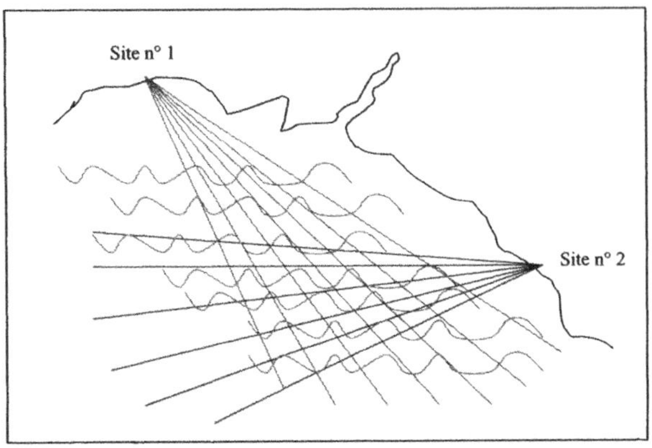

Figure 2.65. *Example of two sites equipped with transmitter–receiver radars. Scanning from a distance and the azimuth enable the determination of coordinates 0x and 0y of the surface currents at the intersection of each beam. It provides a current map of a specfied zone*

As a system of antennas only enables measurement of the Doppler component situated in the axis of the beams (radial component), it is necessary to equip at least two sites to obtain components according to axes $0x$ and $0y$ of the surface currents (see Figure 2.65). In order to avoid measuring the same components, the angle between the emission axes of these sites must be between 30° and 150°. Principally, there are two types of antenna and installations. The first are composed of a whip emission antenna and a receiver antenna in the form of crossed loops. These systems are known as *coastal ocean dynamics applications radar* (CODAR). Compact and automated versions have been commercialized by CODAR Ocean Sensors.

Figure 2.66. *Example of an installation of receiver antennas of a WERA-type system (Courtesy of © SHOM)*

The second are in the form of a network with eight to 16 whip antennas (see Figure 2.66). The antennas are separated by $\lambda/2$ in order to limit the formation of lateral lobes and they are aligned perpendicularly to the direction of observation. Radio pulses are emitted by an emission network, and reflected waves are captured by a reception network. The *WERA* (*WEllen RAdar*) system designed and produced at the University of Hamburg is a working example of this.

CODAR systems allow the coverage of larger oceanic ranges as they are omnidirectional. This presents a problem in the study of waves. WERA systems only cover sectors of 90°, however they can be orientated in specific directions and enable the determination of wave characteristics through the analysis of second-order portions of their spectra.

In the operating mode shown in Figure 2.65, two sites are equipped with a transmitter–receiver couple. Each couple covers a quasi-circular zone whose center is the transmitter–receiver couple, if the two elements are close to each other. Another operating mode exists, which is known as *bi-static*, where the transmitter is placed several kilometers from the receiver. The area sounded out by radar echoes is therefore much bigger, as is density of measurements. The disadvantage with this operating mode is that signals received lose their spatial and temporal coherence. To solve this problem, it is necessary to synchronize the transmitter and receiver with the help of a GPS clock. A receiver can therefore work with transmitters from several other sites.

Usual radar techniques are used in these instruments to measure the distance, speed and azimuth of waves. As is the case of ADCP, the sending of simple pulses does not allow us to obtain sufficient reflected energy and a good signal-to-noise ratio, therefore sequences of pulses are used. These trains are coded with Barker codes (see section 2.5.4.2). As this technique presents limitations, FMCW modulation is often preferred (see section 2.8.1.2), particularly in a bi-static operating mode. The desired phase loop is therefore synchronized each second by a pulse coming from a GPS clock.

Regardless of the modulation technique used, the Doppler resolution δf obtained is inversely proportional to the integration time T of the reflected signal. Radial speed resolution Δv is therefore given by:

$$\Delta v = \lambda \frac{\delta f}{2} = \frac{\lambda}{2T} = \frac{c}{2n\Delta t f_r} \qquad [2.216]$$

where:

– c is speed of light;

– n is the number of samples in the spectrum;

– Δt is the sampling interval; and

– f_r the working frequency of the radar.

For $n = 512$, $\Delta t = 0.5$ s and $f_r = 12.1453$ MHz, $\Delta v = 4.82$ cm/s.

In WERA systems, the beam is orientated by calibration of the amplitude and the phase of signals from each antenna. As these signals are coherent with each other, they are then added. The uncertainty σ_v that is therefore obtained on speed measurement depends once again on T and the width Δf of the Bragg lines:

$$\sigma_v = \frac{\lambda}{4}\sqrt{\frac{\Delta f}{T}} \qquad [2.217]$$

The angular resolution of this beam formation technique is inversely proportional to the aperture D of the radar beam:

$$\delta\varphi = \frac{\lambda}{D} \qquad [2.218]$$

Likewise, in order to obtain an angular resolution of 5° or 0.1 radian, an angular aperture of $10\times\lambda$ is necessary, implying the need for a large surface area in order to install the network.

For CODAR systems, they use a goniometry technique known as "*direction finding*". The in phase and in amplitude signals from two very close antennas are compared at each frequency of the band under analysis. If there are *N* antenna, it is therefore possible to resolve $N - 1$ directions for each frequency. At present, this technique is also used in WERA systems that can simultaneously combine beam formation and *direction finding* techniques.

Several methods of signal treatment have been developed to carry out this direction finding. One of the most frequently used is the *MUltiple SIgnal Classification* (*MUSIC*) method, developed by Schmidt [SCH 86]. It consists of calculating the proper vectors of the interspectral matrix of signals received for a given Doppler frequency, and then, a function which depends on these eigenvectors and the vector direction of the emission sources. The maximum of this function corresponds to the direction of Doppler frequencies measured that radial speeds are associated with. This method allows us to find multiple directions for a given spectral component. The maximum number of directions that can be determined is less than the number of reception antenna minus one (WERA systems) or to the number of independent spectrums that are collected for a given inversion. The higher this number, the lower the uncertainty in the number of directions corresponding to a Doppler shift. For CODAR systems, only three spectrums are used, which renders the determination of the number of directions relatively uncertain.

The frequency bands analyzed can be polluted by other remote sources, particularly relating to conditions of ionospheric propagation. Radar systems are equipped with software that enable automatic selection of the bandwidth presenting the best signal-to-noise ratio for each acquisition. The software also enables the calculation of the rate of interference, and elimiates signals whose pollution level is too high or corrects the interference.

Uncertainties of current measurements that are made largely depend on the resolution of the Doppler spectrum, which is typically between 2 and 5 cm/s. In order to obtain stable estimations, analysis and the averaging of several hundreds of reflections must be done, which limits the spatiotemporal resolution of the installation. If the measurements can be made in ranges of 1–150 km, the spatial resolution goes from 100 m to 3 km.

Their accuracy depends on numerous factors:

– directing errors;

– geometry of antenna or distortions in the sensitivity diagram (it is necessary to measure this diagram to improve the accuracy of measurement);

– various noises and radio interference;

– limitations due to the method of treating the signal used;

– sea states too degraded; and

– nature of the ground surrounding the antenna. This last parameter also limits measuring range.

Its evaluation is made by comparison to current meters, wavemeters or drifting buoys. These intercomparisons are difficult to carry out, as radars cover surfaces of several kilometers squared, over a depth of 0.5–2 m, while current meters make punctual measurements at depths that are often deeper. Wavemeters only deliver information on the waves and floats on currents that are several tens of meters under the surface, at a different timescale. Differences that are recorded between radars and these other methods have error dispersions of about 8–15 cm/s (to one standard deviation).

In order to study sources of error, a simulation was done in 2010 by Laws *et al.* [LAW 10]. It showed that the rms error of a CODAR system in ideal conditions (linear current profiles, strong signal-to-noise ratio, ideal antenna diagram, etc.), could be at a maximum of 1.9 cm/s, that in cases of random sea states this error increased to 2.9 cm/s, and in the case of antenna diagrams with strong distortion it can rise to 12 cm/s.

In order to study the uncertainty of measurements, the results of an intercomparison experiment were published in 2009 by Lipa *et al.* [LIP 09]. Forty-two drifting buoys, anchored at 1 m depth, were dropped into an area covered by six high-frequency CODAR radars in bi-static mode. The data from these instruments were synchronized with the help of GPS clocks, and averaged over 10 mins. The standard deviation obtained between the speeds of buoys and the radial speeds of radars was 8.5 cm/s, while the typical spatial and temporal uncertainties of these radars alone were estimated at between 6.0 and 8.5 cm/s.

Measurements on waves are possible with these systems if we assume that the ocean is homogeneous on the surface observed. For this, we study confined surfaces in directions that have been determined. In 1977, Barrick proposed a method based on the measurement of the spectral powers P_1 and P_2 of the first- and second-order spectrums (see Figure 2.64) in order to obtain efficient wave height. It consists of calculating:

$$H_{rms} = \left[\frac{\lambda}{\sqrt{2}\pi} \frac{\int_{-\infty}^{+\infty} \sigma_2(f) W^{-1}\left(f/_{f_b} \right) df}{\int_{-\infty}^{+\infty} \sigma_1(f) df} \right]^{1/2} \quad [2.219]$$

where:

– f_b corresponds to the Bragg frequency;

– λ to the wavelength of the radar;

– $W(f/f_b)$ to a weighting function; and

– σ_1 and σ_2 to first- and second-order sections.

Application of relation [2.219] enables us to obtain an estimation of wave height between 0.5 and 4.0 m, with a standard uncertainty of 0.14 m.

Obtaining the directional spectrum of wave energy $E(f,\theta)$ is more problematic. A method proposed by Wyatt consists of first estimating their significant wave heights with the help of relation [2.219] (this information enables the calculation of a wave number spectrum), then wind direction (in order to calculate an angular distribution). This one can equally be obtained by studying the signal retrodiffused by high-frequency radars, particularly by establishing the ratio ξ between amplitudes B^+ and B^- of the Bragg peaks:

$$\xi = 10 \ln\left(\frac{B^+}{B^-} \right) = \ln\left[\tan^s\left(\frac{\theta^-}{2} \right) \right] \quad [2.220]$$

where s is a dispersion parameter whose value is often fixed at four. Inversion of relation [2.220] enables an estimation of the value of θ^-, which is the angle between the wind direction and Bragg waves, to a nearest right–left direction ambiguity. This ambiguity can only be dispelled by utilization of two independent measurements.

This information is therefore used to power a directional spectrum model that leads to the calculation of the Doppler spectrum. Differences between this spectrum and the spectrum that is measured are used to modify the directional spectrum for each wave number in a limited range and from specified directions. This calculation is reiterated until the difference between the two Doppler spectra is sufficiently small. The success of this method, when the calculation converges, is based on the way in which corrections are made. It remains more or less experimental, and has frequency limitations.

2.10. Determining the turbidity or sea water's optical properties

Turbidity is a parameter that has been the object of several definitions, and is not presented as an absolute or objective notion. Water turbidity comes from the presence of various types of suspended matter, such as clay, silt, organic and inorganic matter, plankton and other microorganisms. It is therefore presently defined as the optical property, which means that the incident light is scattered and absorbed rather than transmitted in a straight line. In physics, this notion is linked to light diffusion theories that introduce the notion of particle sizes and of distribution of sizes (diffusion of Rayleigh, Mie or diffraction) and absorption. In practice, we can either look to determine the percentage of light absorbed by the medium, and therefore use *transmissiometers*, or look to determine the percentage of scattered light, and therefore use *nephelometers*. These two instruments are used simultaneously when we want to completely characterize the medium. These properties are relative to the wavelength of light.

In order to study the vertical penetration depth of light in water, there is a simpler method called the *Secchi* disc method (named after its inventor). The observer vertically descends a matt white disc around 30 cm in diameter and measures the depth of disappearance on descent and its reappearance when being hauled up. This rudimentary technique, in which accuracy varies to a certain extent according to the observer, surface sea state and lighting, is still used to characterize the penetration depth of light or Secchi depth.

However, turbidity measurements also aim to estimate particle concentrations in the medium or the visibility we can meet. These two notions can be found in the definition of this quantity given by ISO 7027, which states that:

> "turbidity is the reduction of a liquid transparency due to the presence of non-dissolved substances"

Therefore, it is more convenient to talk about the optical properties of sea water. Some of the properties can be observed from space with satellites equipped with spectrometers, such as the SeaWifs (*Sea-viewing Wide Field-of-view Sensors*) launched in 1997 by NASA, which opened the possibility of studying "the color of the ocean". An increase in the concentration of suspended organic and inorganic particles, for example, will cause greater absorption of blue solar wavelengths and will lead to darker coloring of the ocean. Adequate calibration of spatial data collected with the help of sampling or *in situ* measuring instruments, allows us to find the concentration of these particles.

Generally, the optical properties of sea water are classified in two large categories:

– *apparent optical properties* (AOPs) linked to the convolution of solar light with the medium; and

– *inherent optical properties* (IOP) linked to the intrinsic properties of the medium.

AOPs are defined by the notions of radiance or irradiance and are subjected to passive measurements. IOPs are characterized by absorption, transmission or diffusion of light by the medium, and are measured with the help of instruments provided with sources of light.

2.10.1. *Theoretical notions of the optical properties of sea water*

The fraction of daylight that penetrates the ocean can be characterized at all depths by spatial and spectral repartition of a quantity called the *radiance*. Radiance L is defined by the ratio of the elementary energy intensity dI emitted by a source of light onto the elementary surface $d\sigma$ of the solid angle under which an observer views the source. In other words, the radiance of a source is relative to the angle of observation. As the source forms an angle α in relation to the direction of the solid angle of observation, the observer views a surface dS as follows:

$$d\sigma = dS\, cos(\alpha) \qquad [2.221]$$

Energy luminance can therefore be written as:

$$L = \frac{dI}{d\sigma} = \frac{dI}{dS\cos(\alpha)} \qquad [2.222]$$

It is expressed in $Wm^{-2}sr^{-1}$. The energy intensity of a light source can be defined by the ratio of energetic flow dF emitted in a cone with a solid angle $d\omega$.

$$I = \frac{dF}{d\omega} \qquad [2.223]$$

As energy intensity emitted by the source can vary in relation to the wavelength λ of light, radiance also depends on λ. We therefore talk about spectral radiance, $L(\lambda)$. These quantities also vary in the ocean in relation to time and position, but we will limit ourselves to expressing them in relation to λ.

If the observer is now replaced by a photoelectric cell of surface dS', the notion of *irradiance* E or spectral irradiance $E(\lambda)$ is defined in the same way as the ratio between the flow quantity $dF(\lambda)$ intercepted by the cell and its surface dS':

$$E(\lambda)=\frac{dF(\lambda)}{dS'}=\frac{I(\lambda)d\omega}{dS'} \qquad [2.224]$$

With this definition, irradiance is expressed in W m^{-2}. If α' is the incident angle of luminous rays on the photoelectric cell, and r the distance of this to the source, we have:

$$d\omega=\frac{dS'\cos(\alpha')}{r^2} \qquad [2.225]$$

Relation [2.224] can therefore be written in a different form:

$$E(\lambda)=\frac{I(\lambda)\cos(\alpha')}{r^2} \qquad [2.226]$$

We can also define the solid angle $d\omega'$ under which the surface element $d\sigma$, considered to be at distance r, appears on the receiver:

$$d\omega'=\frac{dS\cos(\alpha)}{r^2} \qquad [2.227]$$

From relations [2.222], [2.226] and [2.227], it is possible to express radiance in relation to irradiance:

$$L(\lambda)=\frac{dE(\lambda)}{d\omega'\cos(\alpha')} \qquad [2.228]$$

Expressed in this form, radiance becomes independent of the geometric characteristics of the source. We can therefore apply this notion to a luminous volume, such as water mass that diffuses light, or equally to the canopy of the heavens.

If the medium crossed is absorbent, a fraction $dF(\lambda)$ of the initial luminous flow $F_0(\lambda)$ is absorbed by the elementary water layer dr. If a is defined as being the *absorption coefficient* of the medium (expressed in m^{-1}), it is written as:

$$dF = a\,F\,dr \qquad [2.229]$$

As the energy transported is not the same whatever the wavelength of light, absorption will not be the same whatever the wavelength, and a will also depend on λ. We can therefore define the coefficient of spectral absorption $a(\lambda)$. For the thickness r of the absorbent medium, after integration, equation [2.229] becomes:

$$F(\lambda) = F_0(\lambda)\,e^{-a(\lambda)\,r} \qquad [2.230]$$

This is the approximation or the *Beer-Lambert law.*

If the medium is diffusing, i.e. if it scatters light in all directions, the same reasoning can be applied as previously to define a *spectral scattering* or *total scattering coefficient* $b(\lambda)$, homogeneous to the inverse of a length. If this time $dF(\lambda)$ is the fraction of flow scattered in all directions:

$$dF(\lambda) = b(\lambda)F(\lambda)dr \qquad [2.231]$$

For thickness r of the absorbent medium, after integration we obtain:

$$F(\lambda) = F_0(\lambda)\,e^{-b(\lambda)\,r} \qquad [2.232]$$

However, this relation is only rigorous if no photons are sent back in the direction of the incident flow, and if the photons are not subject to multiple reflections, which is difficult to verify in practice where it is observed that $b(\lambda)$ also depends on direction (see relation [2.236]).

If the medium is absorbent and scatters at the same time, a *spectral attenuation coefficient* $c(\lambda)$ is defined in the same way, so that:

$$c(\lambda) = a(\lambda) + b(\lambda) \qquad [2.233]$$

where $b(\lambda)$ and $c(\lambda)$ are also expressed in m^{-1}. As flow is linked to light intensity by relation [2.223], intensity will also decrease exponentially in an absorbent and scattering medium. As radiance is relative to luminous intensity, by relation [2.222] it will also respond to an exponential relation, so that for a thickness r we have:

$$L_r(\lambda) = L_0(\lambda)\,e^{-c(\lambda)\,r} \qquad [2.234]$$

In fact, c is the sum of a certain number of terms:

$$c(\lambda) = c_0(\lambda) + a_S(\lambda) + a_P(\lambda) + b(\lambda) \qquad [2.235]$$

In this expression:

– $c_0(\lambda)$ is the attenuation coefficient of pure water;

– $a_S(\lambda)$ the absorption coefficient of dissolved substance in water;

– $a_P(\lambda)$ is that of suspended particles; and

– $b(\lambda)$ the total scattering coefficient, which can also be decomposed in $b_w(\lambda)$, mostly due to water molecules and in $b_p(\lambda)$, mostly due to particles.

$c_0(\lambda)$ has been theoretically and experimentally determined with good precision, as in pure water, diffusion is essentially molecular and therefore lends itself to modeling.

Coefficient $a_S(\lambda)$ represents the result of light absorption by dissolved salts ($a_w(\lambda)$) and dissolved organic matter ($a_g(\lambda)$), known as *colored dissolved organic matter (CDOM)*. The CDOM is also known as *Gelbstoff* or *yellow substance.* In ultraviolet light, the presence of aggregates of vegetal origin in water considerably increases the slope of the attenuation coefficient.

For a long time, the absorption coefficient $a_p(\lambda)$ could only be estimated by deduction, knowing the value of $c_p(\lambda)$:

$$c_p(\lambda) = a_p(\lambda) + b_p(\lambda).$$

$a_p(\lambda)$ was the subject of numerous publications, and its value is decomposed in $a_{ph}(\lambda)$, which represents the absorption coefficient of phytoplankton, and in $a_{NAP}(\lambda)$, which represents the absorption coefficient of non-algae particles. $a_{ph}(\lambda)$ variations can be considerable, depending on the biomass contained in the medium.

The principal effect of suspended matter, however, is the scattering of light. Coefficient $b(\lambda)$, which is representative of water turbidity, has been the subject of many studies, but the prediction of its value in relation to the medium remains poorly documented. The value of $b(\lambda)$ would suffice to define the scattering properties of a medium if the scattering was made with the same intensity in all directions, which is not the case. We are therefore led to define a *scattering phase function* $\beta_\theta(\lambda)$:

$$b(\lambda) = 2\pi \int_0^{180°} \beta_\theta(\lambda) \sin(\theta)\, d\theta \qquad [2.236]$$

The integration of particular variation of $\beta_\theta(\lambda)$ in all directions for a given wavelength is known as the *volume scattering coefficient.*

Diffusion of light by particles is related to reflection, refraction and diffraction phenomena. It has been, and continues to be, the subject of numerous theoretical studies, as it is found in several scientific disciplines (astrophysics, aerosol physics, aerology, oceanography, etc.) as well as agri-food (study of wine, beer, etc.) and in biology and medicine. Statistically, particles are assimilated to spheres with diameter d and relative refraction index n. Their size is determined with respect to parameter α, which is defined by the relation:

$$\alpha = \frac{\pi d}{\lambda} \qquad [2.237]$$

The smallest particles ($\pi d \leq \lambda$) respond to the Rayleigh-Gans dipole theory. They offer a certain forward symmetrical diffusion "forward *scattering*" and back diffusion "*backscattering*". Those of intermediate size ($\pi d > \lambda$) respond to Mie's theory (1908) and present backscattering that is weaker when the value of α is high compared to one. The Mie theory has spread to oval and cylindrical particles. For $\alpha < 0.2$, Rayleigh diffusion results in a symmetrical diffusion indicator of the form:

$$\beta_\theta(\lambda) = \beta_{90°}(\lambda)(1 + cos^2\theta) \qquad [2.238]$$

In this case, $\beta_{90°}(\lambda)$ only needs to be measured in order to determine $\beta_\theta(\lambda)$ and to deduct from it the value of $b(\lambda)$ by integration.

Therefore, turbidity can be evaluated by making measurements of the amount of light transmitted or diffused:

– in the direction of propagation (with the help of transmissiometers);

– at 90° from the direction of propagation (with the help of nephelometers); and

– in an angle between 90° and 180° in backscatter.

Standardized nephelometers used for analyzing soft water measure diffusion under a 90° angle in relation to the direction of propagation of the source of light. This equipment is calibrated with Formazine suspensions, whose concentrations are know in mg /1. As the measurement is carried out with a wavelength of 860 nm, the turbidity of water designated for human consumption is measured in Formazine nephelometric units (FNU). However, the unit to be adopted for measurements made at wavelengths different from 860 nm is the nephelometric turbidity unit (*NTU*).

A suspension of Formazine produced by the reaction of hydrazine sulfate $(NH_2)_2H_2SO_4$ (50 mg/l) and hexamethylene-tetramine $C_6H_{12}N_4$ (500 mg/l) in rigorously established conditions produces a turbidity of 40 NTU. By dilation or increasing of these concentrations, it is possible to create standard solutions corresponding to turbidity ranges that we will meet. Despite this calibration technique, it is difficult to link the turbidity and mass concentration of suspended matter (see section 2.11.1.5), as when α increases, (at constant n) the intensities diffused increase and dissymmetries appear, resulting for $\alpha > 30$ in very complex scattering curves. Furthermore, in a natural medium the particles present can have very different sizes, forms and refraction indices (see section 2.10.3).

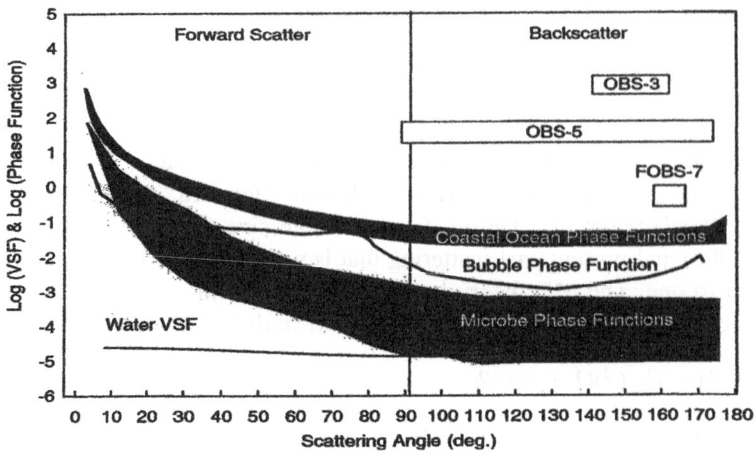

Figure 2.67. *Representation of curves of scattering intensity caused by different sea water constituents relative to the scattering angle. The angular ranges of the operation of nephelometers made by D&A Instruments are superimposed on the diagram (Courtesy of D&A Instrument)*

Finally, it should be noted that Formazine has been declared cancerous and other chemicals already exist, such as the commercial preparation of suspended styrene-divinylbenzene balls, that are being studied as replacements.

2.10.2. *Measurement of apparent optical properties*

These properties are measured *in situ*, usually to calibrate or validate the measurements made by satellites, and to determine the quantity of light absorbed by the ocean or the quantity of light retroreflected. This last one is anisotropic and depends on the constituents of the environment. Its measurement relates to the origin of applications (notably satellites) that focus on "color of water".

Instruments that measure radiance or irradiance must be orientated and kept perpendicular to the surface of the water as much as possible, towards the bottom of the sea and (or) towards the sky, depending on whether absorption or retroreflection is being measured. Radiance introduces the notion of direction of observation. It is obtained using instruments equipped with photoelectric cells topped with a lens that enables a reduction of the entry angle of light (creation of a parallel beam). Only rays perpendicular to the plan of the cell are measured.

Inversely, irradiance is measured with cells topped with a lens that lets rays that come from all directions pass into the plan of the cell. If irradiance that comes from all directions is to be measured, the cell must be placed in a sphere that allows integration in space, and no longer in the plan. We therefore have "irradiance meters" or "scalar radiometers". The light must then be focused on the detector by another lens. These sensors are equipped with filters that allow wavelength bands to pass. As these bands are limited in number, they are known as *multi-spectral instruments*. When the bands are narrow, and about one hundreth of the spectrum, they are known as *hyper-spectral instruments*. These instruments must be calibrated and linked to primary references that are found in national metrology laboratories (such as NIST, NPL, LNE-CNAM). It is the flow or spectral power in the visible range that is calibrated, with a relative uncertainty of about 0.1%. It is by this mean that satellite measurements are also referenced.

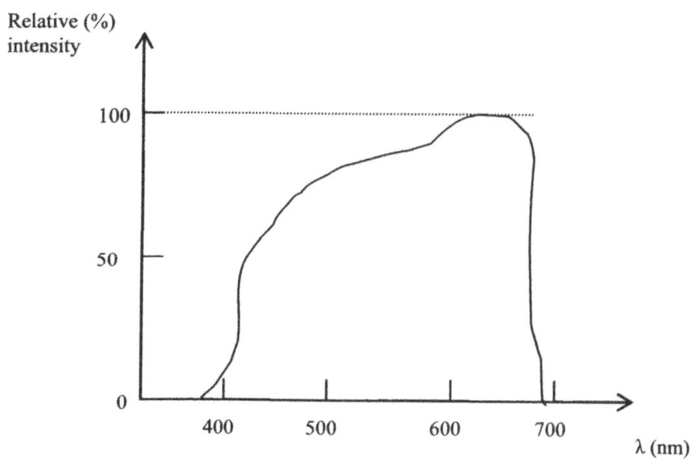

Figure 2.68. *Schematic representation of the response curve in wavelength of a PAR sensor*

Filters used can also be weighted in order to take into account the difference in energy of photons according to their wavelength, and likewise to enable counting of them on a large wavelength range (400–700 nm). These sensors are known as *photosynthetic active radiation* (*PAR*) and used in conjunction with fluorimeters (see section 2.10.5). Directed towards the surface, they measure the quantity of light that locally penetrates the medium. They enable estimation of the *primary production* of phytoplankton (produced uniquely by the interaction of light with molecules) and above all the thickness of the *euphotic layer*, which corresponds to the depth of solar irradiance. It is estimated that there is typically 1% of the surface light to 5 m, but that the light can penetrate up to 50–70 m in the Mediterranean and up to 170 m in clearer seas.

The response of a PAR sensor using the photovoltaic effect must be stronger for elevated wavelengths (red) than for short wavelengths (blue). This effect is obtained with the help of gallium arsenide photodiodes and tinted optical filters.

These are calibrated in μmole m^{-2} s^{-1}. This calibration consists of illumination with a standard source of light (an halogen light of 1,000 W) whose density of photon flow is known. This density flow is expressed in micromoles, by knowing that 1 μmole = 6.023×10^{17} photons. If $E(\lambda)$ is the bulb's spectral irradiance at wavelength λ, the density flow ΔE is expressed by the relation:

$$\Delta E = E(\lambda)\Delta\lambda \qquad [2.239]$$

The number of photons emitted in the spectral interval $\Delta\lambda$ is obtained by multiplying the inverse of the quantum (hc/λ), by relation [2.239], h being the Plank constant and c the speed of light in vacuum. We can therefore integrate this equation in the spectral domain of the bulb (400–700 nm, for example) and devide the result by 6.023×10^{17} to obtain the calibration of the sensor in μmole m^{-2} s^{-1}. If Q is the number of μmole measured per second and per square meter:

$$Q = \frac{1}{6.023\times10^{17}}\left[\frac{1}{hc}\right]\int_{400}^{700}\lambda E(\lambda)\Delta\lambda \qquad [2.240]$$

This absolute calibration can be carried out with a relative uncertainty of ±5%.

2.10.3. *Transmissiometers and measurements of absorption*

The attenuation of the intensity of a luminous beam that propagates in a marine environment is due to scattering and absorption. Transmissiometers are instruments that try to measure these losses, regardless of their origins. Different methods can be used for this. The oldest one consists of projecting a parallel beam of light over a distance of 10–100 cm, with a lens placed at the end in order to focus the beam on a photoelectric cell (see Figure 2.69). This principle is a direct application of relation [2.234].

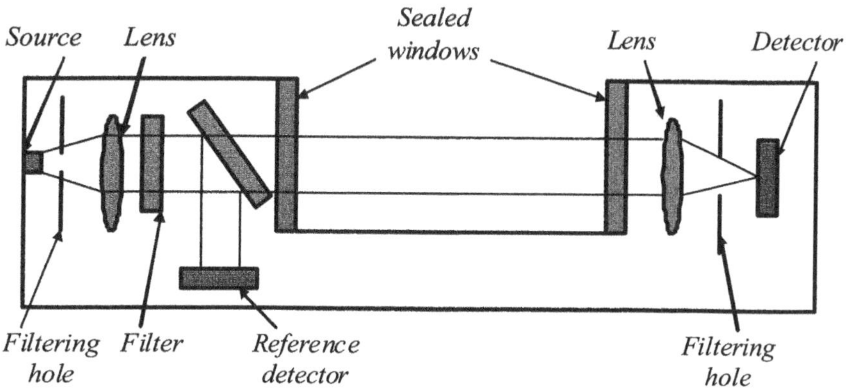

Figure 2.69. *Diagram showing the principle of a transmissometer (Courtesy of Wet Labs, C Star, User's Guide)*

Filtering holes can be positioned after the source and before the receiver in order to limit beam divergence and parasitic reflections. A filter allows selection of the wavelength to be transmitted. As diffusion by particles varies considerably in relation to the wavelength of light, different selective interferential filters are proposed to select an operating wavelength. In certain instruments, several wavelengths can be selected and others can operate in hyperspectrum. A part of the beam is removed by a semi-transparent mirror and sent to a reference detector that enables us to control and stabilize the intensity emitted by the light source; the latter can vary in relation to the temperature of the medium. Voltage measured by this detector also serves as a reference for determining the intensity emitted. The transmitter and the receiver are protected from water and pressure by sealed windows. The principal detector converts the intensity transmitted into voltage.

In the case of C Star transmissometers made by Wet Labs, Inc., the light source is a LED. This component, associated with an oscillator, enables the modulation of light intensity. As the detection assembly possesses a rejection system of frequencies other than those emitted by modulation, the system is less sensitive to ambient light and other sources of signal pollution. All that is required to obtain a signal with an amplitude that is proportional to the light transmitted by the medium is the amplification and demodulation of the signal measured by the detector. The value of the attenuation coefficient, $c(\lambda)$, can be obtained by calibration of the instrument. If d represents the distance between the two sealed windows, relation [2.233] enables us to establish the amplitude Tr of light transmitted:

$$Tr = L / L_0 = e^{-cd} \qquad [2.241]$$

As the response of the instrument is linear, its calibration is carried out over two points. The minimum attenuation value is determined by placing distilled water between the transmitter and receiver. The maximum voltage V_{max} is therefore measured. As the attenuation coefficient of distilled water is known (for example, for λ = 660 nm it is 0.364 m^{-1}), it is possible to precisely determine which attenuation or transmission value V_{max} corresponds to. The maximum attenuation value is given by completely blocking the receiver. The minimum voltage V_{min} is therefore obtained. If V_M is the voltage corresponding to the attenuation that we want to measure, from relation [2.241] we can write:

$$\frac{V_M - V_{\min}}{V_{\max} - V_{\min}} = e^{-c(\lambda)d} \qquad [2.242]$$

From relation [2.242], the value of $c(\lambda)$ expressed in m^{-1} can be extracted:

$$c(\lambda) = -\frac{1}{d}\ln\left(\frac{V_M - V_{\min}}{V_{\max} - V_{\min}}\right) \qquad [2.243]$$

The linearity of these instruments can also be verified and eventually corrected by placing Indian ink solutions of different concentrations in the path of the beam. Temperature dependency must also be verified and eventually corrected prior to the first utilization by placing the instrument in a temperature-controlled bath. In order to show the possible variations in the attenuation of the medium, its beam can be blocked. We are therefore able to track variations of V_{min} at different stages of response to temperature.

Finally, when purchasing an instrument it is important to effectively choose the length of its optical path d in relation to the turbidity conditions of the medium. The path must be long enough that the volume of measurement represents a good statistical average of the surrounding environment, and so that the instrument has sufficient sensitivity in low turbidity environments. It must also be short enough that multiple reflections linked to the turbidity of the environment do not return light into the beam. The optimal value of d is considered to have been obtained if it responds to the inequality:

$$c(\lambda)^{-1} \leq d \leq 3\, c(\lambda)^{-1}.$$

Beyond three $c(\lambda)^{-1}$, there is a probability not equal to zero that diffused photons will be subjected to more than three multiple reflections (the approximation of the Beer-Lambert law is therefore no longer valid) and will return in detected flow, thereby introducing a bias. Inversely, if $d < c(\lambda)^{-1}$, the instrument will not be sufficiently sensitive or able to correctly follow variations in turbidity of the environment.

While transmissometers measure the attenuation of a beam, it can be necessary to determine how much is due to *molecular absorption* and to the components of the medium. The total attenuation coefficient $a_t(\lambda)$ is given by equation [2.234]: $a_t(\lambda) = a_S(\lambda) + a_P(\lambda)$. $a_s(\lambda)$ is the sum of the absorption of water and the dissolved salts that form it ($a_w(\lambda)$), and absorption by the CDOM, ($a_g(\lambda)$). The principle of these measurements is based on the fact that if the absorption spectrum of certain molecules is known and unique, it is possible to estimate these terms separately. Thus, values of $a_P(\lambda)$ (for non-algal contribution) and $a_g(\lambda)$ can be approximated.

There are different types of instruments used to measure these coefficients. One of them is known as a reflecting tube for absorption measurements. It consists of using a source with a large diameter and redirecting all light diffused by the environment towards a photo-electric detector, which also has a large diameter. From relations [2.232] and [2.233], it is possible to extract the value of $a_t(\lambda)$. If ϕ_0 represents flow emitted by the source, $\phi_T(d)$ the flow of light transmitted over distance d, and $\phi_R(d)$ the flow retroreflected over the same distance, $a_t(\lambda)$ is given by the relation:

$$a_t(\lambda) = -\frac{1}{d}\ln\left(\frac{\varphi_T(d) - \varphi_R(d)}{\varphi_0}\right) \qquad [2.244]$$

This concept is based on the hypothesis that reflection in the tube is made without loss and that no photon is reflected towards the source. In reality, it is estimated that only around 80–90% of photons reach the detector. The absorption coefficient measured is therefore always higher than the real absorption coefficient, and it is necessary to apply empirical corrections to the measurements in order to infer that absorption represents around 5% of this loss. Multi-spectral or hyper-spectral versions of this type of instrument are available.

Another type of instrument, commercialized by Hobilabs and called "a-beta", is based on the measurement of the angular diffusion coefficient, as defined by relation [2.235]. It uses a divergent light source that illuminates a scattering surface with an angle θ_0. A detector placed in the same plane as the source measures the light scattered at an angle θ_d. The source and the detector are situated behind a window at a distance d from the scattering surface. $a_t(\lambda)$ is therefore obtained from a complex development:

$$a_t(\lambda) = -\frac{1}{2z}\left[\ln(k) - \ln\left(V_d - V_d^0\right)\right] - \chi\beta_\theta(\lambda) \qquad [2.245]$$

where:

– $\bar{z}$ represents the mean optical path between the source and the detector;

– k a constant characteristic of the instrument;

– V_d voltage measured;

– V_d^0 the dark voltage of the detector; and

– χ a scale factor of $\beta_\theta(\lambda)$ valid for the sum of angles θ_0 and θ_d.

Coefficients and unknown variables of relation [2.245] are obtained by the least squares technique and several calibration points.

Finally, another process based on the utilization of an integrating sphere was developed in the 1990s, which is called the *point source integrating cavity absorption meter* (PSICAM). The sphere is loaded with sea water and illuminated by a point source. The light produced is subject to multiple reflections, which provokes an elongation of its optical path and favors its absorption. It has been shown that the utilization of certain wavelengths enables us to use this instrument to measure the absorption due to the CDOM and micro-algae.

2.10.4. *Nephelometers and turbidity sensors*

Nephelometers are instruments used to characterize the dispersion of light by particles. This scattering is not homogeneous in all directions. Relation [2.235] gives the definition of the volume scattering coefficient $b_p(\lambda)$. According to this definition, nephelometers should therefore, strictly speaking, enable the measurement and integration of intensity scattered in all directions between 0° and 180°. Certain laboratory prototypes enable this, such as the instrument developed by Lee and Lewis in 2003 or the Hydrobeta produced by Hobilabs. However, they are not easily transposable to make *in situ* measurements, and their measuring range is often limited to 10 mg/l of suspended matter.

In order to facilitate the estimation of the volume scattering coefficient and enable a comparison of measurements, studies have been initiated to determine the angle values characteristic of medium scattering. It has been observed that most scattering phase functions $\beta_\theta(\lambda)$ present a common angle for which the volume scattering coefficient is about equal to $b_b(\lambda)$, which is the backscattering coefficient. These studies have led to different commercial achievements where scattering is measured according to one or two discrete angles. These instruments allow the measurement of particle concentrations in ranges that can range from zero to several hundreds of mg/l. In order to attempt to homogenize these measurement results, a standard (NF EN 27027 April 1994 or ISO 7027 from 1990) was created focusing on the analysis of the quality of soft water in relation to its turbidity. It specifies the conditions under which the nephelometric measurement must be carried out (see Figure 2.70) in accordance with:

– wavelength;

– spectral band $\Delta\lambda$ of the source (60 nm);

– the angle θ between the source and the detector (90 ±2.5°); and

– aperture of the beam of measurement (20°).

Turbidity sensors generally give an indication of the concentrations of suspended particles from the measurement of optical scattering centered on an axis at 90° from the beam's axis. Their calibration enables the provision of absolute values of scattered flow relative to the part of the scattering phase function under observation. For most sea water, light backscaterred by particles between 90 and 180° is proportional to the signal viewed by satellites that observe water color.

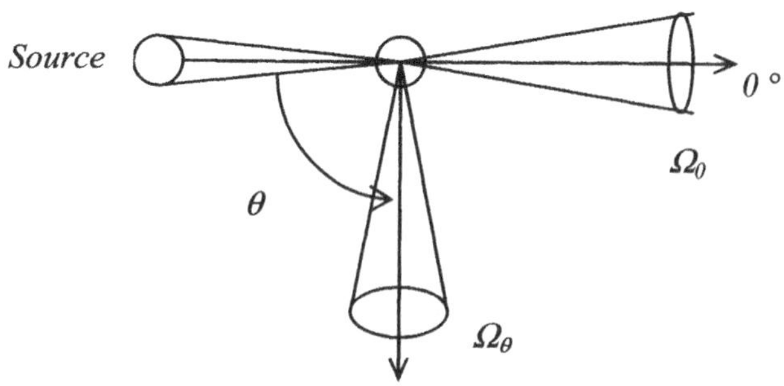

Figure 2.70. *Representation of light cones transmitted and scattered as they must be measured according to standard ISO 7027*

In compact and inexpensive commercial products, the light transmitter and detector are placed in the same plan, and the voltage that they deliver is proportional to the integral of the backscattered light in the illumination cone generally situated between 100 and 170° (see Figure 2.71). Once calibrated in NTU, they enable the measurement of concentrations up to 100 mg/l with resolutions that can reach 100 µg/l. These good resolutions enable them to be sensibly used to measure the relative variations in concentrations, the scattering in their aperture angle or, by extrapolation, scattering in the 90–180° spectrum.

In the field of oceanography it seems, according to Ivanoff [IVA 75], that measurements made according to angles of less than 90° enable us to obtain a good correlation with the volume scattering coefficient for particles $b_p(\lambda)$, on the condition that we deduct the part due to the molecular scattering from the measured scattering:

$$\beta_{p\theta}(\lambda) = \beta_\theta(\lambda) - \beta_{w\theta}(\lambda) \quad [2.246]$$

where $\beta_\theta(\lambda)$ corresponds to what is measured by the nephelometer. Certain commercial products therefore use angles of less than 90°, as the difficulty is not taking into account the quota due to the light transmitted. In fact, although water molecules contribute little to the value of the volume scattering coefficient, they contribute in a much larger way to the backscattering coefficient value, particularly in the blue zone. It has been shown that relation [2.246] must also be applied when we have angles between 90 and 160°.

As instruments for *in situ* measurements must be compact and low consumers of energy, the light source most often used is a LED. This component allows the emission of light pulses in which the reflected part is more easily detected by a photodiode. It is expected that the intensity $I_b(\lambda)$ measured will be proportional to the backscattering coefficient of particles $b_{bp}(\lambda)$. In fact, in relation [2.235], $\beta_\theta(\lambda)$ must be multiplied by a weighting function $w(\theta)$, which depends on the illuminated volume and aperture cone of the detector to determine what the device is really measuring. We have:

$$I_b(\lambda) = K\, b_{bp}(\lambda) = \int_0^{180°} \beta_\theta(\lambda) w(\theta) \sin(\theta)\, d\theta \quad [2.247]$$

Following work done by Oishi [OIS 90], Boss and Pegau [BOS 01] showed that when $b_{bp}(\lambda)$ was determined from a single angle close to 120° (exactly 117°), the ratio $\beta_\theta(\lambda)/b_b(\lambda)$ is constant, at around 0.15 sr^{-1}, and insensitive to the molecular contribution of water. They also showed that the ratio $\chi_\theta = b_b(\lambda)/(2\pi\beta_\theta(\lambda))$ is constant and worth 1.1 to ±4% for $\theta = 117°$. From these observations, it is therefore possible to use χ in order to find the value of $b_{bp}(\lambda)$, with the relation:

$$b_{bp}(\lambda) = 2\pi\chi\beta_{p\theta}(\lambda) \quad [2.248]$$

where $\beta_{p\theta}(\lambda)$ is obtained by relation [2.246] (with $\beta_{w\theta}(\lambda) = 0$ if $\theta = 117°$). It is, however, necessary to take into account the characteristics of the detector and therefore to determine $w(\theta)$ before using formula [2.248]. $w(\theta)$ is calculated from calibration with polystyrene beads 2 µm in diameter. The procedure to be followed is described in a document edited by NASA in 2002 [NAS 02].

If we then want to find the volume backscattring coefficient $b_b(\lambda)$ from relation [2.248], the backscattering coefficient of pure water $b_{bw}(\lambda)$ needs to be added. It is obtained from a relation published by Morel [MOR 74]:

$$b_{bw}(\lambda) = (0.0022533(\lambda/500)^{-4.23})/2 \qquad [2.249]$$

where λ is expressed in nm and $b_{bw}(\lambda)$ in m^{-1}. According to Boss and Pegau, the contribution of water to measurements made at different angles of 117 ± 3°, can considerably bias the calculation of $b_b(\lambda)$ from measurement at a single angle.

If we want to estimate turbidity, calibration of the nephelometer with Formazin solutions enables us to make the value of output voltage correspond to a value expressed in NTU. According to the range of turbidity explored and modes of lighting and detection, it is possible that the proportional coefficient K will not be constant. In other words, it is possible that the relation between the voltage V_b obtained on output from the instrument and the value of I_b of intensity detected will not be linear. In this case, calibration must be carried out on several points in order to determine linearization coefficients $c_1, c_2, \ldots c_n$ so that:

$$V_b = c_1 + c_2 I_b + \ldots + c_n I_b^{(n-1)} \qquad [2.250]$$

Despite their calibration, various factors can influence the response of nephelometers. In 2000, Sutherland *et al.* determined an equation that allows the quantification of luminous flow F diffused by a small water sample with a volume V relative to the specific mass concentration C of suspended particles, incident lighting E, effectiveness of diffusion Q_b, diameter of particles D and their mass density ρ [SUT 00]:

$$F = \frac{3}{2}\frac{VCEQ_b}{\rho D} \qquad [2.251]$$

If V and E can be fixed, the other parameters strongly depend on the characteristics of the medium, which are generally unknown at the time of measurement.

Finally, there are some laboratory nephelometers, such as the HACH 2100 AN, whose respective repeatability and accuracy characteristics have been evaluated to be ±0.1 NTU to 4.5 NTU and 96% of the meter reading at 2.48 NTU, which enables an appreciation of the relative performances of measurements that can be reached with this type of instrument.

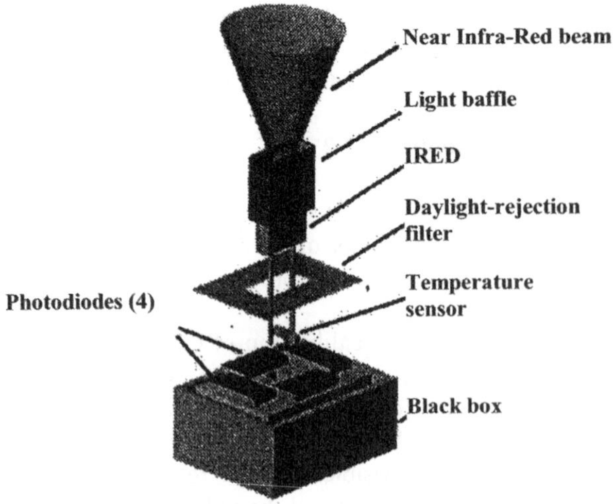

Figure 2.71. *Representation of the operation of the OBS-3 scatterometer produced by D&A Instruments. The LED transmitter and the detector are in the same plan. Wavelength chosen is situated in the infra-red spectrum in order to limit the propagation distance and the influence of daylight. Detection is made in the cone between 140 and 160° (see Figure 2.67) (D&A Instrument Company [DOW 06])*

2.10.5. *Fluorimeters*

A considerable portion of particles suspended in the ocean is of vegetal or organic origin. *Plankton* is defined as the set of pelagic organisms likely to be carried away by ocean currents. There is a difference between animal plankton or *zooplankton* and vegetal plankton or *phytoplankton*. The latter is made up of unicellular algae, which are the first link in the food chain (primary production). The development of these algae occurs by photosynthesis: the solar energy absorbed triggers a complex process leading to the formation of organic matter and to the release of oxygen. A specific molecule, chlorophyll *a*, converts luminous energy into chemical energy. Other pigments are present in the various classes of phytoplankton (chlorophyll *b*, *c*, fucoxanthine, peridinine, zeaxanthine, etc.) They also capture light energy, but at different wavelengths to chlorophyll *a*, to which they transfer this energy.

Concentrations of *phytoplankton* are correlated with dynamic oceanic structures, notably thermal fronts. It is therefore an ocean *biomarker*. It has, however, been shown and acknowledged that relation:

$$a_p(\lambda) = a_{ph}(\lambda) + a_{NAP}(\lambda)$$

is generally dominated by the term $a_{ph}(\lambda)$, which represents absorption by phytoplankton, and that $a_p(\lambda)$ can be estimated by the measurement of $a_{ph}(\lambda)$.

Research has therefore been conducted to attempt to determine the relations between the absorption coefficient $a_p(\lambda)$ of particles [2.234], and their concentration in type *a* chlorophyll pigments, noted *<chl>*, that are found in phytoplankton. This concentration is linked to the seasonal production of phytoplankton, chlorophyll *a* being the most important photosynthetic pigment in plants. Relations of the form [2.252], where c_1 and c_2 are empirical constants, are used to determine the value of $a_p(\lambda)$ under certain conditions of concentration in chlorophyll pigments. This concentration is determined *in situ*, with the help of *fluorimeters*.

$$a_p = c_1 <chl>^{c_2} \qquad [2.252]$$

Fluorimeters are not only used to determine particle concentrations in chlorophyll *a*. They can also be used to determine the rate of *dissolved organic carbon (DOC)*. DOC is found in river water or water that has had high phytoplanktonic activity that is subsequently degraded. It is a component of yellow substances, or CDOM. Its level enables the identification of water masses whose origins are particular. It has been shown that the concentrations of CDOM are correlated with *<chl>* in open water and that the term $a_g(\lambda)$ can also be expressed in relation to *<chl>*. This correlation is weaker in coastal areas due to the input from rivers, and fluorimeters specific to CDOM measurements have been developed and commercialized.

Fluorescence is a property of certain molecules that emit light rays in all directions when they are subjected to the action of incident light rays. After having absorbed a photon, theses molecules return to a state of intermediary excitation and emit a photon in returning to their initial energy state. The photon emitted has lower energy than that absorbed. Therefore, emitted wavelengths are redshifted in relation to incident wavelengths, according to *Stokes'* law, which states that the *maximum absorption of the light spectrum that excites a molecule corresponds to a wavelength that is always lower than that of maximum emission.*

Fluorescence, with phosphorence, is a photoluminescence phenomenon. Contrary to phosphorescence, however, fluorescent emission is independent of exterior conditions (temperature, pressure, etc.). Its duration, which reflects the mean time spent by the molecule in an intermediary excitation state, is around a nanosecond, while in the case of phosphorescence this duration is around a micro or millisecond.

The energetic efficiency of fluorescence ρ_e can be defined according to relation [2.253], where W_f is the energy re-emitted by fluorescence and where W_a is not the incident energy but the energy absorbed by the substance.

$$\rho_e = W_f / W_a \qquad [2.253]$$

The calculation of this efficiency allows determination of the wavelength of maximum excitation by a layout of the curve $\rho_e = f(\lambda)$. In oceanography, the measurement of this ratio allows us to determine which energy quantity can be converted by phytoplankton. This information is useful for the evalution of the effectiveness of fluorescent pigments and the health of ecosystems.

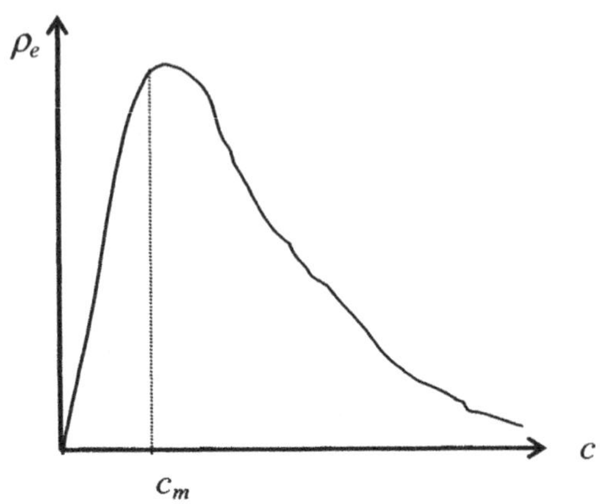

Figure 2.72. *Representation of an energetic efficiency and fluorescent molecule concentration curve. For concentration c_m this goes through a maximum, called the fluorescence optimum*

In addition to wavelength, fluorescent intensity I depends on a certain number of factors including:

– the intensity of incident flow, I_i;

– the geometry of the system – notably the thickness l of the sample crossed that affects absorption of the incident flow by the medium and therefore the lighting of the last layers;

– the molar absorption coefficient $\varepsilon(\lambda)$ of the substance that fluoresces (expressed in l mol^{-1} cm^{-1}). This expresses the effectiveness with which a molecule absorbs light in a given medium. This medium is generally a solvent. Its value can be determined;

– its concentration C in fluorescent substances (expressed in mol l^{-1}). If we start with an infinite dilution, the fluorescent flow is zero. If we increase the concentration, it also increases up to a maximum point, called the optimum fluorescence (Figure 2.72). For even higher concentrations, it decreases as the increase in shocks between molecules leads to a deactivation of fluorescence and a decrease in yield. This phenomenon also causes a decrease in the mean duration τ of the emission of fluorescence, or the *half-life of fluorescence*. However, we can note that the ratio ρ_e / τ is *constant*, regardless of the concentration. This property enables the identification of substances without ambiguity.

In fact, the absorbance $A(\lambda)$ of a sample follows the Beer-Lambert law:

$$A(\lambda) = \ln\left(\frac{I_i}{I}\right) = \varepsilon(\lambda) lC = OD \qquad [2.254]$$

where OD corresponds to the optical density of the sample that can be measured.

Relation [2.254] therefore shows that a measurement of I in a determined wavelength enables to us retrieve the concentration of fluorescent substance via some constants.

The absorption and emission of photons are very quick processes ($\approx 10^{-15}$s). After the excitation of a high number of molecules by very brief pulses, the intensity of the light re-emitted decreases following an exponential law, with a characteristic time τ corresponding to the lifespan of molecules in an excited state:

$$I = k_r \, I_0 \, e^{-t/\tau} \qquad [2.255]$$

Here, I_0 corresponds to the number of molecules that are excited when $t = 0$ s and k_r, at a constant of the radiative de-excitation speed. k_r is defined from the *quantum yield of fluorescence* ϕ, which corresponds to the fraction of excited molecules that return to a fundamental state after having emitted fluorescence photons:

$$\phi = k_r \tau \qquad [2.256]$$

I is the impulse response of the system if excitation is from a very short pulse.

Fluorometers used in oceanography exploit different principles that have differing levels of sophistication. The simplest ones operate in a *stationary mode*, and use a single wavelength specific to the molecule whose concentration we want to measure. Their measurement principle responds to relation [2.254]. Numerous commercial products exist, for exmaple the Aquatracka fluorimeter III made by Chelsea Instruments (see Figure 2.73). It is equipped with a Xenon flashlight that emits white light whose rays are filtered in the blue zone around 430 ±105 nm in order to excite chlorophyll *a* molecules. Light emitted by the sample is detected perpendicularly to the direction of excitation, with the help of a photomultiplier in front of which a second filter centered in the red spectrum around 670 ±30 nm is placed.

To avoid taking fluctuations of light intensity of the flashlight into account, a reference detector measures the intensity emitted by the flashlight, and the instrument calculates the ratio between the light intensity of the fluorescence and that of the flashlight. The ratio between these two signals results in voltage V, which then passes into a logarithmic amplifier, which releases a voltage proportional to 10^V. In-factory calibration with the help of distilled water, acetone and chlorophyll *a* solutions allows the determination of <*chl a*> in µg/l:

$$<chl\ a> = slope\left[\frac{10^{V} - 10^{V_B}}{10^{V_1} - 10^{V_A}}\right] + offset \qquad [2.257]$$

where:

– V_B is the voltage (in Volts) obtained with a pure water solution (this corresponds to the correction of electronic offset of the instrument); and

– V_I to voltage obtained by a pure acetone solution that corresponds to a null chlorophyll concentration and therefore to a second offset correction of the instrument.

The quantity in brackets is *a priori* without unity, as a ratio between voltages is created. However, as V_I is measured for a concentration of 1 µg/l, it is implied that this ratio corresponds to a variation of 1 µg/l. This comes down to multiplying it by the coefficient of a slope that is initially equal to 1 µg/l. To render this coefficient different to 1, Seasoft software produced by Sea Bird Electronics, Inc., provides the possibility of adding another offset correction to data to adjust the instrument to real concentrations measured *in situ*. These slope and offset coefficients also allow the calibration of the instrument without preparing a solution at 1 µg/l, so that acetone and distilled water solutions. It must be noted that the chlorophyll *a* molecule is not soluble in water, and that it is necessary to use pure acetone to create solutions of different concentrations.

Figure 2.73. *The Aquatracka fluorimeter III produced by Chelsea Instruments (© SHOM)*

The opinions and practices relating to the calibration of chlorophyll *a* fluorimeters differ according to various laboratories. If calibration from solutions of chlorophyll *a* is rigorous and allows quantification of the measurement uncertainty (which is about 2%, relatively, in the best cases), the comparison of measurements made with instruments calibrated in this way and the analysis of samples made *in situ* often show big differences.

This analysis is carried out according to a well-defined protocol. First of all, to be compared the samples must be taken from water layers that are sufficiently homogeneous. Samples must be kept away from the Sun as they are photosensitive. Phytoplanktonic cells are separated from sea water by filtering (see Arar *et al.* [ARA 97]). The filter used is then soaked in 90% acetone in order to extract pigments without altering them. Analyses are made, preferably with the help of a laboratory spectrofluorimeter whose coefficient *K* is determined by a calibration curve with the help of a standard range of chlorophyll *a* solutions. <*chl a*> is obtained with an equation developed by Lorenzen:

$$<chl\ a> = K\frac{R_{\max}}{(R_{\max}-1)}(F_{NA}-F_A)\left(\frac{v}{1000V}\right) \quad [2.258]$$

where:

– R_{max} is the maximum acidification ratio. $R_{max} = \frac{1}{n}\sum\left(\frac{F_{NA}}{F_A}\right)$ and n is the number of samples used to obtain the calibration curve;

– F_{NA} is the fluorescence amplitude of a non-acidified sample and F_A is that of an acidified sample.

– V is the volume of filtered water (in l); and

– v is the volume of extraction solvent (in ml).

However, the ratio $\left(\frac{v}{1000V}\right)$ is not to be taken into account when solutions are prepared from pure chlorophyll *a*, as no extraction is done.

It must be noted that this analysis can also be done with the help of a spectrophotometer. It is therefore necessary to know the absorption coefficient of the molecule being studied. For example, it is 87. 67 l g^{-1} cm^{-1} for chlorophyll *a*, 51.36 l g^{-1} cm^{-1} for chlorophyll *b* and 166 l g^{-1} cm^{-1} for fucoxanthin. As the spectrophotometer is adjusted to the emission wavelength of the pigment (664 nm for chlorophyll *a*), it provides an absorbance value that just needs to be multiplied by 1,000 and divided by the absorption coefficient of the molecule in order to find the concentration in mg/l.

This analysis protocol shows that laboratories measurements come directly from pigments, while measurements *in situ* come from living plankton. In living plankton, pigments are therefore not directly exposed to light, and the incident light can be subject to absorption and diffusion before reaching the chlorophyll *a* molecules. These factors depend on the size and structure of phytoplankton, and also on chloroplasts that contain cells with pigments. This is called "the package effect" and it is dependant on the wavelength used. In order to compare the laboratory analysis with *in situ* measurements, it is therefore necessary to combine fluorescence measurements with absorption and scattering measurements to correct the latter.

Temperature can also slightly influence *in situ* measurements. First, it must be ensured that the fluorimeter used has a response independent of the medium temperature by carrying out temperature variations in a calibrating bath (see section 1.2.1), with a blocked measuring cell. When this response is known, *in situ* measurements must then be referenced to one temperature, as the quantum yield of fluorescence ϕ presents dependence on the temperature according to the relation:

$$1/T \approx ln(1/\Phi - 1) \qquad [2.259]$$

Likewise, variation of ±10°C around 20°C can cause maximal error of ±10% in the value of fluorescence measured.

Taking into account all of these effects, certain manufacturers of fluorimeters recommend using phytoplankton cultivation, particularly of species close to those that will be found in the medium in which measurement will be made, in order to calibrate the instruments. Cultivating plankton is hard to do. Buying sufficient quantities to calibrate across an entire measuring range is a costly process and this technique can present measurement uncertainties that are difficult to estimate; phytoplankton possesses photo-adaptation capacities and a change in temperature or luminosity can affect the fluorescence of the cultivation. The room must therefore be dimly lit and the set (instrument plus cultivation) must be placed in thermal equilibrium. Furthermore, measurements carried out in laboratories have shown that at identical chlorophyll *a* concentrations, certain species of plankton emit *in vivo* fluorescence that is three times higher than others.

Measurements of the amplitude of fluorescence made *in situ* therefore remain difficult to quantify, and are often used in relative terms to simply determine the profile of the medium. Intercomparison of measures made with instruments made by different companies remains difficult to interpret, as there is no standard procedure for the calibration or conception of fluorimeters.

Chlorophyll *a* is not the only pigment to be detected by fluorescence, and several manufacturers produce multi-spectral fluorimeters to detect the presence of secondary pigments, such as fluorescein, phycoerythrin, or CDOM (see Table 2.11). These instruments are frequently used to quantify environments of species of plankton or dissolved organic matter.

The question of the best method for calibrating these fluorimeters therefore arises. Those dedicated to phycoerythrin are generally calibrated with the help of solutions of rhodamine (which is soluble in ethanol). Similarly, as CDOM is a substance that cannot be produced industrially, CDOM fluorimeters are generally calibrated with the help of perylene solutions (soluble in ethanol) or quinine sulphate (which is diluted in sulphuric acid), as perylene is a carcinogen. For this, we need to establish concentration ratios taking into account the quantum yield of fluorescence of these molecules defined by relation [2.256], the optical density (OD) of the sample in relation to that of a reference sample, OD_r, and the ratio of their refractive indices n. For equal intensities of fluorescence, I and I_r, , we have:

$$\varphi = \varphi_r \frac{OD}{OD_r} \frac{n^2}{n_r^2} \qquad [2.260]$$

Combined with relation [2.254] and by taking into account the ratio of molar masses, m to m_r, we obtain:

$$C = C_r \frac{\varphi_r}{\varphi} \frac{\varepsilon_r(\lambda)}{\varepsilon(\lambda)} \frac{m}{m_r} \frac{n^2}{n_r^2} \qquad [2.261]$$

As the numerical values of this equation can be determined or found in the literature, it is therefore possible to establish concentration ratios and calibrate these fluorimeters in µg/l with substances other than those measured.

	Excitation (nm)	Emission (nm)
Chlorophyll a	470 ± 30	685 ± 10
Chlorophyll b	460 ± 20	650 ± 10
Phycoerythrin and rhodamine WT	510 ± 40	580 ± 20
Uramine and fluorescein	475 ± 30	530 ± 30
CDOM and oil, UV colorants, quinine sulphate	360 ± 12	450 ± 40
Pheophytin a	410 ± 20	667 ± 10
Phycocyanin	610 ± 20	685 ± 10

Table 2.11. *Central wavelengths and excitation and emission bandwidths of principal pigments. It must be noted that when the bandwidths of instruments used to detect these pigments cover each other, it is impossible to discriminate between them. Light detected is the sum of all wavelengths emitted*

When it is necessary to characterize the properties of phytoplankton photosynthesis, and particularly its effectiveness, it is possible to exploit relation [2.255] with instruments that enable us to control the intensity and duration of excitation and to observe the decrease in fluorescence in micro or milliseconds. These fluorimeters operate in *pulse repetition* or *time resolved* mode. Several techniques that are distinguished by the way in which a pulse of light is applied and the emission is measured can be implemented. These include: pulse amplitude modulation, the pumping during probe, fluorescence induction and recovery and fast repetition rate fluorescence.

The latter is put into practice in fluorimeters used *in situ*. The light source they use is generally a Xenon flashlight (polychromatic source) in front of which a filter is placed that selects a wavelength band. This flashlight generates very brief flashes at a frequency of 100–200 kHz. The intensity re-emitted is measured by a photodetector that is equipped with an optical filter that selects the wavelength or wavelengths of the substances being studied. The fluorescence signal obtained is the convolution product of the excitation signal with the kinematic decrease in fluorescence determined by equation [2.255]. We can express it as:

$$F_n = F_0 + \left(F_m - F_0\right)\left[1 - e^{-E_n \sigma_{ps}}\right]\left\{\sum_{j=0}^{n-1} e^{-jE_n \sigma_{ps} + \frac{\Delta t}{\tau}}\right\} \qquad [2.262]$$

where:

– F_n represents the intensity of fluorescence obtained after n^{th} flashes;

– F_0 is the minimum signal of fluorescence;

– F_m is the maximum;

– σ_{ps} the section of absorption of the photosystem;

– E_n the intensity of the flash n and Δt the delay between flashes.

Finally, the resolved time mode can also be exploited in laboratory fluorimeters that enable measurement of the value of τ, which is a parameter characteristic of molecular kinetics. This measurement enables the detection of substances in complex mediums and the evaluation of their concentrations. It is then possible to distinguish traces of petrol or volatile polycyclic composites from traces of CDOM.

2.11. Determining various physicochemical properties

The study of exchanges between the ocean and atmosphere and that of the impact of human activity on the ocean have brought about the development of sensors in recent years whose role is to quantify several of the chemical parameters of sea water. Until now, these physicochemical properties were essentially determined from samples made during campaigns at sea, and then analyzed in laboratories using analytical chemistry and spectroscopic methods. Certain phenomena remained "invisible" with these discrete sampling methods, however, which led to the development of instruments dedicated to the *in situ* determination of specific chemical constituents. These instruments enable us to avoid the problems surrounding the conservation of samples and have the advantage of being able to sample the ocean environment at a higher frequency for easier deduction of its variability. They are beginning to be integrated into most oceanographic instruments: CDT profilers, drifting buoys, etc. The constituents detected today are principally dissolved oxygen, carbonates and nutrients; however, the door remains open for the detection of other molecules.

2.11.1. *Notions of the chemical parameters of sea water*

In the chemistry of sea water, two categories of elements are distinguished: *dissolved elements* and *particles*. The difference between the two is made in an arbitrary way according to a simple rule: every composite retained by a filter with holes 0.45 µm in diameter is considered particular and what passes through is considered to be in dissolved form.

2.11.1.1. *Major constituents and traces of elements*

Sea water solution is constituted of 96.5% pure water. In the remaining 3.5%, there are 11 principal elements present in the form of cations or anions, due to the power of dissolution of water. These are the ions Na^+, K^+, Mg^{2+}, Ca^{2+} and, Cl^-, SO_4^{2-}, HCO_3^-, B_r^- and F^-. The origin of these constituents is said to essentially be linked to erosion and "leaching" of the Earth's crust by oceans and from sandstorms from deserts. These constituents represent 99.9% of the total mass of dissolved substances. The eleventh element is boron, which is found in the form of boric acid (H_3BO_3). We consider these to be major constituents, as their concentration is above 1 mg/kg. We then have HCO_3, CO_3^- and H_3BO_3 molecules whose concentrations vary according to the pH and alkalinity of water (see Table 2.12). HCO_3^- and CO_3^- originate from the dissolution of atmospheric CO_2 or from primary organic producers.

The concentration of chlorides and bromides can be regularly evaluated in laboratories by methods of measuring out. The analysis of other constituent contents is more complicated. It is acknowledged that the relative proportion of these principal salts remains sensitively constant at the surface, from one ocean to another if coastal areas are excluded, as river input can modify this composition. This principle of constant proportions is known as *Dittmar's law* or *Marcet's principle*. However it is acknowledged, for example, that deep water values of Ca^{2+} are 0.3% higher than those at the surface due to the dissolution of $CaCO_3$ in deep water. Other soluble or secondary elements that are found in sea water are present in trace amounts and account for less than 0.01% of the total mass of dissolved salts.

All of these salts are *elements or natural tracers said to be conservative* of water mass; variations in their oceanic concentrations are uniquely linked to mixing phenomena. These variations are globally estimated by conductivity measurements that have led to the definition and quantification of the salinity of sea water (section 1.2.3).

In 2008, Millero *et al.* reviewed the reference composition of standard sea water [MIL 08], as defined in section 1.2.3.3. Its composition is given in Table 2.12. The sum of the concentrations of principal constituents, given in this table, is equivalent to the value of S_R defined in section 1.2.3.1. This reference composition is valid for a pH of 8.1. This can evolve over time, notably in relation to the absorption of CO_2. There are also *local anomalies of composition* that remain a major source of errors in the calculation of the physical properties of sea water. They can also induce errors in measurement. Thus, when salinity measurements are made at two different temperatures with a conductimeter in the same water sample containing composition anomalies, it is possible to find different salinity values, as the PSS-78 was produced from measurements made in samples of standard sea water.

Elements	*Concentration in g/kg*
Sodium: Na^+	10.78145
Magnesium: Mg^{2+}	1.28372
Calcium: Ca^{2+}	0.41208
Potassium: K^+	0.39910
Strontium: Sr^{2+}	0.00795
Chloride: Cl^-	19.35271
Sulphate: SO_4^{2-}	2.71235
Bicarbonate: HCO_3^-	0.10481
Carbonate: CO_3^{2-}	0.01434
Bromide: Br^-	0.06728
Borate: $B(OH)_4^-$	0.00795
Fluoride: F^-	0.00130
OH^-	0.00014
$B(OH)_3$	0.01944
CO_2	0.00042

Table 2.12. *Composition of standard sea water and the concentrations of its principal constituents. The total sum of dissolved salts gives a total concentration of 35.16504 g/kg here, which corresponds to the reference salinity S_R*

In addition to these principal constituents, nearly all chemical elements have been detected as traces in sea water. Their concentration varies from 1 mg/kg to 10^{-10} mg/kg. The sum of these elements accounts for less than 0.01% of the total salts present. However, taking into account the volume of the ocean, their total mass can be very high. They have a biological role as they are used by phytoplankton, but in concentrations that are too high (in cases of pollution) they can have a very negative impact.

2.11.1.2. *Nutritive substances and dissolved organic matter*

The marine environment is one that contains living organisms. Microorganisms that are found there constitute phytoplankton or zooplankton (section 2.10.4). They feed off the *nutrients* contained in water. These nutrients are local anomalies of composition, as defined above. They are essentially nitrates NO_3^- (0–420 µg/l), phosphates PO_4^{3-} (0–100 µg/l), silicates $S_i(OH)_4$ (0–5,000 µg/l) and ammonium NH_4^+ (0–50 µg/l). Nitrates, phosphates and silicates are non-conservative elements of water mass and typical waters, as their concentrations depend on the presence of living organisms. Their measuring out is made usually with onboard colorimeters or in laboratories using sea water samples; however analyzers and *in situ* measurement processes have been developed and are starting to be routinely used.

Silicates are essential constituents of the “skeletons” of numerous marine organisms. Phosphates and nitrates are direct products of living matter. Silicates are present in dissolved form (and therefore in the category of secondary elements) or in the form of suspended particles. Nitrogen or phosphorous-based compounds are also present, either in suspended or solution form. They each follow their own transformation and assimilation cycle in vegetal and marine environments.

Deep water is reputedly rich in nutrients as they recuperate organic matter from higher ocean layers that are biologically active, or because they come themselves from water layers close to the surface. It is possible to date water samples in relation to their nutrients’ content, hence the importance of the measurements made of these elements.

Marine dissolved organic matter (DOM) can be of an external (due to pollution, river input and atmosphere) or internal (a result of photosynthesis) origin. Much of DOM is of internal origin. It is composed of sugars, amino acids, fatty acids, hydrocarbons and vitamins. Its mean concentration is 266 µg/l.

2.11.1.3. *Dissolved gases*

When sea water is in contact with air, it disolves gases contained in the air. As dry air is essentially composed of nitrate (at 78.08%), oxygen (at 20.95%), argon (at 0.93%) and carbon gas (at 0.03%), these elements are also present in dissolved form in sea water. For oxygen, nitrate and rare gases in general, *Henry’s law* shows that, at equilibrium, the solubility of gas in solution or aqueous gas is directly proportional to its partial pressure in gaseous state. If this is higher than that of the same gas in liquid, there is a gas-to-liquid transfer and vice versa. In other words, the concentration c of each dissolved gas in sea water is proportional to its partial pressure p in the atmosphere:

$$c = \alpha p \qquad [2.263]$$

α being the saturation or solubility coefficient. It depends on gas, but also on temperature and water salinity. It is defined as the volume of gas that can be absorbed per unit of solution volume, under standard temperature and pressure conditions, when the partial pressure of the gas in question is equal to a normal atmosphere (1013.25 mbar). It is expressed in mole kg^{-1} atm^{-1} or in mol l^{-1} atm^{-1}. Generally, the solubility of gas decreases when temperature or salinity increases. It is therefore stronger in distilled water than in sea water.

The dissolved air nitrate and oxygen can be used as tracers. Water mass at the surface that is in contact with the atmosphere loads itself with gas. Under the effects of seasonal cooling, the water density increases and it descends, taking the dissolved gases with it, meaning that their content therefore become an indicator of the "age" of displacement of the water mass.

The situation is more complex for carbon gas. One part becomes dissolved according to Henry's law (giving aqueous CO_2) and another part reacts with sea water to give carbonic acid H_2CO_3, which is a weak acid as its molecules are not entirely disassociated in water and the disassociation reaction is reversible. H_2CO_3 is decomposed in the following way:

$$CO_2 + H_2O \leftrightarrow H_2CO_3$$

$$H_2CO_3 \leftrightarrow H^+ + HCO_3^-$$

$$HCO_3^- \leftrightarrow 2\ H^+ + CO_3^{2-}$$

None of these components can be measured directly. In fact, the system of carbon dissolution can only be described by measuring temperature, salinity, pressure and:

– The *fugacity of CO_2* or *$f\ CO_2$*. Fugacity is determined as being "the partial pressure of CO_2 in the air, which is in equilibrium with sea water, knowing that CO_2 is not a perfect gas". [SCH 09] Its unit of measurement should be the pascal (Pa), but it is generally expressed in μatmosphere (μatm). At the surface, its measuremnt range is 250–550 μatm, but it can reach 2,000 μatm at great depths. In order to detect the influence of human activity, it is necessary to take measurements with an accuracy of 1 μatm in air and 2 μatm in water.

In this definition, the notion of *partial pressure* of CO_2 or *pCO_2* intervenes. pCO_2 is defined as being the equivalent of the molar fraction xCO_2 of CO_2 in the air, which is in equilibrium with water, multiplied by the total pressure of equilibrium:

$$pCO_2 = xCO_2 \times (p_{atm} - p_{H2O}) \quad [2.264]$$

pCO_2 depends on the temperature and salinity of water, and its value can be estimated by algorithms from the measurement of these quantities (see [COP 88]). xCO_2 corresponds to molar fraction, and is expressed in μmol/mol.

– *DIC or dissolved inorganic carbon*, which is the sum of the dissolved CO_2 concentration, carbon dioxide, bicarbonates and carbonates:

$$DIC = [CO_2\ (aq)] + [H_2CO_3] + [HCO_3^-] + [CO_3^{2-}] \quad [2.265]$$

It is expressed in µmol/kg and its concentrations classically go from 1,800 to 2,300 µmol/kg. In order to detect the influence of human activity, it is necessary to make measurements with an accuracy of 1 µmol/kg. It has been shown that conductivity was barely sensitive to variations in DIC, which is not the case for S_A. A variation of 100 µmol/kg is equivalent to $\delta S_A \approx 0.006$ g/kg while $\delta S_p \approx 0.000\,7$.

– pH, which is the measurement of acidity or bascity of a solution. We have:

$$pH = -\log_{10}([H^+]) \qquad [2.266]$$

according to the definition provided by Sörensen [SÖR 09]. $[H^+]$ represents the concentration of hydrogen ions, or protons, in solution. For pure water, pH = 7. Adding an acid will cause the pH to decrease and adding a base will increase it. The pH can be expressed in different scales: full scale, that of the National Bureau of Standards (NBS), that of sea water or the free scale. The pH of oceans varies generally between 7.8 and 8.3, and in the study of the carbon cycle it is necessary to measure it with an accuracy of ± 0.002 *pH* units.

– The *total alkalinity* (*TA)* of sea water, which is defined as being "the number of hydrogen moles in ionic form that is equivalent to the excess of proton donors in a 1 kg sample" [SCH 09]. Therefore, the measurement of TA implies that we take into account the contribution of all ions present in sea water. If, to simplify, we only take the most abundant species into account, TA can be written as:

$$TA = [HCO_3^-] + 2\,[CO_3^{2-}] + [B(OH)_4^-] + [OH^-] - [H^+] \qquad [2.267]$$

It will be necessary to add the contribution of ions of nutrients to this expression if their concentration is not negligible. $[B(OH)_4^-]$ can be obtained from salinity for water whose composition is close to that of standard water. The *alkalinity of carbonates AC* can be calculated with:

$$AC = AT - [B(OH)_4^-] - [OH^-] + [H^+] \qquad [2.268]$$

AT is expressed in µmol/kg or in µEqu/kg, i.e. microequivalents per kg. It can also be expressed in µmol/l, but its value therefore depends on the density of sea water. Oceanic ranges classically go from 2,000 to 2,500 µmol/kg and it is necessary to estimate this to an accuracy of ±1 µmol/kg in order to detect the anthropogenic influence.

Equations [2.264] to [2.268] constitute a system of four equations with four unknowns that can be resolved to determine their values. However, if at least two of these parameters are measured, it is possible to determine the other two from equilibrium constants by knowing the temperature, pressure and salinity.

Dissolved oxygen is one of the best documented oceanic parameters, as since the 19th Century there have been methods to measure it, and understanding its behavior is simpler. A quota of this oxygen is consumed by organic matter during degredation. Another quota aids the respiration of living aquatic organisms. Plant life present in the medium produces oxygen via photosynthesis. On this basis, it is an indicator of water quality and a non-conservative tracer of water mass. The amount present, however, is associated with that of certain nutrients (phosphates in particular), which means that it can be used to become a partially conservative tracer.

Solubility of oxygen in sea water can be estimated at normal atmospheric pressure from different algorithms if we know the temperature and salinity of the medium (see [GAR 92]), but the existence of sensors and sufficiently precise measuring techniques that can be used at sea has resulted in dissolved molecular oxygen becoming a parameter that is studied regularly and in detail. This has enabled the assessment of the age of deep water, among other things. Calibration of these sensors is done from samples taken at different depths using the *Winkler method.*

In order to carry out these analyses, first of all, we have to prevent the samples from bubbling in order to limit exchanges with ambient air. The Winkler method consists of oxidizing iodine molecules that are present in the form of iodide. Sea water is mixed beforehand with a manganese (II) solution that is precipitated with a strong base:

$$Mn^{2+} + 2\ OH^- \rightarrow MN(OH)_2 \downarrow \qquad [2.269]$$

Dissolved oxygen fixes itself to the precipitate, thereby causing higher oxidation to manganese (III):

$$Mn(OH)_2 + \tfrac{1}{4}\ O_2 + \tfrac{1}{2}\ H_2O \rightarrow Mn(OH)_3 \downarrow \qquad [2.270]$$

From this reaction, the medium is acidified in the presence of iodide, the manganese becomes bivalent and the precipitate dissolves:

$$Mn(OH)_3 + I^- + 3\ H^+ \rightarrow Mn_2^{\ +} + {}^3/_2\ H_2O + \tfrac{1}{2}\ I_2 \qquad [2.271]$$

Iodide ions are therefore oxidized in turn and are transformed into iodine (I_2). In the presence of excess iodide ions at the start of the reaction, however, equilibrium establishes itself:

$$I_2 + I^- \rightarrow I_3^- \qquad [2.272]$$

The quantity of iodine released (in the form of I_2 and I_3^-) during mixing is proportional to the concentration of dissolved oxygen. Half a mole of dissolved oxygen will release one mole of I_2. The amount of I_2 can then be measured using a sodium thiosulphate solution. The oxygen contents must be expressed in moles of O_2 per kilogram of solution; however, in practice it is often expressed in mg l^{-1} or in ml l^{-1}. The change from ml to moles can be done if we know the molar volume of gaseous oxygen. In normal conditions of temperature and pressure, it is 22.390l. It is possible to convert measurements from ml l^{-1} into mol kg^{-1} (unit of *molality concentration*) if we devide ρ by 22.390, ρ being the density of the sea water analyzed.

The Winkler method presents an alternative that consists of replacing the measurement of thiosulphate by measuring the coloration linked to the iodine at a wavelength of 466 nm, with the help of a visible spectrophotometer. The wavelength 466 nm corresponds to that for which the molar absorptivities of I_2 and I_3 molecules are equal. As measuring out using thiosulphate is a source of error, this test has the advantage of no longer needing to resort to this measure out. Moreover, the use of pump and flow cell allows us to limit iodine volatilization, human handling and to increase the speed of sample analysis, notably during the calibration of sensors.

This method enables us to obtain measurement uncertainties of 0.1–0.25 µmol/kg (to 1σ). However, although it is considered as a reference for some authors, it tends to overestimate the concentration of dissolved O_2.

2.11.1.4. *Transient tracers*

In addition to principle elements, trace elements, nutrients and dissolved gas, the ocean contains a certain number of other components that can be classified as *transcient tracers*. They are also usually gas composites that are soluble in water. Those of anthropogenic origin have progressively penetrated into the ocean via exchange at the interface between air and water. They are taken to greater depth in cold water zones by thermohaline circulation, which makes them interesting tracers.

Among these components, we can find radioactive elements, such as potassium 40, rubidium 87 and uranium 238, as well as their products of disintegration and transformation. Potassium 40 is the most widely spread. Lighter radioactive elements also exist such as tritium, which is a heavy hydrogen isotope that disintegrates into helium 3 by emitting β rays, and carbon 14, which disintegrates into nitrogen 14 by also emitting β rays. The origin of these elements can be natural or anthropogenic. The tritium–helium 3 couple is used to evaluate the time taken by a water mass to pass from the surface to a certain depth. For this, we use the disintegration period of tritium, which is 12.3 years. The time after which the number of atoms N of the element being considered is reduced to half its initial value N_0 is called the radioactive period T. The law of radioactive decrease having the form:

$$N = N_0 e^{-\lambda t} \qquad [2.273]$$

where λ is the radioactive constant. The radioactive period of the element is calculated from:

$$T = Log\left(\frac{2}{\lambda}\right) \qquad [2.274]$$

The disintegration period of carbon 14 is also used to date deep water mass and to decuce its flow speed.

Finally, sea water is also composed of freons and chlorofluoromethanes (CFM) that can be classified as transient tracers. Like all of the components present in air, these molecules are found in water and can be used to date water mass from the measurement of the ratio of two components, called F_{11} and F_{12}. They have the advantage of not having natural sources. These molecules have been produced from industrial production since 1928. However, only water masses formed between 1950 and 1980 can be dated with sufficient accuracy. Other components, such as CCL_4 or F_{113}, must be analyzed to retrieve data from periods before 1950 and after 1980. The analysis of this is more delicate. If great caution has been undertaken, F_{11} and F_{12} can be measured aboard oceanographic ships. This is one of the other advantages of their measurement.

2.11.1.5. *Suspended matter*

In addition to dissolved elements, suspended matter is also found in sea water. While the distinction between these two categories is arbitrary, as there are no real limits, it is considered that particles larger than 1 nm are "suspended matter" (SM). Between 1 nm and 0.1 µm, SM is detectable with an electronic microscope. From 0.1 µm it is visible with a classic microscope. SM concentration can be measured in a vacuum, with the help of a filtration membrane that can retain particles bigger than 0.45 µm. However, as the membrane becomes blocked, it is likely that smaller particles will be retained. These membranes are dried and weighed before (P_1) and after filtration (P_2). The difference in weight enables us to determine the mass of matter retained per volume of filtered water (V):

$$SM(mg/l) = \frac{P_2 - P_1 - B}{V} \qquad [2.275]$$

In this relation, B is a correction linked to the loss of weight of the filter when the solution is poured.

Determining these concentrations is particularly important in studies that cover water transparency, photosynthetic production, river or wind input, and nutrient sources for plankton and fauna in general.

Certain studies also attempt to determine the size distribution of particles, as this enables us to understand the light dispersion in water. This is possible with the help of electronic microscopes and energy dispersive spectrometers, as shown by Groundwater *et al.* [GRO 12] with the help of these instruments and image processing software, it is possible to determine the average area A of particles and from this to determine their equivalent spherical diameter D:

$$D = 2\sqrt{\frac{A}{\pi}} \qquad [2.276]$$

Finally, it must be recalled that suspended matter generally has a negligible influence on thermal and electrical properties as well as on the density of sea water. Their size leads to the scattering of optical and acoustic waves. In optics, they reduce the propagation of light (see section 2.10 for more information). In acoustics, they favor the operation of a good number of instruments, particularly those for which an inbound wave is expected (see section 2.5.2 for more information). This is one of the reasons why acoustics has had greater success underwater than optics.

2.11.2. *In situ measurement of dissolved oxygen*

The measurement of concentrations of dissolved oxygen enables us to trace the physical processes of advection, scattering and mixing, but also the biogeochemistry activities of photosynthesis, breathing and remineralization of the medium. This parameter is therefore very sensitive to changes in climatic conditions.

2.11.2.1. *Polarographic or Clark sensors*

Oxygen concentration is traditionally measured in gas and liquids with the help of polarographic sensors, also known as membrane sensors or Clark sensors (named after its inventor, Leland Clark). These elements are constituted of two electrodes: one in silver that serves as an anode and another in platinum or gold that serves as a cathode (see Figure 2.274). These electrodes are placed into an electrolyte consisting of potassium chloride in gel form. This gel is in contact with the exterior environment via a Teflon membrane. The electrodes are subjected to a fixed potential difference of 0.8 V. In the case of sea water, dissolved oxygen diffuses according to the value of its partial pressure across the membrane and it is electrochemically reduced at the cathode. The potential applied means that the cathode is negatively charged in relation to the anode. Its value is chosen so that only oxygen is reduced. Platinum acts like a catalyst in this reaction. The current that circulates between the electrodes is proportional to the quantity of oxygen that diffuses across the membrane, which is itself relative to partial pressure. At the cathode, we have the reaction:

$$4\,H^+ + 4\,e^- + O^2 \rightarrow 2\,H_2O \qquad [2.277]$$

The complete reaction gives:

$$4\,Ag + 4\,Cl^- + 4\,H^+ + O^2 \leftrightarrow 4\,AgCl + 2\,H_2O \qquad [2.278]$$

As this set consumes oxygen in water proportionally to the partial pressure of this element, it is necessary for the measurement to be made in a moving medium or with liquid circulation to avoid the impoverishment of the boundary layer that is created at the membrane that can produce false measurements.

These sensors generally contain a temperature probe that enables the variations in the speed of oxygen scattering and solubility that are relative to this parameter to be compensated for. In order to calculate the concentration of dissolved oxygen, it is neceassary to also determine the pressure and salinity of the medium. For this, O_2 sensors are often fixed on CDT profilers and pressure, conductivity and temperature sensors are used to process the data. Therefore the problem that arises relates to the alignement of the response times of these sensors with that of the O_2 sensor.

The tension of the membrane is important if we are to obtain a high quality sensor, as tension determines the thickness of the electrolyte layer that separates it from the cathode. This is what sets its response time. As the electrolyte serves to reduce oxygen, its lifespan is limited and it is necessary to replace it periodically. The periodicity of this replacement is relative to the flow of reduced oxygen. This flow depends on the diameter of the cathode. A low diameter will enable us to increase the periodicity, but as consequence the sensor will have a reduced sensitivity.

Maintenance of the membrane is important in the assessment of the accuracy of measurement. It is necessary to avoid all surface deposits, as this could modify permeability. Sensors must therefore be meticulously cleaned in soft water or with a soft surfactant product such as Triton x-100 at 0.1% (like conductivity cells) after each usage *in situ*. They must then be covered with a case to protect them from drifts and polluted atmospheres, or stored in a distilled water circuit.

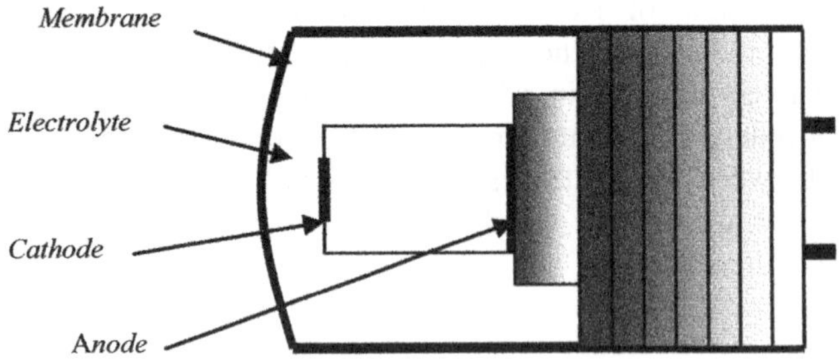

Figure 2.74. *Representation of a dissolved oxygen sensor (Courtesy of Thermo Russell)*

2.11.2.2. *Calibration of dissolved oxygen polarographic sensors*

The relation that exists between dissolved oxygen and the current I_O that circulates between the electrodes involves the value of I_O, the temperature of the medium T, its salinity S, its pressure p, and the internal temperature of the sensor T_0 that approximately corresponds to that of its membrane. In order to obtain the oxygen concentration OX, expressed in ml/l, these parameters must be linked to the OX_{sat} value of water saturation in oxygen. OX_{sat} itself depends on T, S and p. In 1970, *Weiss* established a relation that enables us to calculate the value of OX_{sat} in relation to these parameters:

$$OX_{sat}(ml/l) = e^{\left[a_1 + a_2\left(\frac{100}{T}\right) + a_3 \ln\left(\frac{T}{100}\right) + a_4\left(\frac{T}{100}\right) + S\left(b_1 + b_2\left(\frac{T}{100}\right) + b_3\left(\frac{T}{100}\right)^2\right)\right]} \qquad [2.279]$$

with:

$a_1 = -173.4292$ $\qquad$ $b_1 = -0.033096$

$a_2 = 249.6339$ $\qquad$ $b_2 = 0.014259$

$a_3 = 143.3483$ $\qquad$ $b_3 = -0.00170$

$a_4 = -21.8492$

The Weiss relation is a general relation that results in the dependence of the solubility coefficient of gas α, defined by equation [2.263], on the temperature and salinity of water. T is expressed in Kelvin in this relation ($T = t(°C) + 273.15$) and it is valid for normal atmospheric pressure. It can be applied to other gases dissolved in water, to the nearest coefficients $a_{i(1\leq i\leq 4)}$ and $b_{j\,(1\leq j\leq 3)}$. In the case of oxygen, to convert the values of OX_{sat} into mg/l they must be multiplied by 1.4276.

Temperature		***Salinity***			
t (°C)	***T (K)***	**0**	**10**	**25**	**35**
-2	271.15	10.82	10.10	9.10	8.49
0	273.15	10.22	9.54	8.61	8.05
5	278.15	8.93	8.36	7.57	7.09
10	283.15	7.89	7.41	6.73	6.32
15	288.15	7.05	6.63	6.05	5.69
20	293.15	6.35	5.99	5.48	5.17
25	298.15	5.77	5.45	5.00	4.73
30	303.15	5.28	4.99	4.60	4.35
35	308.15	4.85	4.60	4.25	4.03

Table 2.13. *Values of sea water saturation with oxygen OX_{sat} expressed in ml/l at normal atmospheric pressure, calculated per unit volume from the Weiss relation [WEI 70]*

A first relation, established by Millard, [MIL 82] enables us to calculate *OX*:

$$OX = (slopexI_O + offset)OX_{sat}e^{\left[t_{memb}(T+w_T(T_0-T))+p_{memb}P\right]} \qquad [2.280]$$

It involves:

– the *slope* and *offset* factors that enable correction of the value of current provided by the sensor in relation to the calibrated values of dissolved oxygen;

– factors t_{memb} and p_{memb}, which enable the correction of variations in the permeability of the membrane in relation to temperature and pressure; and

– w_T, which is a weighting factor of the difference between the temperature of the sensor and the exterior medium.

A second relation created by Owens and Millard [OWE 85] takes the response time τ of the membrane into account:

$$OX = \left[Slope\left(I_O + \tau\frac{dI_O}{dt}\right) + offset\right]OX_{sat}e^{\left[t_{memb}(T+w_T(T_0-T))+p_{memb}P\right]} \qquad [2.281]$$

The values of t_{memb}, p_{memb}, w_T and τ can be obtained from the manufacturers of dissolved O_2 sensors. On an indicative basis, for a Teflon® membrane that is 0.5 mm thick, t_{memb} = -0.033, $p_{memb} = 1.5\times10^{-4}$, τ = 2 s. The value of w_T then depends on the manufacturer: 0.67 for a Beckman sensor and 0.85 for a YSI sensor.

Slope and *offset* coefficients are determined with the help of two solutions:

– one solution does not contain dissolved oxygen, and the other is saturated with oxygen; or

– with the help of sea water samples in which the different concentrations of O_2 are determined using the Winkler method (section 2.11.1.3).

A sodium sulfur solution (Na_2SO_3) can be used to obtain the "zero" of the sensor. It is also possible to place the sensor in an inert gas stream (nitrogen or argon) in order to create a local medium that is oxygen deprived. Relation [2.279] enables us to calculate the conditions necessary to obtain water saturated with oxygen. In practice, the same result can be obtained by airing the medium with the help of an aquarium pump. As measurements are rarely made at normal atmospheric pressure, p_0 = 101,325 Pa, it is necessary to multiply the ratio OX/OX_{sat} by a corrective factor called A_{ns}, which depends on the atmospheric pressure of the day p_{atm}:

$$A_{ns} = \left(\frac{p_{atm}}{p_0}\right)\left[\frac{\left(1-\frac{p_{H2o}}{P_{atm}}\right)}{\left(1-\frac{p_{H2o}}{P_0}\right)}\right] \qquad [2.282]$$

with:

$$p_{H2O} = e^{\left[\left(-216961\left(\frac{1}{T}\right)-3840.7\right)\left(\frac{1}{T}\right)+16.4754\right]} \qquad [2.283]$$

However, the most reliable technique for carrying out calibration remains that devised by Winkler. For this, a bath in which the oxygen concentration remains stable during measurements and sampling is needed. This condition is fulfilled if its temperature is regulated and known to at least 0.1°C, and stirred properly in order to obtain equilibrium between the air and the dissolved gas concentration. After the analysis of samples taken at different stages of temperature, the *slope* and *offset* coefficients can be determined with the help of a least squares technique.

It is possible to calibrate these sensors with an accuracy to 2% of the value of oxygen saturation. When correct maintenance is ensured, this accuracy can be preserved over about 1,000 hours of operation.

In 2010, Edwards *et al.* [EDW 10], from Sea Bird Electronics, improved the accuracy of the Clark cell sensor (SBE 43 produced by the same company) by carrying out a dynamic study of its behavior in the sea. Their study showed that the factor τ of relation [2.281] depends on temperature T and pressure P. Pressure has an effect on the plasticity of the membrane polymer and on the components ratio present in amorphous and crystalline forms. In order to correct these effects, they calculated a τ normalized at 20°C at normal atmospheric pressure (τ_{20}), and their measurements showed that τ followed an exponential response of the following type:

$$\tau = \tau_{20} e^{[D1P + D2(T-20)]} \qquad [2.284]$$

where $D1$ and $D2$ are constants determined during these measurements. Only τ_{20} is to be determined by calibration in order to characterize each sensor.

They also studied the hysteresis of their sensor, which induced differences of 2–5% between the descent and ascent at pressures greater than 1,000 dbar. In order to correct this effect, they successfully experimented with a digital filtering equation of the same type as that used to align the response time of SBE 4 conductivity sensors with that of SBE 3 temperature sensors (see [EDW 10]). Using this equation, differences of 10 µmol/l have been reduced to 1µmol/l.

2.11.2.3. *Optodes*

With the aim of compensating for the disadvantages of electrode sensors, another technique for measuring dissolved oxygen was developed in 1996. The possibility of effectuating material deposits using sol–gel techniques cleared the way for the arrival of "*optodes*" or "optrodes". The prefix "opt"was added infront of the radical "ode", similar to the word "electrode", the principle of measurement this time being based on optical techniques.

Sol–gel techniques are used in the production of coverings that have specific properties. In our case, these are optical properties that we seek to exploit through the addition of thin layers on supports allowing the transmission of light. It is necessary to control the thickness of the refractive index of these layers during their production. Two types of technology allow this:

– the physical deposit in a vacuum; and

– the deposit of layers in liquid phase or a "solution–gelatin" process known as sol–gel.

The sol–gel process is particularly well adapted to deposits on glass. It is relatively simple and inexpensive to develop compared to the technology used to create physical deposits in a vacuum. It needs reactive precursors such as alkoxides or metallic salts and can decompose in two principal stages: hydrolysis or the action of water and condensation that enables the formation of a covalent network. Reactive salts can have a base of silicon, titanium, zirconium, ruthenium, etc. The "sol" is constituted of solid elements dispersed in a liquid. Its stability and homogeneity determine whether it is suitable or not for industrial use. After condensation, it evolves towards the formation of a system of high molecular complexity that imprisons the solvent, or "gel". The constitution of this gel is relative to the sequence of aging, drying and thermal treatment developed. The optical and mechanical properties of the deposit made will depend on the mode of sol–gel synthesis used.

In the case of dissolved O_2 sensors, the deposit can be made at the extremity of an optical fiber, or simply at a glass window. It must be hydrophobic. Photophore substances such as ruthenium are used. Their role is to generate light by fluorescence (for more information about fluorescence, see section 2.10.5). They fluoresce when they are illuminated by a monochromatic source in blue spectrum (around 475 nm). According to Stoke's law, under the effect of blue radiation, the reactive salt imprisoned in the deposit fluoresces by sending a wavelength situated the red spectrum (around 600 nm). However, when molecules of excited reactive salts come into contact with oxygen molecules, a non-radiative energy transfer takes place, provoking the loss of fluorescence, which is measured as a reduction of its yield. This phenomenon, described by Kautsky in 1931 [KAU 31] is known as "quenching" or the dynamic extinction of fluorescence. The degree of quenching is proportional to the frequency of collisions and therefore to the concentration of O_2, or more precisely its partial pressure p_{O2}. The decrease in intensity I of fluorescence can be quantified with the *Stern-Volmer* linear equation:

$$\frac{I}{I_0} = 1 + kp_{O2} \qquad [2.285]$$

where I_0 represents the intensity of fluorecence with $p_{O2} = 0$. Its value depends on the optical assembly used and the intensity of the source. k is the Stern-Volmer extinction constant, which depends on the chemical composition of the deposit. The concentration of oxygen molecules that penetrates the matrix of the deposit is different to that of the ambient medium. It is therefore necessary to correct equation [2.285] with a term s_o, which represents the solubility coefficient of oxygen in the matrix. We therefore have:

$$\frac{I}{I_0} = 1 + ks_o p_{O2} \qquad [2.286]$$

Furthermore, the temperature of the sample influences the frequency of collisions between oxygen molecules and the molecules of active substrates, and as a result the intensity of fluorence and its duration. Coefficients k and s_o therefore depend on temperature. Equations [2.285] and [2.286] can only be used at constant or relatively stable temperature. Tests have shown that the effect of temperature can be reproduced if the composition of the deposit remains chemically stable. Also, terms I_0 and ks_o can be modeled with relations of the form:

$$I_0 = a_0 + b_0 t + c_0 t^2 \qquad [2.287]$$

$$ks_o = a_1 + b_1 t + c_1 t^2 \qquad [2.288]$$

where a_0, b_0, c_0, a_1, b_1 and c_1 are coefficients to be determined by calibration.

If we experimentally trace the curve $I/I_0 = f(p_{O2})$, we realize that it can be approximated by a second-order relation. This nonlinear effect has been attributed to heterogeneities of the molecular matrix of the deposit, which modifies the probability of collisions between oxygen molecules and reactive salt molecules. It is therefore necessary to replace equation [2.286] by:

$$\frac{I}{I_0} = 1 + Ap_{O2} + Bp_{O2}^2 \qquad [2.289]$$

to take into account the nonlinear effects of the B coefficient. In this relation, A, which is equivalent to coefficient ks_o from equation [2.286] and B depend on temperature. The value of A can be determined experimentally from equation [2.288] and that of B by the anologue relation:

$$B = a_2 + b_2 t + c_2 t^2 \qquad [2.290]$$

where a_2, b_2 and c_2 are coefficients obtained by calibration.

Calibration must therefore be carried out on both the temperature and the oxygen concentration. To find the value of the coefficients from equation [2.286], two calibrated oxygen solutions are needed: one at 0% and the other at the highest value that we wish to measure. Equation [2.289] needs a minimum of three equidistant solutions that cover the range of exploration. Experiments have shown that the salinity of water does not have any influence on the measurement results and therefore does not affect the calibration of probes.

As these probes are sensitive to the partial pressure of oxygen, it is possible to calibrate them in gas atmospheres. Henry's law, which is expressed by relation [2.263], enables the conversion of partial pressure into concentration. However, this law must be corrected for gases that are highly soluble in water. The Weiss relation

[2.279] enables calculation of the value of oxygen saturation in water in relation to its temperature and salinity, and:

$$X = e^{\left[-66.7354+\frac{8747.55}{T}+24.5426\ln\left(\frac{T}{100}\right)\right]} \qquad [2.291]$$

enables calculation of the molar fraction X of oxygen that can be found in water relative to absolute temperature T at normal pressure.

Finally, according to the type of probe used, the diffusing or reflecting properties of the medium can have an influence on the intensity of the fluorescence signal measured. In fact, the emission of wavelengths obtained by fluorescence is omnidirectional and only the part of the flow situated in the aperture cone of the detector is measured. According to the reflecting properties of the medium, a higher intensity can be returned to the cone and cause errors in measurement. To avoid this problem, it is recommended that calibration of the probes is carried out in a medium where composition is as close to that of the real medium of measurement as possible.

The accuracy of measurements that can be reached with this technology is usually expressed in *molarity* (number of moles per unit of volume). Measurements made by probes with a range that varies from 0 to 500 µmol/l or 0 to 40.7 ppm are currently accurate to 3–6% (while the objective fixed by the WHP is closer to 1%). The resolution of measurements can go from 0.002 ppm to 0.02 ppm at ambient temperature, which corresponds to the detection limit of probes.

To evaluate this accuracy, Ushida *et al.* [USH 07] carried out calibration experiments *in situ* with 3,830 optodes produced by Aanderaa Data Instruments AS. The concentrations of O_2 $[O_2]$, expressed in µmol/l, measured by this optode are obtained by a polynomial:

$$[O_2] = C_0 + C_1 P + C_2 P^2 + C_3 P^3 + C_4 P^4 \qquad [2.292]$$

where P is phase shift (in °) measured in the detected signal. As explained previously, coefficients C_x (x = 0 to 4) depend on temperature t (°C), and it is necessary to calculate them with the help of another polynomial:

$$C_x = C_{x0} + C_{x1}\, t + C_{x2}\, t^2 + C_{x3}\, t^3 \qquad [2.293]$$

As pressure p decreases the sensitivity of the response by around 4%/1,000 dbar, this effect is taken into account by adopting a calculation equation with a corrected concentration $[O_2]_c$:

$$[O_2]_c = [O_2]\,(1 + C_pP/1{,}000) \qquad [2.294]$$

where C_p is a compensation coefficient of pressure. Its value, which should have been 0.04, has been optimized by measurements and adjusted to 0.032.

As equation [2.291] proposed by Aanderaa is not a product of the theory discussed earlier, and as its coefficients are difficult to determine, Ushida *et al.* opted for another form of relation, based on the recognition of the variation time of decreases in fluorescence, τ:

$$[O_2] = \frac{\left(\frac{\tau_0}{\tau} - 1\right)}{k} = \frac{\left(\frac{P_0}{P_c} - 1\right)}{k} \qquad [2.295]$$

where τ_0 is the time it takes for $[O_2] = 0$, which corresponds to phase shift P_0. k is the Stern-Volner constant and P_c is the phase shift obtained after calibration from raw values of shift P_r: $P_c = c_1 + c_2P_r$. P_0 depends on temperature, as does parameter k, hence: $P_0 = c_3 + c_4t$ and $k = c_5 + c_6t + c_7t^2$. The number of coefficients c_i ($i = 0$ to 6) to be determined is thus reduced. Equations [2.294] and [2.295], *a priori*, give comparable results.

As measurements have shown that the response time of these optodes is not constant in relation to temperature (it increases below 10°C), they calculated a relation to compensate for this delay in an empirical manner and align it with that of the temperature sensor of the CDT profiler used to acquire the measurements (SBE 9 by Sea Bird Electronics). Values of P_r have been advanced by calculating times T_a so that: $T_a = 25e^{-0.13t}$ for $t < 16.3$°C and $T_a = 3$ for $t \geq 16.3$°C. With these compensations, the differences between descent and ascent profiles made at 1.2 m/s have been reduced to around 1 µmol/kg.

Taking into account the problems of calibration and the influence of particles, optodes have a certain number of advantages in relation to electrode sensors:

– They do not consume oxygen. There is therefore no need to "air" the medium, as is the case for electrode sensors. They are adapted to static measurements, in mooring for example.

– Their upkeep is reduced (the electrolyte does not need to be changed), which leads to their characteristics having a good stability.

– The drift in their characteristics over time is lower than that of membrane sensors. This was shown by data from ARGO floats (ARGO Workshop, May 2011 [REI 11]). However, deployments over nearly three years by Sea Bird Electronics showed very weak drifts of about 1 µmol/kg/year for their SBE 43 sensor (see also [MAR 07]).

These advantages are ofset by a certain number of inherent disadvantages of this technology, namely:

– the accuracy of measurements can be affected by the presence of components other than oxygen in the medium that are equally likely to lead to the extinction of fluorescence;

– the sensitive deposit can be degraded by certain mediums (biofouling, abrasion) and by prolonged exposure to light. The probe must therefore be returned to the manufacturer for its substrate to be reloaded; and

– ambient light can also influence the accuracy of measurement. In order to protect against this effect, it is necessary to place probes in protective sheaths.

2.11.3. *In situ measurement of dissolved carbon*

As described in section 2.11.1.3, the system of carbon dissolution can only be described by measuring temperature, salinity, pressure and at least two of the following elements: fugacity of CO_2 (fCO_2), *DIC*, pH and *TA*. This enables the evaluation of the acidification of oceans and its repercussions on marine biology, and also the quantification of CO_2 exchanges between the ocean and the atmosphere or the evaluation of absolute salinity.

2.11.3.1. *Measuring* fCO_2

Fugacity of CO_2 cannot be approached directly in liquid phase. Either a measurement of xCO_2 mole fraction must be made in the air, equilibrated beforehand using the sea water sample to be analyzed, or a measurement must be made using a dye indicator pH after the probe has been equilibrated with the same sample of sea water.

The first method requires a sampling and water filtering system positioned at the bow of the boat, a dry air pumping system with a condenser, four reserves of reference CO_2 (from 0 to 500 ppm) enabling calibration of the infra-red analyzer (generally a non-dispersive spectrometer). Filtered water is then sent to an equilibrator and the gas xCO_2 present in this chamber is measured by the infra-red analyzer. Periodically, the xCO_2 in atmospheric air is also measured by this analyzer. This device is conditioned so that it can be installed in research or commercial vessels, as the latter make regular journeys and therefore can take periodic samples of certain ocean radials. The sensitive element of this assembly is the equilibrator, which can include different technologies and is available in different volumes. High volume equalibrators have a longer response time, but they are less sensitive to fouling and have higher thermal inertia. They require a higher water flow rate so that their temperature, T_{equi}, will be close to that of the medium at the time of analysis. Temperature must be measured in the equilibrator at the time of the spectrometric measure, as for atmospheric pressure p_{atm}, in order to calculate corrections. According to Henry's law, when equilibrium is reached the fugacity of CO_2 in gaseous space is proportional to its concentration in the circulating water. Its value is calculated from the equation:

$$fCO_2 = pCO_2 e^{\left\{\frac{\left[\beta + 2(1 - xCO_2)^2 \delta\right] p_{atm}}{RT_{equi}}\right\}} \quad [2.296]$$

where R is the constant of perfects gases (R = 82.0578 cm^3 atm/mol), and β and δ are parameters that are expressed in cm^3/mol and depend on T. For CO_2, their values are calculated with the following equations:

$$\beta = -1636.75 + 12.0408T - 3.27957 \times 10^{-2} T^2 + 3.16528 \times 10^{-5} T^3 \quad [2.297]$$

$$\delta = 57.7 - 0.118T \quad [2.298]$$

pCO_2 is obtained with relation [2.263], which involves xCO_2 and water vapor pressure pH_2O, which depends on T and on salinity S:

$$pH_2O = e^{\left[24.4543 - 67.4509\left(\frac{100}{T}\right) - 4.8489 \ln\left(\frac{T}{100}\right) - 0.000544S\right]} \quad [2.299]$$

The sensitivity of fCO_2 to temperature being 15 µatm/°C, it is necessary to determine it to an accuracy better than 0.01°C in order to respond to the constraint of 1 µatm, which is indispensable if we want to detect anthropogenic influence. Likewise, the pressure must be known to within ±0.2 mbar, which remains difficult. Another source of error in this measurement resides in the fact that absorption rays of CO_2 in the infra-red spectrum are very close to that of water. The air from the equilibrator has 100% humidity. It is therefore necessary to either dry it before analysis or to measure the absorption rays by another means, and to carry out corrections for the proportion of water molecules (xH_2O). However, this device is a reference for *in situ* measurements of fCO_2.

The achievement of CO_2 equilibrium in air and sea water can also be achieved through the use of a membrane, therefore enabling the development of lighter devices that can be set up on moorings or small carriers (buoys, gliders, etc.). The determination of the partial pressure of CO_2 is achieved with a spectrophotometer, by measuring the maximum wavelength of absorbance in a colored solution, thymol blue, in sea water. CO_2 contained in air is put into pressure equilibrium with this solution through a silicon membrane. Thymol blue is an indicator of pH. It changes color in relation to the acidity or basicity of the solution, whose CO_2 concentration varies. The spectrophotometer enables the quantification of this variation in coloration and therefore absorbance of the solution.

The "zero" is obtained by making a measurement using a non-absorbant wavelength. As the colorant can degrade over time and cause drifts in measurement, it is necessary to have a third wavelength corresponding to the isobestic point of the colorant in acid and basic form so that these corrections can be applied to the measurements. These sensors can be calibrated in laboratories with the help of standard gases or by comparison to Li-Cor non-dispersive reference spectrometers. Finally, as the membrane is sensitive to biofouling, it must be protected by adding mercuric chloride to the colorant or by a choration device, or with the use of other methods such as those presented in section 2.2.5.

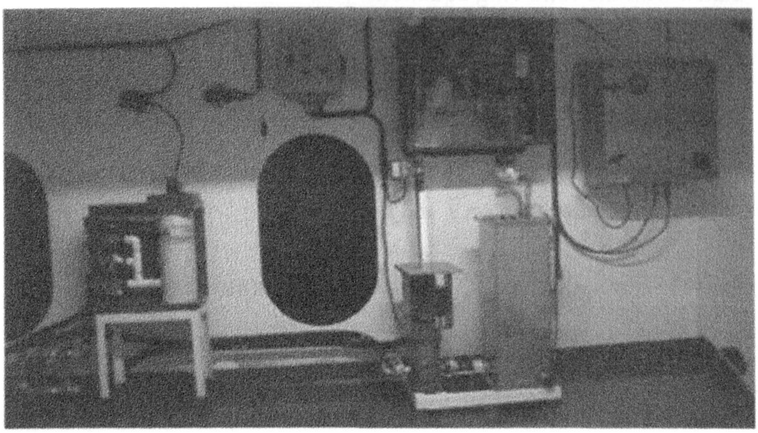

Figure 2.75. *Measuring equipment of the CO_2 mole fraction (xCO_2) allowing the calculation of fCO_2. Set up on Monte Olivia (a boat in south Hamburg). The two boxes fixed to the wall on the right contain the analysis device. To the left, we can see an SBE 21 thermosalinometer (Courtesy of the Institute for Research and Development)*

2.11.3.2. *Measurement of dissolved inorganic carbon and total alkakinity*

Dissolved inorganic carbon (*DIC*) is the sum of concentrations of dissolved/aqueous CO_2, carbonic acid, bicarbonates and carbonates defined by equation [2.265]. At present, its can only be determined in laboratories from sea water samples by:

– *Potentiometry*: a sea water sample whose mass is known and is titrated with a strong acid in a sealed container. During measurement, this container must be kept at a constant temperature.

– *Coulometry*: this technique enables us to obtain a more accurate measurement than the previous one. The sample is acidified with phosphoric acid, which enables us to extract the dissolved CO_2 in gas form. The gas concentration can be measured by coulometry or with the help of a non-dispersive infra-red analyzer. These measurements must once again be made at a constant temperature.

Measurement of total alcalinity corresponds to the titration of the number of hydrogen moles in ion form. As for DIC, this measurement can only be made in laboratories from sea water samples by:

– *potentiometery*: the sample is titrated with hydrochloric acid, with the help of a pair of electrodes in which potential difference is measured. This measurement, which can take 10 to 20 mins, is made at a constant temperature while the sample is constantly stirred; or

– by *spectrophotometry*.

2.11.3.3. *Measuring pH*

A pH value describes the activity of hydrogen ions in an aqueous solution. If this activity is high, the solution is acidic and its pH is below 7, while if it is low, the solution is basic, with a pH above 7 and the hydroxide ions have a higher activity.

As the activity of an ion is not directly measurable, we measure it via potentiometry, as is the case for *TA*. The accuracy of this technique is limited (±0.01 pH), however, and it is not recommended to follow the pH of sea water. pH can be measured also with spectrophotometry. The spectrophotometric technique was recently the object of research for the development of a measuring instrument ([AßM 11]). It needs use of a colorimetric indicator with a known *dissociation coefficient* or pK_a. Thymol blue is used in certain applications, as variations of its pK_a in relation to temperature are known in a large temperature range. For sea water, the purple m-Cresol or mCP is preferred, as its variation in the pH range of 7.4 to 9.0 is better adapted to the range in variation of the pH of sea water (7.5 to 8.5). Its pK_a has been established in relation to temperature in the $293 < T < 303$ K range and to salinity in the range $30 < S < 37$:

$$pK_a = \frac{1245.69}{T} + 3.8275 + 0.00211(35 - S) \qquad [2.300]$$

This relation was validated over a larger temperature range ($278 < T < 308$ K) in 2011 by Liu *et al.* [LIU 11] for purified mCP, however, Aßmann *et al.* opted to use a non-purified mCP and temperature regulation of the measurement vat around 25°C, with the help of Peltier effect modules to remain within the validity area of formula [2.300].

The pH of sea water is generally determined using the total pH scale to take into account all hydrogen ions, whose concentration is determined by:

$$[H^+] = [H^+]_F + (1 + 1 / K_S) \approx [H^+]_F + [H_2SO_4^-] \quad [2.301]$$

in which $[H^+]_F$ represents the concentration of hydrogen ions released and K_S the dissociation constant of hydrogen sulfate ions, $H_2SO_4^-$. The pH value is then calculated with the help of relation [2.266].

mCp also dissociates and it is its second dissociation reaction that intervenes in the pH values of sea water. It gives:

$$HI^- \Leftrightarrow I^{2-} + H^+ \quad [2.302]$$

mCp is present in a low quantity in the sea water sample, but the total concentration of hydrogen ions is obtained by:

$$pH_T = pK_a + \log_{10}\left(\frac{\left[I^{2-}\right]}{\left[HI^-\right]}\right) \quad [2.303]$$

As the I^{2-} and HI^- ions have different absorption spectrums, their concentrations ($[I^{2-}]$ and $[HI^-]$) can be evaluated by determing the ratios of molar extinction e_i ($i = 1$ to 3) for two distinct wavelengths, 434 and 578 nm, and in calculating the absorbance ratio $R = A_{578}/A_{434}$ from the measurement of A_{578} and A_{434} with a spectrophotometer. Relation [2.302] therefore becomes:

$$pH_T = pK_a + \log_{10}\left(\frac{R - e_1}{e_2 - Re_3}\right) \quad [2.304]$$

In relation [2.303], $e_1 = 0.00691$, $e_2 = 2.2220$ and $e_3 = 0.1331$.

The system developed by Aßmann *et al.* can only be installed on boats (not on floats for example), taking into account its weight and volume. The indicator is injected with the help of a peristaltic pump in a pipeline that constantly powers a coil with sea water. This coil is thermostated in order to carry the liquid at 25°C, and the mixing indicator–sea water enters through a thermally-isolated analysis cell. An optical fibre illuminates the mixture with the help of a white LED, and another fiber placed in alignment, 1 cm away measures this light in order to send it to a Hamamatsu spectrophotometer (CDD 2,048 pixels). The peristaltic pump, temperature regulation and the spectrophotometer are controlled by a microprocessor. Short-term repeatability of the measurements has been evaluated at ±0.0007 pH. Long-term repeatability is harder to estimate. The accuracy of measurement has been evaluated to 0.0081 pH, in comparison to a reference pH meter.

Recently, autonomous devices have been created by a team at the University of Montana. These can be fixed to moorings, and apply the principles of measurements previously discussed. These devices are called SAMIs (Submersible Autonomous Moored Instruments). There is a SAMI pCO_2 operating with the help of bromothymol blue that enables us to determine the fCO_2, and a SAMI-pH that enables us to find pH with the help of mCP.

2.11.4. *In situ measurement of some other components*

Instruments to measure several other components other than dissolved oxygen and CO_2 are under development for *in situ* measurements. These developments are interesting because they provide the possibility of accessing "high frequency" variations of the medium. This is the case for heavy ions such as Cu^{2+}, Pb^{2+}, Cd^{2+} or Hg^{2+} whose trace presence in the form reveals the degrees of pollution in certain areas or the presence of aromatic hydrocarbons due to oil pollution at sea. It is also useful for the analysis of nutrients, silicates, nitrates, nitrate whose evaluation is necessary to determine the absolute salinity or biology of the medium, and for the study of dissolved iron or sulphites from hydrothermal sources. *In situ* detection and analysis present several advantages: the absence of sampling avoids the alteration to the sample studied; the measurement can be made with temporal and spatial sample frequencies that are a lot higher; and the process can be automated. We can therefore manufacture surveillance systems.

2.11.4.1. *Detection of heavy ions and organic pollutants*

The detection of these components best carried out using optical techniques such as optodes, whose principle is described in section 2.11.2.3 and a well known phenomenon: *the Raman effect.*

In 1928, Raman showed that, besides radiation corresponding to incident wavelengths λ_I (Rayleigh scattering) the light diffused by a liquid contains rays of different wavelengths that are characteristic of the molecule studied. These rays are generated by the inelastic scattering of several photons off molecules (of about $1/10^8$) and are of a much lower intensity (up to 100 times). The energy of photons diffused in this way is equal to the variation of vibrational energy the molecules are subjected to during "the shock" and it is characteristic of the molecule in question, which enables us to identify it. The frequencies observed represent the difference between the frequency of incident radiation and that of absorbed radiation. They are differentiated from fluorescent rays as they can be generated, regardless of incident wavelength, to the nearest shade. The effectiveness of Raman scattering of substances decreases in relation to λ_i^4. Therefore, it is better to work with short incident wavelengths, knowing that they still favour fluorescence. Rays linked to the fluorescence of numerous pigments in sea water present real problems in the analysis of Raman spectrums.

The large majority of dissolved components (Na^+, Cl^-, Mg^{2+}, etc.) cannot be detected by this effect due to their presence in very weak concentrations. Water itself has a relatively weak Raman signature (see Figure 2.277) but its rays, which are always present in analyses, can easily be distinguished from the responses of other components. The Raman spectrum of standard sea water has been known since 1964, thanks to the studies carried out by *Walfaren*. These studies showed the possibility of detecting ion sulfates (liaison S-O at 981 nm).

More recent studies have shown that the influences of temperature, pressure and species of molecules are negligible on shifts and sulfate concentrations measured by the Raman effect, which is promising for technical developments in this field. These developments are very interesting because Raman spectroscopy enables us to make measurements with a relatively high sampling rate ($\approx$ 1 sample/10 s), in samples of liquids, solids and gases without analytical modification of the components. It even enables us to detect the composition of the rocks that make up the seabed through the recognition signatures of carbons and sulfides in hydrothermal zones. As the deep ocean is barely lit by solar light, it is an ideal medium for the exploitation of this technique.

In order to improve sensitivities of measurement using the Raman effect, and to move from the detection of traces to exploitable signals, sol–gel deposit techniques in thin layers on glass supports have been developed. When applied to the detection of aromatic hydrocarbon, they enable us to obtain increases of about 10^6 to 10^8 in the amplitude of Raman rays, by the absorption of target molecules in a rough metallic layer. This technique is known as surface-enhanced Raman scattering, as the presence of nanostructured metal in the form of thin layers or nanoparticles electromagnetically enhances the signal. This effect can be further amplified through chemically functionalizing the substrate.

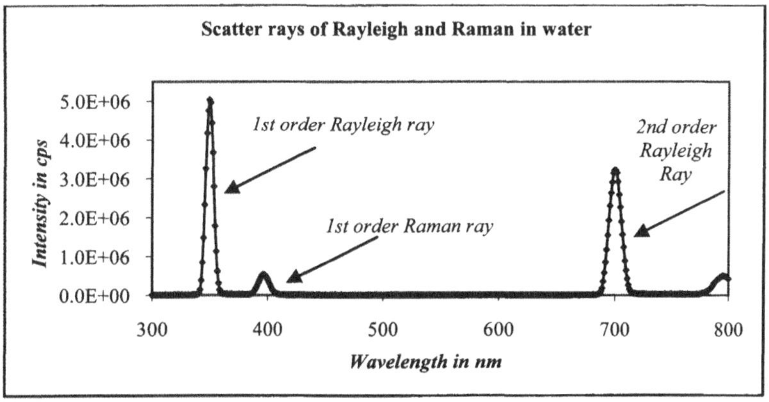

Figure 2.76. *Spectrum of scatter rays obtained with the help of a spectrophotometer in a sample of distilled water. This enables us to demonstrate the difference in amplitude between Raman rays and those linked to molecular scattering*

2.11.4.2. *Detection of nutrients and other dissolved elements*

Analyzers that can be used *in situ* have been developed to detect and measure the concentrations of nutrients and other dissolved elements. Their development is based on the transposition of measurement techniques present in chemistry laboratories, such as electrochemistry, spectrophotometry or analysis of continuous flow followed by colorimetric detection in submergible instruments. Concentration ranges being explored go from 0.1 to 40 μmol/l for offshore nitrates (200 μmol/l at the coast), are 160 μmol/l for silicates and 5 μmol/l for phosphates.

Continuous flow analysis is one of the first methods to have been exploited in these analyses. Since the mid 1960s, Technicon AutoAnalyser analyzers have been mounted onto boats and linked to sea water pumping systems. These devices separate water flow to send it to different colorimeters dedicated to the measurement of different nutrients. They segment the water with air bubbles in order to sequence the analysis. The water samples are mixed with a reagent that is specific to each substance under analysis to form a molecule that can be detected by colorimetry.

A colorimeter is an instrument that enables the measurement of absorbance or the percentage of light transmitted when a light beam, whose spectral band has been selected, crosses the sample. The establishment of the calibration curve of the instrument enables us to link percentage absorbance to concentration. This method of measurement, which takes into account the characteristics of the equipment (spectral source band, optical characteristics of the cell where measurement is made), requires the utilization of solutions containing known concentrations of the molecule under analysis. It is equally possible to determine unknown concentrations of a molecule by using the Beer-Lambert law, as long as the molar absorption coefficient of the substance analyzed is known (see [2.254]). The measurement can therefore equally use fluorimetry (see section 2.10.5.).

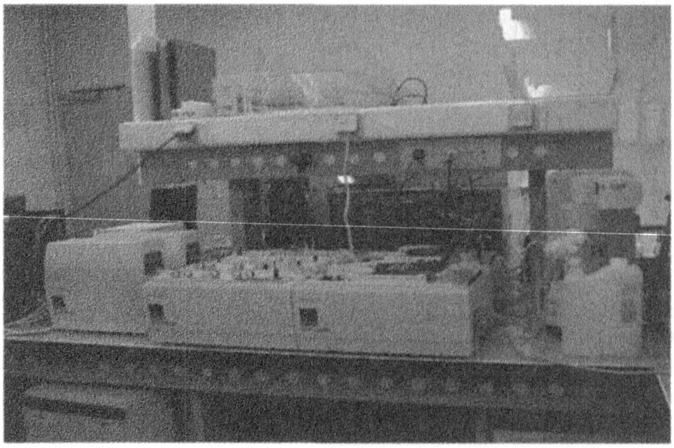

Figure 2.77. *The Laboratory Technicon AutoAnalyser colometric analysis chain produced by Bruän & Lubbe, consists of four measurement channels for the measurement of silicate, phosphates, nitrates and nitrites (Courtesy of © SHOM)*

In order to carry out *in situ* measurements, this method has been modified by replacing sequencing with air bubbles, by a technique called flow injected analysis. In the system, developed by Daniel *et al.* in 1995 [DAN 95], a pump propels water into a collector tube where it is mixed with the reagent. A system of rotating vanes enables the selection of different reagents, which enables the analysis of several molecules using the same device. Another vane enables injection of the mixture into the measurement cell. This cell is illuminated, at an extremity, by an optic fiber and the light transmitted across the sample is collected by another fiber that sends it to a photodetector. This instrument enables the analysis of nitrates and nitrites. Nitrate concentrations are determined from their reduction with copper cadmium, which transforms them into nitrite. A colored complex is obtained by mixing nitrites with sulphanilamide and N-(1-naphthyl) ethylenediamin dihydrochloride. This instrument enables the analysis of 40 samples per hour with accuracy estimated to be to 1% and a detection limit of 0.45 µmol of NO_3 over a measuring range of 0–150 µmol of NO_3. The measuring range can be adapted by varying the length of the cell being analyzed.

An analyzer of the same type was previously developed by Johnson *et al.* in order to measure dissolved silicates and sulphites. This analysis technique was taken up again in 2009 by Vuillemin *et al.* [VUI 09], in order to create an instrument to analyze dissolved iron (Fe (II) + Fe (II+III)) and sulfites ($H_2S + HS^- + S^{2-}$), capable of descending to 6,000 m, with detection limits of 0.3 and 0.1 μmol, respectively. This instrument can be adapted to measure silicates and ammonium and phosphates by fluorimetry.

The principle faults of colorimeters developed to make *in situ* measurements are their high consumption of electricity and their size and weight, which renders them inappropriate for integration in autonomous platforms. Their consumption of reagents is also a problem. The use of reagents can be overcome through the use of electrochemistry. These techniques have the advantage of potentially being able to be integrated in silicon substrates in the manufacture of integrated sensors. They can make voltametric, coulometric, potentiometric or amperometric measurements, with the help of electrodes inserted into the solution being analyzed.

The voltametric method can be used to measure silicates and phosphates, by anodic oxidation of molybdbenum:

$$Mo + 4\ H_2O \rightarrow MoO_4^{2-} + 8\ H^+ + 6\ e^- \qquad [2.305]$$

The products of this oxidation form a complex with silicates to produce a silico-molybdic molecule in a solution of pH = 1.5, which can be detected by cyclic voltametry:

$$P0_4^{3-} + 12\ MoO_4^{2-} + 28\ H^+ \rightarrow H_4P(Mo_{12}O_{40}) + 12\ H_2O \qquad [2.306]$$

Likewise for phosphates, in a solution with pH = 1, a phospho-molybdic complex is formed:

$$Si(OH)^4 + 12\ MoO_4^{2-} + 24\ H^+ \rightarrow H_4Si(Mo_{12}O_{40}) + 12\ H_2O \qquad [2.307]$$

This technique requires us to use or produce four electrodes:

– a platinum electrode or counter electrode situated in front of a membrane that prevents the crossing of H^+ions;

– a reference Ag/AgCl electrode situated behind the membrane;

– a molybdbenum electrode used as an anode; and

– a gold working electrode.

Cyclic voltametry applies an electric potential between the reference electrode and the working electrode and measures the current that circulates between the working electrode and the counter electrode. The current that circulates is said to be faradic, as it is produced by a redox reaction that occurs after application of the potential. In fact, potential $V_{au} - V_{Ag/AgCl}$ is increased in a linear way. The current, which has a negative starting value, increases suddenly following the application of a specified potential to a maximum level corresponding to the maximum oxidation or reduction of substances. Then, potential is inversed and we witness the reduction or progressive oxidation of species formed previously until the solution returns to the initial current (see Figure 2.278). The rapidity of this cycle and the intensity *i* of the current measured are relative to the scattering coefficient of species between electrodes. *i* is governed by *Cottrell's law*:

$$i = nFSC\sqrt{\frac{D}{\pi t}} \qquad [2.308]$$

where:

– n represents the number of electrons exchanged per mole of substance;

– F is the Faraday constant, which has a value of 96,485 C/mol;

– S is the exchange surface of the electrode;

– C is the reducing substance concentration expressed in mol/cm^3;

– D is the scattering coefficient of this substance in cm²/s; and

– t the time in s.

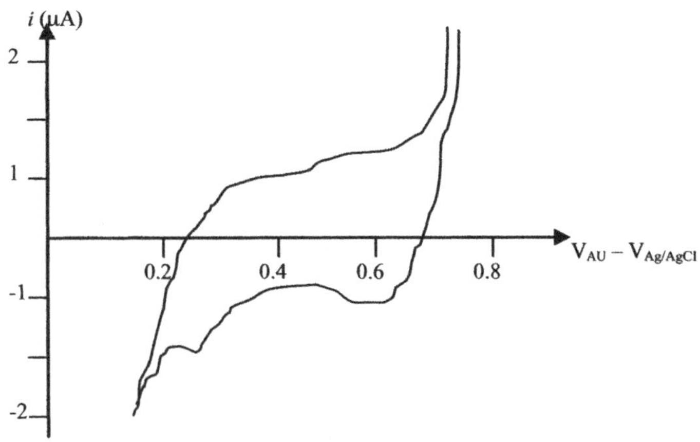

Figure 2.78. *Schematic representation of the current diagram that can be obtained from a phospho-molybdic complex subjected to a potential variation of 200 mV/s*

The efforts to integrate this technique for the analysis of silicon substrates has led to the development of ultra microelectrodes. Cottrell's law therefore takes a significantly different form:

$$i = 4nFDCr \qquad [2.309]$$

where r is the ultra microelectrode ray, which can be smaller than 20 µm. This relation enables us to see that if we measure i and manage to determine the value of D, it is possible to find the absolute value of concentration C if we manage to determine r with sufficient accuracy. This is the path to be explored to make sensors that do not require calibration with reference substances. The relative expected uncertainty relating to C is several per cent and the limit of detection is about 1 µmol.

The need to have sensors with a better accuracy, a lower detection limit and higher long-term stability has encouraged the development of instruments using *in situ* spectrophotometry. Several problems arise with regards to their production:

– most dissolved inorganic components absorb light at very short wavelengths – below 280 nm in the ultraviolet range – and it is difficult to work with these wavelengths;

– this ultraviolet signal is largely dominated by the response of bromides Br^-, nitrates and organic matter (CDOM); and

– these spectrums are superimposed on very narrow bands (see Figure 2.279).

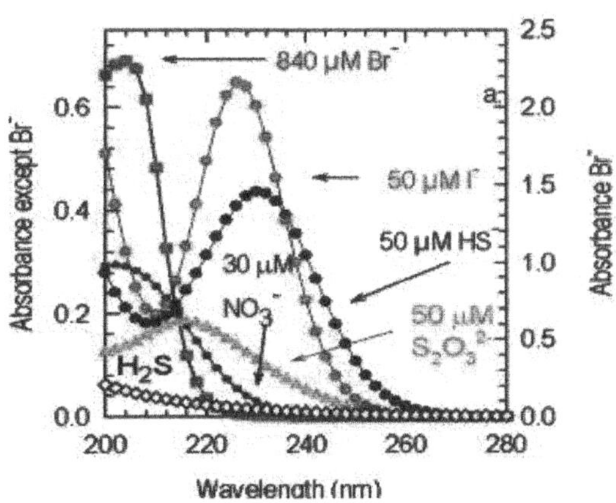

Figure 2.79. *Absorption spectrums of principal components of sea water that absorb in the 200–280 nm band (Courtesy of K.S. Johnson, MBARI)*

The absorption spectrums of these components are well known, and their molar absorbencies, $\varepsilon_{\lambda,j}$, can be determined from laboratory measurements. It is therefore possible, from the Beer-Lambert law [2.254] to find the concentrations C_j of each component j. It casts off absorption linked to CDOM, i.e. the dissolved organic matter. It responds to an exponential relation of the same type as relation [2.252]. In order to find the coefficients of this equation that best correspond to the medium where measurements are made, it is necessary to use nonlinear least-square regression algorithms. These algorithms are time-consuming and have risks of divergence. Therefore, the solution adopted by Holm *et al.* in 1997 [HOL 97] consists of modeling with the help of a second-order polynomial relative to wavelength, λ, which constitute for the measurement of nutrients, a background noise. This simplification was validated by measurements at sea. The absorbency spectrum of sea water, measured using an ultraviolet spectrophotometer, can therefore be modeled with the following equation:

$$A(\lambda) = l\left[\sum_j \varepsilon_{\lambda,j} C_j + e + f\lambda + g\lambda^2\right] \qquad [2.310]$$

where l is the length of the measured cell, and e, f and g are coefficients to be determined to cast off the absorption of CDOM. Knowing the absorption spectrums for each component, we need to deconvolute $A(\lambda)$, knowing that the first term represents the sum of the absorption spectrums of these components, and the others, the fluctuates of the base line.

This is how the ISUS and SUNA nitrate sensors produced by SAtlantic operate. The ISUS uses a deuterium flashlight as a light source. The light is transmitted by an optical fiber to a measuring cell that is 1 cm in length. Another optical fiber collects the light that has crossed the sample and returns it to a spectrophotometer. The entry of the light in the measurement volume is made through a quartz window. The absorption spectrums of principal sea water components (see Figure 2.280) are programmed into the instrument, and an iterative process adjusts these spectrums until they correspond to the spectrum measured. The concentration of nitrates is therefore deducted. If the concentration of CDOM is greater than 30 mg/l, a correction is applied according to equation [2.310]. As molar absorption coefficients of principal ions, and particularly that of Br^-, vary with temperature, it is necessary to measure these coefficients in order to apply corrections during deconvolution. For this, the ISUS sensor is usually connected to a CTD profiler that exploits this parameter. This correction in temperature also serves to compensate for instrumental drift (flashlight, detector) linked to this quantity. Likewise, in order to produce measurements to an accuracy increased from 2 µmol/l to 0.4 µmol/l, the ISUS sensor must exploit the practical salinity calculated by the CTD profiler. Nitrate concentration measurements are increased in the same fraction as the error that exists between the salinity estimated from its measurements, and that calculated by the profiler. Finally, the accuracy shown by this instrument is 2 µmol/l across a range of 2,000 µmol/l, or 10% of the reading. Its precision is 0.5 µmol/l and its possible drift is 0.05 µmol/hour of operation. However, its essential advantages are the absence of a reagent, and that it can reach a sampling frequency of 1 Hz, so it can be used to carry out concentration profiles up to a depth of 1,000 m.

Instruments called *biosensors* are also under development to enable the detection of nutrients (sulfites, phosphates, nitrates) related to organic pollutants, such as pesticides and certain metals. Their principle of measurement is based on the usage of bacteria, enzymes, animal or plant tissue, etc., which are gathered on a membrane from the element to be detected. Their transformation or resulting emission produces a substance or reaction that can be measured with physical transducers. For example, Unisense has recently developed and commercialized a biosensor whose element is sensitive to denitrifying bacteria. In the presence of nitrates, NO_3^-, this bacterium generates NO_2 molecules that can be detected by a specific chemical electrode. The sensitivity of the biosensor varies from 5 to 800 M of NO_3, according to the range chosen, and its response time is 30 s. Its lifespan, however, is only five days. Transduction can also be carried out by surface-plasmon resonance (SPR). SPR consists of measuring the variations of the refraction index at the interface between a metal and the medium being studied. It enables the study of interactions between a receiver present on a metallic surface and the target molecule that we are looking to quantify.

These instruments have undeniable advantages: they are selective, they offer good sensitivity, they are barely sensitive to interference, and they lend themselves to miniaturization. Their development is therefore promising.

2.12. Bibliography and further reading

2.12.1. *Measuring temperature*

[BEN 72] BENNETT A.S., "The calibration of thermistors over the temperature range 0°-30°C", *Deep-Sea Research*, vol. 19, pp. 157-163, 1972.

[BIA 04] BIANCHI A.M., FAUTRELLE Y., ETAY J., *Transferts Thermiques*, Presses Polytechniques et Universitaires Romandes, 2004.

[BRO 88] BROWN N.L., "New generation CTD system", *IEEE Journal of Oceanic Engineering*, vol. 13, no. 3, pp. 129-134, 1988.

[BUL 90] INTERNATIONAL TEMPERATURE SCALE OF 1990, *Bulletin du Bureau National de Métrologie*, n° 79, Comité Consultatif de Thermométrie (CCT), 1990.

[GEN 84] GENT A.E., "Pressure effects on NBIS CTD temperature and conductivity sensors", in *Proceedings of the Marine Technology Society STD-84 Conference and Workshop*, pp. 85-90, 1984.

[GOU 10] GOURETSKI V., RESEGHETTI F., "On depth and temperature biases in bathythermograph data: development of a new correction scheme based on analysis of a global ocean database", *Deep-Sea Research,* I, vol. 57, pp. 812-833, 2010.

[GRE 80] GREGG M.C., MEAGHER T.B., "The dynamic response of glass rod thermistors", *Journal of Geophysical Research,* vol. 85, no. C5, pp. 2779-2786, 1980.

[GRE 82] GREGG M.C., SCHEDVIN J.C., HESS W.C., MEAGHER T.B., "Dynamic response calibration of the Neil Brown conductivity cell", *Journal of Physical Oceanography,* vol. 12, pp. 720-742, 1982.

[GRE 85] GREGG M.C., HESS W.C., "Dynamic response calibration of Sea-Bird temperature and conductivity probes", *Journal of Atmospheric and Oceanic Technology*, vol. 2, pp. 304-313, 1985.

[HAR 01] HAREN H.V., GROENEWEGEN R., LAAN M., KOSTER B., "A fast and accurate thermistor string", *Journal of Atmospheric and Oceanic Technology*, vol. 18, pp. 256-265, 2001.

[KIZ 08] KIZU S., ONISHI H., SUGA T., HANAWA K., WATANABE T., IWAMIYA H., "Evaluation of fall rates of the present and developmental XCTDs", *Deep Sea Research*, I, vol. 55, pp. 571-586, 2008.

[KIZ 10] KIZU S., SUKIGARAL C., and HANAWA K., "Comparison of the fall rate and structure of recent T-7 XBT manufactured by Sippican and TSK", *Ocean Science Discussion*, vol. 7, pp. 1811–1847, 2010.

[LAR 96] LARSON N., PEDERSON A.M., "Temperature measurements in flowing water: viscous heating of sensor tips", *Proceedings of the 1st IGHEM Meeting*, Montreal, Canada, June 1996, from www.seabird.com.

[LEW 84] LEWIS E.L., "Optimal design of CTD sensor heads for salinity determination", in *Proceedings of the Marine Technology Society STD-84 Conference and Workshop*, pp. 23-27, 1984.

[MAG 84] "MANGUM B.W., Stability of small industrial platinum resistance thermometers", *Journal of Research of the National Bureau of Standards,* vol. 89, pp. 305-316, 1984.

[MAG 82] MANGUM B.W. and EVANS G.A., "Investigation of the stability of small platinum resistance thermometers", *Temperature, its Measurement and Control in Science and Industry*, vol. 5, pp. 795-801, 1982.

[PED 84] PEDERSON A.M., "A modular high resolution CTD system with computer-controlled sample rate", *Marine Technology Society STD Conference and Workshop*, February 27-29, San Diego, 1984.

[REV 09] REVERDIN G., MARIN F., BOURLES B., L'HERMINIER P., "XBT temperature errors during French research cruises (1999-2007)", *Journal of Atmospheric and Oceanic Technology,* vol. 26, pp. 2462-2473, 2009.

[RUD 07] RUDNICK D.L., KLINKE J., "The underway conductivity-temperature-depth instrument", *Journal of Atmospheric and Oceanic Technology,* vol. 24, pp. 1910-1923, 2007.

[SCH 05] SCHMITT R.W., MILLARD R.C., TOOLE J.M., WELLWOOD W.D., "A double-diffusive interface tank for dynamic-response studies", *Journal of Marine Research*, vol. 63, pp. 263-289, 2005.

[STE 68] STEINHART J.S., HART S.R., "Calibration curves for thermistors", *Deep-Sea Research,* vol.15, pp. 497-503, 1968.

[TRU 83] TRUMP C.L., "Effects of ship's roll on the quality of precision CTD data", *Deep Sea Research*, vol. 30, 11A, pp. 1173-1183, 1983.

[USH 07] USHIDA H., OHYAMA K., OZAWA S., FUKASAWA M., "In situ calibration of the SeaBird 9plus CTD thermometer", *Journal of Atmospheric and Oceanic Technology*, vol. 24, pp. 1961-1967, 2007.

[WIL 09] WILLIS J.K., LYMAN J.M., JOHNSON G.C., "In situ data biases and recent ocean heat content variability", *Journal of Atmospheric and Oceanic Technology*, vol. 26, pp. 846-852, 2009.

[WMC 94] WMC/IOC IGOSS TASK TEAM ON QUALITY CONTROL OF AUTOMATED SYSTEMS, Calculation of New Depth Equations for Expendable Bathythermographs using a Temperature-error-free Method, UNESCO Technical Papers in Marine Science, No. 67, UNESCO, 1994.

2.12.2. *Measuring conductivity*

[AND 05] ANDO K., MATSUMOTO T., NAGAHAMA T., UEKI I., TAKATSUKI Y., KURODA Y., "Drift characteristics of a moored conductivity-temperature-depth sensor and correction of salinity data", *Journal of Atmospheric and Oceanic Technology*, vol. 22, pp. 282-291, 2005.

[ARC 03] ARCHER C., BAPTISTA A., LEEN T.K., "Fault detection for salinity sensors in the Columbia estuary", *Water Resources Research*, vol. 39, no. 3, 1060, pp. 1-10, 2003.

[COO 10] COOK M., "Copper-alloy biofouling control on marine water-monitoring systems", *Sea Technology*, October, pp. 19-23, 2010.

[COW 98] COWLING M.J., HODGKIESS T., KERR A., PARR A., SMITH M., BEVERIDGE C., "Biofouling of oceanographic sensors – is there a solution?", *Oceanology International 98 – The Global Ocean*, vol. 1, pp. 203–214, 1998.

[DEL 10] DELAUNAY L., COMPÈRE C., LEHAITRE M., "Biofouling protection for marine environmental sensors", *Ocean Science*, vol. 6, pp. 503-511, 2010.

[FES 98] FESTY D., LE BRAS S., CLEGG M., LACOTTE N., LEHAITRE M., MENLOVE R., SEBASTIAO P, "Bioflim prevention on optics by chlorine compound generation on Tin oxide coating", *Ocean'98, IEEE Conference*, Nice, Sept 28–Oct 1, 1998.

[GIL 86] GILES A.B., MC DOUGALL T.J., "Two methods for the reduction of salinity spiking of CTD", *Deep-Sea Research*, vol. 33, no. 9, pp. 1253-1274, 1986.

[GRE 82] GREGG M.C., SCHEDVIN J.C., HESS W.C., MEAGHER T.B., "Dynamic response calibration of the Neil Brown conductivity cell", *Journal of Physical Oceanography*, vol. 12, pp. 720-742, 1982

[HOR 10] HORIUCHI T., WOLK F., MACOUN P., "Long-term test of a new conductivity and temperature sensor", *Sea Technology*, pp. 37-40, 2010.

[JOH 07] JOHNSON G.C., TOOLE J.M., LARSON N.G., "Sensor correction for Sea-Bird SBE 41CP and SBE 41 CTDs", *Journal of Atmospheric and Oceanic Technology*, vol. 24, pp. 1117-1130, 2007.

[LEM 03] LE MENN M., Capteurs de conductivité en océanographie: état de l'art, Communication orale n° 05 de l'Atelier Expérimentation Instrumentation INSU, IFREMER, METEO FRANCE, Brest, January 28-29, 2003.

[LEM 04] LE MENN M., Etalonnage de thermosalinomètres dans le cadre du projet CORIOLIS, Communication orale S7-02 de l'Atelier Expérimentation Instrumentation INSU, IFREMER, METEO FRANCE, Paris, March 23-24, 2004.

[LUE 90] LUECK R.G., "Thermal inertia of conductivity cells: theory", *Journal of Atmospheric and Oceanic Technology*, vol. 7, pp. 741-755, 1990.

[LUE 90] LUECK R.G. and PICKLO J., "Thermal inertia of conductivity cells: observations with a Sea Bird cell", *Journal of Atmospheric and Oceanic Technology,* vol. 7, pp. 756-768, 1990.

[MEN 09] MENSAH V., LE MENN M., MOREL Y., "Thermal mass correction for the evaluation of salinity", *Journal of Atmospheric and Oceanic Technology*, vol. 26, no. 3, pp. 665–672, 2009.

[MIN 05] MINEART G.M., CROUT R.L., "Technologies for global observations of ocean constituents", *Marine Technology Society Journal*, vol. 39, no. 3, pp. 36-48, 2005.

[MOR 94] MORISON J., ANDERSEN R., LARSON N., D'ASARO E., BOYD T., "The correction of thermal-lag effects in Sea-Bird CTD data", *Journal of Atmospheric and Oceanic Technology*, vol. 11, pp. 1151-1164, 1994.

[PED 79] PEDERSON A.M., GREGG M.C., "Development of a small *in-situ* conductivity instrument", *IEEE Journal of Oceanic Engineering*, vol. OE-4, no. 3, July, pp. 69-75, 1979.

[SCH 55] SCLICHTING H., *Boundary Layer Theory*, Pergamon Press, 1955.

[STR 85] STRIGGOW K., DANKERT R., "The exact theory of inductive conductivity sensors for oceanographic application", *IEEE Journal of Oceanic Engineering*, vol. OE-10, no. 2, April pp. 175-179, 1985.

[WOL 05] WOLK F., LI H., "Self-cleaning sensors for long-term moorings", *Sea Technology*, pp. 43-49, 2005.

2.12.3. *Measuring pressure*

[PAR] www.paroscientific.com/appnotes.htm

[VIS] www.vishay.com/company/press/releases/2006/061219mmhs/

2.12.4. *Measuring velocity*

[DUS 93] DUSHAW B.D., WORCESTER P.F., CORNUELLLE B.D., "On equation for the speed of sound in seawater", *Journal of the Acoustical Society of America,* vol. 93, no. 1, pp. 255-275, 1993.

[EAT 96] EATON G., DAKIN D.T., "Miniature time of flight sound velocimeter offers increased accuracy over sing-around technology and CTD instrumentation", *Oceanology International 96 Proceedings*, Brighton, UK, March 5-8, 1996.

[MAC 73] MACKENZIE K.V., "Calibrations of oceanographic research velocimeters", *The Journal of the Acoustical Society of America*, vol. 53, no. 3, pp. 869-875, 1973.

[SWE 06] SWEENEY A.D., CHADWELL C.D., HILDEBRAND J.A., "Calibration of seawater sound velocimeter", *IEEE Journal of Oceanic Engineering*, vol. 31, no. 2, pp. 454-461, 2006.

[YOS 92] YOST W.T., CANTRELL J.H., KUSHNICK P.W., "Fundamental aspects of pulse phase-locked loop technology-based methods for measurement of ultrasonic velocity", *The Journal of the Acoustical Society of America*, vol. 91, no. 3, pp. 1456-1468, 1992.

2.12.5. *Measuring current*

[ALD 99] ALDERSON S.G., CUNNINGHAM S.A., "Velocity errors in Acoustic Doppler Profiler measurements due to platform attitude variations and their effect on volume transport estimates", *Journal of Atmospheric and Oceanic Technology*, vol. 16, pp. 96-106, 1999.

[BEV 81] BEVIR M.K., O'SULLIVAN V.T., WYATT D.G., "Computation of electromagnetic flowmeter characteristics from magnetic field data", *Journal of Physics D: Applied Physics*, vol. 14, pp. 373-388, 1981.

[BRU 83] BRUMLEY B.H., CABRERA R.G., DIENES K.L. and TERRAY E.A., "Performance of a broad-band acoustic Doppler current profiler", *Journal of Oceanic Engineering*, vol. 16, no. 4, pp. 402-407, 1983.

[CAR 97] CARUSO M.J., "Applications of magnetoresistive sensors in navigation systems", *Sensors Actuators*, vol. 1220, pp. 15-21, 1997.

[DIL 11] DILLON J., ZENDEL L., HAY A.E., "Asymptotic properties of an autocorrelation coefficient for coherent Doppler sonar", *Journal of Atmospheric and Oceanic Technology*, vol. 28, pp. 966-973, 2011.

[DIL 12] DILLON J., ZENDEL L., HAY A.E., "Simultaneous velocity ambiguity resolution and noise suppression for multifrequency coherent Doppler sonar", *Journal of Atmospheric and Oceanic Technology*, vol. 29, pp. 450-463, 2012.

[DOH 10] DOHAN K., MAXIMENKO N., "Monitoring ocean currents with satellite sensors", *Oceanography*, vol. 23, no. 4, pp. 94-103, 2010.

[EDE 07] EDELSTEIN A., "Advances in magnetometry", *Journal of Physics: Condense Matter*, vol. 19, pp. 1-28, 2007.

[GIL 00] GILBOY T.P., DICKEY T.D., SIGURDSON D.E., YU X., MANOV D., "An intercomparison of current measurements using a vector measuring current meter, an acoustic Doppler current profiler and a recently developed acoustic current meter", *Journal of Atmospheric and Oceanic Technology*, vol. 17, pp. 561-574, 2000.

[HOV 96] HOVDENES J., "The RMC9 – A unique new instrument for measuring ocean currents and other oceanographic parameters", *Ocean 96 MTS/IEEE Conference Proceedings*, Florida, September 23-26, vol. 1, pp. 293-295, 1996.

[JOY 89] JOYCE T.M., "On in situ "calibration" of shipboard ADCPs", *Journal of Atmospheric and Oceanic Technology*, vol. 6, pp. 169-172, 1989.

[LEM 07] LE MENN M., LE GOFF M., "A method for absolute calibration of compasses", *Measurement Science and Technology*, vol. 18, pp. 1614-1621, 2007.

[LHE 84] LHERMITTE R., SERAFIN R., "Pulse-to-pulse coherent Doppler sonar signal processing techniques", *Journal of Atmospheric and Oceanic Technology*, vol. 1, no. 4, pp. 293-308, 1984.

[MAY 07] MAYER D.A., VIRMANI J.I., WEISBERG R.H., "Velocity comparisons from upward and downward acoustic Doppler current profilers on the West Florida shelf", *Journal of Atmospheric and Oceanic Technology*, vol. 24, pp. 1950-1960, 2007.

[PIE 99] PIERCE S.D., BARTH J.A., SMITH R.L., "Improving acoustic Doppler profiler accuracy with wide-area differential GPS and adaptative smoothing of ship velocity", *Journal of Atmospheric and Oceanic Technology*, vol. 16, pp. 591-596, 1999.

[POL 89] POLLARD R., READ J., "A method for calibrating shipmounted Acoustic Doppler Profilers and the limitations of gyro compasses", *Journal of Atmospheric and Oceanic Technology*, vol. 6, no. 6, pp. 859-865, 1989.

[PRI 79] PRIMDHAL F., "The fuxgate magnetometer", *Journal of Physics E: Scientific Instruments*, vol. 12, pp. 241-253, 1979.

[SCH 62] SCHVERCLIFF J.A., *The Theory of Electromagnetic Flow-measurement*, Cambridge University Press, 1962.

[SIE 10] SIEGEL E., LOHRMANN A., "Eliminating the blanking distance for acoustic Doppler current profilers", *Sea Technology*, pp. 33-36, 2010.

[TEL 94] TELEDYNE R.D., Technical Manual DR/SC-BBADCP, December, R.D. Teledyne Instruments, 1994.

[TES 06] TESSIER C., LURTON X., LE GALL Y., LE HIR P., La mesure de turbidité par ADCP: processus, méthodologies et limitations, Communication Orale n° 76 AEI-2006, IFREMER, Brest, January 31 – February 1, 2006.

[THE 86] THERIAULT K.B., "Incoherent multibeam Doppler current profiler performances: I-estimate of variance", *IEEE journal of Oceanic Engineering*, vol. OE-11, pp. 7-15, 1986.

[THU 10] THURNHERR A.M., "A practical assessment of the errors associated with full-depth LADCP profiles obtained using Teledyne RDI workhorse acoustic Doppler current profilers", *Journal of Atmospheric and Oceanic Technology*, vol. 27, pp. 1215-1227, 2010.

[ZED 96] ZEDEL L., HAY A.E., CABRERA R., LOHRMAN A., "Performance of a single-beam pulse-to pulse coherent Doppler profiler", *IEEE Journal of Oceanic Engineering*, vol. 21, no. 3, pp. 290-297, 1996.

[ZED 08] ZEDEL L., "Modeling pulse-to-pulse coherent Doppler sonar", *Journal of Atmospheric and Oceanic Technology*, vol. 25, pp. 1834-1844, 2008.

[ZED 10] ZEDEL L., HAY A.E., "Resolving velocity ambiguity in multifrequency, pulse-to-pulse coherent Doppler sonar", *IEEE Journal of Oceanic Engineering*, vol. 35, no. 4, pp. 847-851, 2010.

2.12.6. *Measuring time and frequencies*

[BAN 07] BANERJEE P., SUMAN, SURI A.K., CHATTERJEE A., BOSE A., "A study on the potentiality of the GPS timing receiver for real time applications", *Measurements Science and Technology*, vol. 18, pp. 3811-3815, 2007.

[FRE 78] FRERKING M.E., *Crystal Oscillator Design and Temperature Compensation*, New York, Van Nostrand, 1978.

[KAP 96] KAPLAN E.D., *Understanding GPS: Principles and Application*, LARSON L., (Ed.), Artech House Publishers, 1996.

[KIT 07] KITCHING J., "Time for better receiver", *GPS World*, pp. 52-57, November 2007.

[LEW 91] LEWANDOWSKI W., THOMAS C., "GPS time transfer", *IEEE Special Issue on Time*, vol. 79, no. 7, 1991.

[MOU 05] MOUDRAK A., KONOVALTSEV A., FURTHNER J., HAMMENFARH J., BAUSH A., DEFRAIGNE P., BEDRICH S., SCHOTH A., "Interoperability on time, GPS-Galileo offset will biais position", *GPS World*, pp. 24-32, March 2005.

[YAN 11] YANG S.H., LEE C.B., HEO M.B., LEE E., LEE S.J., PARK C., "Analysis of a clock-aided global navigation satellite system (GNSS)", *Measurements Science and Technology*, vol. 22, pp. 1-7, 2011.

[YEH 09] YEH T-K., HWANG C., XU G., WANG C.S., LEE C.C., "Determination of global positioning system (GPS) receiver clock errors: impact on positioning accuracy", *Measurements Science and Technology*, vol. 20, pp. 1-7, 2009.

2.12.7. *Measuring distance*

[BOT 97] BOTTON S., DUQUENNE F., EGELS Y., EVEN M., WILLIS P., *GPS, Localisation et Navigation*, Hermes, 1997.

[CRE 06] CREEL T., DORSEY A.J., MENDICKI P.H., LITTLE J., MACH R.G., RENFRO B.A., "New, improved GPS. The legacy accuracy improvement initiative", *GPS World*, March, pp. 20-31, 2006.

[DIG 07] VAN DIGGELEN F., "GNSS accuracy. Lies, damn lies, and statistics", *GPS World*, pp. 26-32, January 2007.

[FEN 05] FENG Y., "Future GNSS performance. Predictions using GPS with a virtual Galileo constellation", *GPS World*, pp. 46-52, March 2005.

[GLE 07] GLEASON S.T., "Reflecting on GPS. Sensing land and ice from low earth orbit", *GPS World*, October, 44-49, 2007.

[HEI 06] HEINRICHS G., WINKEL J., DREWES C., MAURER L., SPRINGER A., STUHLBERGER R., WICPALEK C., "To locate a phone or PDA GNSS/UMTS prototype for mass market applications", *GPS World*, January, pp. 20-27, 2006.

[HOF 97] HOFMANN-WELLENHOF B. *et al.*, *GPS Theory and Practice*, Fourth revised Edition, Springer Verlag, Vienna, New York, 1997.

[JEN 11] JENSEN A.B.O., MITCHELL C., "GNSS and the ionosphere. What's in store for the next solar maximum", *GPS World*, February, pp. 40-48, 2011.

[KAP 96] KAPLAN E.D. (ed.), *Understanding GPS Principles and Applications*, Artech House, Boston, London, 1996.

[KAZ 94] KAZANTSEV V.N., RESHETNEV M.F., KOSLOV A.G., CHEREMISIN V.F., "Overview and design of the GLONASS System", *Proceedings of the International Conference on satellite Communications*, vol. II, Moscow, Russia, pp. 207-216, 1994.

[LAN 99] LANGLEY R.B., "The integrity of GPS", *GPS World*, vol. 10, no. 3, pp. 60-63, 1999.

[MAT 05] MATTOS P.G., "A single-chip GPS receiver and the steps to mass-market Galileo", *GPS World*, October, pp. 24-31, 2005.

[MCD 02] MC DONALD K., "The modernisation of GPS: plans, new capabilities and future relationship to Galileo", *Journal of Global Positioning Systems*, vol. 1, no. 1, pp. 1-17, 2002.

[MIL 09] MILBERT D., "Improving dilution of precision. A companion measure of systematic effects", *GPS World*, Nov., pp. 38-47, 2009.

[MOE 09] MOERNAUT G.K.J., ORBAN D., "GNSS antennas. An introduction to bandwidth, gain pattern, polarization and all that", *GPS World*, Feb., pp. 42-48, 2009.

[ODO 03] O'DONNELL M., WATSON T., FISHER J., SIMPSON S., BRODIN G., BRYANT E., WALSH D., "Galileo performance", *GPS World*, vol. 14, no. 6, pp. 38-45, 2003.

[RÜE] RÜEGER J.M., *Electronic Distance Measurement, an Introduction*, Fourth Edition, Springer.

2.12.8. *Measuring sea level*

[HAJ 04] HAJJ G.A., ZUFFADA C., "CHAMP and SAC-C atmospheric occultation results and intercomparisons", *Journal of Geophysical Research*, vol. 109, pp. D06109, 2004.

[IOC 03] IOC, Workshop on new Technical Developments in Sea and Land Level Observing Systems, Intergovernmental Oceanographic Commission, Workshop Report No. 193, Paris, October 14-16, 2003.

[IOC 06] IOC, Manuals and Guides 14 Manual on sea level Measurement and Interpretation, JCOMM Technical report no. 31, WMO/TD no. 1339, UNESCO, 2006.

[KAT 05] KATO T., TERADA Y., ITO K., HATTORI R., ABE T., MIYAKE T., KOSHIMURA S., NAGAI T., "Tsunami due to the 5 September 2004 of the Kii peninsula earthquake, Japan, recorded by a new GPS buoy", *Earth Planets Space*, vol. 57, pp. 297-301, 2005.

[LER 04] Le Roy R., Mesures Marégraphiques, Document SHOM GU2003-010 du 14/01/2004.

[MAR 02] MARTIN-NEIRA M., COLMENAREJO P., RUFFINI G., SERRA C., "Altimetry precision of 1 cm over a pond using the wide-lane carrier phase of GPS reflected signal", *Canadian Journal of Remote Sensing*, vol. 28, no. 3, pp. 394-403, 2002.

[MEY 05] MEYER T.H., ROMAN D.R., ZILKOSKI D.B., "What does *Height* really mean? Part II: Physics and Gravity", *Surveying and Land Information Science*, vol. 65, no. 1, pp. 5-15, 2005.

[SIM 06] SIMON B., "Références de hauteur en hydrographie", *Annales Hydrographiques*, vol. 773 (3), 1-12pp, 2006.

[WAT 04] WATSON C., WHITE N., COLEMAN R., CHURCH J., MORGAN P., GOVIND R., "TOPEX/poseidon and Jason-1: absolute calibration in Bass Strait, Australia", *Marine Geodesy*, vol. 27, no. 1-2, pp. 107-132, 2004.

[WOO 96] WOODWORTH P.L., VASSIE J.M., SPENCER R., SMITH D.E., "Precise datum control for pressure tide gauges", *Marine Geodesy*, vol. 19, pp. 1-20, 1996.

2.12.9. *Measuring state of sea*

[ARD 06] ARDHUIN F., Quelles mesures pour la prévision des états de mer en zone côtière?, Communication n° 55 de l'Atelier Expérimentation et Instrumentation, IFREMER, Brest, January 31-February 1, 2006.

[BAR 72] BARRICK D.E., "First-order theory and analysis of MF/HF/VHF scatter from the sea", *IEEE Transactions on Antennas and Propagation*, vol. AP-20, no. 1, pp. 2-10, 1972.

[BAR 77] BARRICK D.E., "Extraction of wave parameters from measured HF sea echo Doppler spectra", *Radio Sciences*, vol. 12, pp. 415-424, 1977.

[BEN 10] BENDER L.C., GUINASSO JR. N.L., WALPERT J.N., "A comparison of methods for determining significant wave heights-Applied to a 3-m discus buoy during hurricane Katrina", *Journal of Atmospheric and Oceanic Technology*, vol. 27, pp. 1012-1028, 2010.

[CHA 97] CHAPMAN R.D., GRABER H.C., "Validation of HF radar measurements", *Oceanography*, vol. 10, no. 2, pp. 76-79, 1997.

[CRO 55] CROMBIE D.D., "Doppler spectrum of sea echo at 13.56 Mc/s", *Nature*, vol. 175, pp. 681-682, 1955.

[GRA 97] GRABER H.C., HERON M.L., "Wave height measurements from HF radar", *Oceanography*, vol. 10, no. 2, pp. 90-92, 1997.

[HAR 05] HARIGAE M., YAMAGUCHI I., KASAI T., IGAWA H., NAKANISHI H., MURAYAMA T., IWANAKA Y., SUKO H., "Abreast of the waves. Open-sea sensor to measure height and direction", *GPS World*, pp. 16-26, 2005.

[HER 99] HERBERS T.H.C., ELGAR S., GUZA R.T., "Directional spreading of waves in the nearshore", *Journal of Geophysical Research*, vol. 104, no. 4, pp. 7683-7693, 1999.

[IRI 06] IRISH J.L., WOZENCRAFT J.M., CUNNINGHAM A.G., GIROUD C., "Nonintrusive measurement of ocean waves: Lidar wave gauge", *Journal of Atmospheric and Oceanic Technology*, vol. 23, pp. 1559-1572, 2006.

[JAN 07] JANSSEN P.A.E.M., ABDALLA S., HERSBACH H., BIDLOT J.R., "Error estimation of buoy, satellite and model wave height data", *Journal of Atmospheric and Oceanic Technology*, vol. 24, pp. 1665-1677, 2007.

[KAV 31] KAUTSKY H., HIRSCH A., "Neue Versuche zur Kohlensäureassimilation", *Naturwissenschaften*, vol. 19, pp. 964, 1931.

[KOH 03] KOHUT J.T., GLENN S.M., "Improving HF radar surface current measurements with measured antenna beam patterns", *Journal of Atmospheric and Oceanic Technology*, vol. 20, pp. 1303-1316, 2003.

[LAW 10] LAWS K., PADUAN J.D., VESECKY J., "Estimation and assessment of errors related to antenna pattern distortion in CODAR seasoned high-frequency radar ocean current measurements", *Journal of Atmospheric and Oceanic Technology*, vol. 27, pp. 1029-1043, June 2010.

[LIP 86] LIPA B.J., BARRICK D.E., "Extraction of sea state from HF radar sea echo: mathematical theory and modeling", *Radio Science*, vol. 21, no. 1, pp. 81-100, 1986.

[LIP 09] LIPA B., WHELAN C., RECTOR B., NYDEN B., "HF radar bistatic measurement of surface current velocities: drifter comparison and radar consistency checks", *Remote Sensing*, vol. 1, pp. 1190-1211, 2009.

[LON 63] LONGUET-HIGGINS M.S., CARTWRIGHT D.E., SMITH N.D., "Observation of the directional spectrum of sea waves using the motion of a floating buoy", *Ocean Wave Spectra*, Prentice-Hall, pp. 111-136, 1963.

[PAD 97] PADUAN J.D., GRABER H.C., "Introduction to high-frequency radar: reality and myth", *Oceanography*, vol. 10, no. 2, pp. 36-39, 1997.

[RIU 02] RIUS A., APARICIO J.M., CARDELLACH E., MARTIN-NEIRA M., CHAPRON B., "Sea surface state measurement using GPS reflected signals", *Geophysical Research Letters*, vol. 29, pp. 23, 2002.

[ROR 00] RORBAEK K., ANDERSEN H., "Evaluation of wave measurements with an acoustic Doppler current profiler", *Proceeding of Oceans 00*, pp. 287-290, IEEE, New York, 2000.

[REY 04] REY V., CERTAIN R., DREVARD D., MEURET A., PIAZZOLA J., "Mesures de houles partiellement stationnaires côtière et littorale", *Communication aux Journées AUM/AFM*, Brest, September 2-3, 2004.

[SCH 86] SCHMIDT R.O., "Multiple emitter location and signal parameter estimation", *IEEE Transactions on Antennas and Propagation*, vol. AP-34, pp. 276-280, 1986.

[SIE 06] SIEGEL E., PEDERSEN T., MAATJE J., "Real-time directional wave measurements", *Sea Technology*, pp. 10-14, February 2006.

[TEA 97] TEAGUE C.C., VESECKY J.F., FERNANDEZ D.M., "HF radar instruments, past to present", *Oceanography*, vol. 10, no. 2, pp. 40-44, 1997.

[TER 97] TERRAY E.A., GORDON R.L., BRUMLEY B.H., "Measuring wave height and direction using upward-looking ADCPs", *Proceeding of Oceans 97*, pp. 287-290, IEEE, New York, 1997.

[TRI 09] TRIZNA D.B., "A bistatic HF radar for current mapping and robust ship tracking", *Sea Technology*, pp. 29-33, 2009.

[WMO 76] WMO, Handbook on Wave Analysis and Forecasting, No. 446, WMO, Geneva, Switzerland, 1976.

[WYA 97a] WYATT L.R., "The ocean wave directional spectrum", *Oceanography*, vol. 10, no. 2, pp. 85-89, 1997.

[WYA 97b] WYATT L.R., LEDGARD L.J., ANDERSON C.W., "Maximum likelihood estimation of the directional distribution of 0.53 Hz ocean waves", *Journal of Atmospheric and Oceanic Technology*, vol. 14, pp. 591-603, 1997.

2.12.10. *Measuring turbidity and optical properties of sea water*

[ARA 97] ARAR E.J., COLLINS G.B., In vitro Determination of Chlorophyll a and Pheophytin a in Marine and Freshwater Algae by Fluorescence, Method 445.0, National Exposure Research Laboratory, Office of Research and Development, US EPA, 1997.

[BAR 78] BARTZ R., ZANNEVELD J.R.V., PAK H., "A transmissiometer for profiling and moored observation in water", *Ocean Optics*, vol. 60, pp. 102-108, 1978.

[BOS 01] BOSS E., PEGAU W.S., "Relationship of light scattering at an angle in the backward direction to the backscattering coefficient", *Applied Optics*, vol. 40, pp. 5503-5507, 2001.

[DIC 97] DICKLEY T.D., FRYE D., JANNASCH H.W., BOYLE E., KNAP A.H., "Bermuda sensor system testbed", *Sea Technology*, pp. 81-86, 1997.

[DOW 06] DOWNING J., *Twenty-five Years with OBS Sensors: The Good, the Bad and the Ugly*, Continental Shelf Research, 2006.

[EIS 93] EISMA D., *Suspended Matter in the Aquatic Environment*, Springer Verlag, 1993.

[IVA 75] IVANOFF A., *Introduction à l'Océanographie, Propriétés Physiques et Chimiques des eaux de Mer*, volumes 1 and 2, Vuibert, 1975.

[JEF 97] JEFFREY S.W., MANTOURA R.F.C., WRIGHT S.W., "Phytoplankton pigments in oceanography", *Monographs on Oceanographic Nethodology*, UNESCO, 1997.

[LEA 00] LEATHERS R.A., DOWNES T.V., DAVIS C.O., "Analysis of a point-source integrating-cavity absorption meter", *Applied Optics*, vol. 39, no. 33, pp. 6118-6127, 2000.

[LEE 03] LEE M., LEWIS M.R., "A new method for the measurement of the optical volume scattering function in the upper ocean", *Journal of Atmospheric and Oceanic Technology*, vol. 20, pp. 563-571, 2003.

[LOR 66] LORENZEN C.J., "A method for continuous measurement of in vivo chlorophyll concentrations", *Deep Sea Research*, vol. 13, pp. 223-227, 1966.

[MEI 03] MEISTER G. *et al.*, "Comparison of spectral radiance calibrations at oceanographic and atmospheric research laboratories", *Metrologia*, vol. 40, pp. S93-S96, 2003.

[MIL 01] MILES R.B., LEMPERT W.R., FORKEY J.N., "Review article: Laser Rayleigh scattering", *Measurements Science and Technology*, vol. 12, pp. R33-R51, 2001.

[MOO 09] MOORE C., BARNARD A., FIETZEK P., LEWIS M.R., SOSIK H.M., WHITE S., ZIELINSKI O., "Optical tools for ocean monitoring and research", *Ocean Science*, vol. 5, pp. 661-684, 2009.

[MOR 74] MOREL A., "Optical properties of pure water and pure sea water", in: JERLOV N.G., NIELSON E.S. (eds), *Optical Aspects of Oceanography,* pp. 1-24, 1974, http://hartree.univ.szczecin.pl/~shendu/i/Morel_OAO_74.pdf.

[PEG 02] PEGAU S., ZANEVELD J.R.V., MITCHELL B.G., MULLER J.M., KAHRU M., WIELEND J., STRAMSKA M., Ocean Optics Protocols for Satellite Ocean Color Sensor Validation, Rev. 4, Vol. IV: Inherent Optical Properties: Instruments, Characterisations, Field Measurements and Data Analysis Protocols, NASA/TM-2003-211621/rev4, Vol. IV, 2002.

[PEÑ 99] PEÑALOZA M.M.A., "Deriving the basic cell-reciprocal integrating nephelometer equation and its use for calibration purpose: a comprehensive approach", *Measurements Science and Technology*, vol. 10, pp. R1-R15, 1999.

[SUT 00] SUTHERLAND T.F., LANE P.M., AMOS C.L., DOWNING J., "The calibration of optical backscatter sensors for suspended sediment of varying darkness level", *Marine Geology*, vol. 162, pp. 58 –597, 2000.

[ZAN 94] ZANEVELD J.R.V., KITCHEN J.C., MOORE C., "The scattering error correction of reflecting-tube absorption meter", *Ocean Optics XII, Proc. SPIE*, vol. 2258, pp. 44-55, 1994.

2.12.11. *Measuring chemical parameters*

[AßM 11] AßMANN S., FRANK C., KÖRTZINGER A., "Spectrophotometric high-precision seawater pH determination for use in underway measuring systems", *Ocean Science*, vol. 7, pp. 597-607, 2011.

[AMI 04] AMINOT A., KÉROUEL R., Hydrologie des Écosystèmes Marins. Paramètres et Analyses, IFREMER, direction de l'Environnement et de l'Aménagement littoral, 2004.

[ATK 95] ATKINSON M.J., THOMAS F.I.M., LARSON N., TERRILL E., MORITA K., LIU C.C., "A micro-hole potentiostatic oxygen sensor for oceanic CTDs", *Deep Sea Research*, vol. 42, no. 5, pp. 761-771, 1995.

[BEW 04] BEWER P.G. *et al.*, "Development of a laser Raman spectrometer for deep-ocean science", *Deep Sea Research Part I*, vol. 51, pp. 739-753, 2004.

[BRI 98] BRICAUD A. *et al.*, "Variation of light absorption by suspended particles with chlorophyll *a* concentration in oceanic (case1) waters: analysis and implications for bio-optical models", *Journal of Geophysical Research*, vol. 103, no. C13, pp. 31,033-31,044, 1998.

[BUL 89] BULLISTER J.L., "Chlorofluorocarbons as time-dependent tracers in the ocean", *Oceanography*, vol. 2, no. 2, pp. 12-17, 1989.

[CAR 66] CARRITT D.E., CARPENTER J.H., "Comparison and evaluation of currently employed modifications of the Winkler method for determining dissolved oxygen water", *Journal of Marine Research*, vol. 24, no. 3, pp. 286-318, 1966.

[COP 88] COPIN-MONTEGUT C., "A new formula for the effect of temperature on the partial pressure of CO2 in seawater", *Marine Chemistry*, vol. 25, pp. 29-37, 1988.

[CUL 91] CULBERSON C.H., "Dissolved Oxygen", *WHP Operations and Methods*, Woods Hole Oceanographic Institution, Woods Hole, MA, USA, July 1991.

[DAN 95] DANIEL A., BIROT D., BLAIN S., TRÉGUER P., LEÏLDÉ B., MENUT E., "A submersible flow-injection analyser for the in-situ determination of nitrite and nitrate in coastal waters", *Marine Chemistry*, vol. 51, pp. 67-77, 1995.

[DEG 95] DEGRANDPRE M.D., HAMMAR T.R., SMITH S.P., SAYLES F.L., "In situ measurements of seawater pCO_2", *Limnology and Oceanography*, vol. 40, pp. 969-975, 1995.

[DON 97] DONEY S.C., JENKINS W.J., BULLISTER J.L., "A comparison of ocean tracer dating techniques on a meridional section in the eastern North Atlantic", *Deep Sea Research,* vol. 44, no. 4, pp. 603-626, 1997.

[DUN 02] DUNNE J.P., DEVOL A.H., EMERSON S., "The oceanic remote chemical/optical analyser (ORCA) – an autonomous moored profiler", *Journal of Atmospheric and Oceanic Technology*, vol. 19, pp.1709-1721, 2002.

[EDW 10] EDWARDS B., MURPHY D., JANZEN C., LARSON N., "Calibration, response and hysteresis in deep-sea dissolved oxygen measurements", *Journal of Atmospheric and Oceanic Technology*, vol. 27, pp. 920-931, 2010.

[GAR 92] GARCIA H.E., GORDON L.I., "Oxygen solubility in seawater: better fitting equations", *Limnology and Oceanography*, vol. 37, no. 6, pp. 1307-1312, 1992.

[GED 01] GEDDES C.D., "Optical halide sensing using fluorescence quenching: theory, simulation and applications – a review", *Measurements Science and Technology*, vol. 21, pp. R53-R88, 2001.

[GRO 12] GROUNDWATER H., TWARDOWSKI M.S., DIERSSEN H.M., SCIANDRA A., FREEMAN S.A, "Determining size distributions and composition of particles suspended in water: a new SEM-EDS protocol with validation and comparison to other methods", *Journal of Atmospheric and Oceanic Technology*, vol. 29, pp. 433-449, 2012.

[HOL 97] HOLM T.R., KELLEY W.H., SIEVERS L.F., WEBB D.L., "A comparison of ultraviolet spectroscopy with other methods for the determination of nitrate in water", *Spectroscopy*, vol. 12, pp. 38-45, 1997.

[HOO 01] HOOD E.M., MERLIVAT L., "Annual to interannual variations of pCO_2 in the northwestern Mediterranean sea: high frequency time series data from CARIOCA buoys (1995-1997)", *Journal of Marine Research*, vol. 59, pp. 113-131, 2001.

[HOR 72] HORNE R.A., *Water and Aqueous Solutions. Structure, Thermodynamics and Transport Processes*, Wiley-Interscience, 1972.

[JAL 07] JALUHSE L., LEITO I., "Model-based measurement uncertainty estimation in amperometric dissolved oxygen concentration measurement", *Measurements Science and Technology*, vol. 18, pp. 1877-1886, 2007.

[JOH 86] JOHNSON K.S., BEEHLER C.L., SAKAMOTO-ARNOLD C.M., "A submersible flow analysis system", *Analytical Chemistry Acta*, vol. 179, pp. 245-257, 1986.

[JOH 10] JOHNSON M.T., "A numerical scheme to calculate temperature and salinity dependent air-water transfer velocities for any gas", *Ocean Science*, vol. 6, pp. 913-932, 2010.

[KAR 98] KARTENSEN J., TOMCZAK M., "Age determination of mixed water masses using CFC and oxygen data", *Journal of Geophysical Research*, vol. 103, no. C9, pp. 18599-18610, 1998.

[KOP 10] KOPISKE E., MUNDERLOH K., SCHWALFENBERG S., "UV spectrometer ProPS conduct marine research field trials", *Sea Technology*, pp. 63-66, 2010.

[KRO 99] KRONFELDT H.D., SCHMIDT H., "Submersible fiber-optic sensor system for coastal monitoring", *Sea Technology*, pp. 51-55, 1999.

[LAB 04] LABASQUE T., CHAUMERY C., AMINOT A., KERGOAT G., "Spectrophotometric Winkler determination of dissolved oxygen: re- examination of critical factors and reliability", *Marine Chemistry*, vol. 88, pp. 53-60, 2004.

[LAC 07] LACOMBE M., GARÇON V., COMTAT M., ORIOL L., SUDRE J., THOURON D., LE BRIS N., PROVOST C., "Silicate determination in seawater: toward a reagentless electrochemical method", *Marine Chemistry*, vol. 106, pp. 489-497, 2007.

[LAM 06] LAMBECK P.V., "Integrated optical sensors for the chemical domain", *Measurements Science and technology*, vol. 17, pp. R93-R116, 2006.

[LIU 11] LIU X., M.C. PATSAVAS, R. H. BYRNE, "Purification and characterisation of meta-cresol purple for spectrophotometric seawater pH measurements", *Environmental Science and Technology*, vol. 45, pp. 4862-4868, 2011.

[MAR 07] MARTINI M., BUTMAN B., MICKELSON M., "Long-term performance of Aanderaa optodes and sea-bird SBE 43 dissolved-oxygen sensors bottom mounted at 32 m in Massachusetts bay", *Journal of Atmospheric and Oceanic Technology*, vol. 24, pp. 1924-1935, 2007.

[MIL 82] MILLARD R.C., "CTD calibration and data processing techniques at WHOI using the 1978 practical salinity scale", *Proceedings Int. STD Conference and Workshop*, La Jolla, Marine Technical Society, pp.1-19, 1982.

[MIL 93] MILLARD R.C., *CTD Oxygen Calibration Procedure, WHP Operations and Methods*, WHP, July 1993.

[MIL 08] MILLERO F.J., FEISTEL R., WRIGHT D.G., MCDOUGALL T.J., "The composition of standard Seawater and the definition of the reference-composition salinity scale", *Deep-Sea Research*, I, vol. 55, pp. 50-72, 2008.

[OIS 90] OISHI T., "Significant relationship between the backward scattering coefficient of sea water and the scatterance at 120°", Appl. Opt., 29, pp. 4658-4665, 1990.

[OWE 85] OWENS W.B., MILLARD R.C., "A new algorithm for CTD oxygen calibration", *Journal of Physical Oceanography*, vol. 15, pp. 621-631, 1985.

[PAW 10] PAWLOWICZ R., "A model for predicting changes in the electrical conductivity, practical salinity, and absolute salinity of seawater due to variations in relative chemical composition", *Ocean Science*, vol. 6, pp. 361–378, 2010.

[PAW 11] PAWLOWICZ R., WRIGHT D.G., and MILLERO F.J., "The effects of biogeochemical processes on oceanic conductivity/salinity/density relationships and the characterization of real seawater", *Ocean Science*, vol. 7, pp. 363–387, 2011.

[PEG 02] PEGAU S., ZANEVELD J.R.V., MITCHELL B.G., MULLER J.M., KAHRU M., WIELEND J., STRAMSKA M., Ocean optics protocols for satellite ocean color sensor validation, revision 4, volume IV: Inherent optical properties: Instruments, characterizations, field measurements and data analysis protocols, NASA/TM-2003-211621/Rev-4Vol.IV, May, 2002.

[PIE 09] PIERROT D., NEILL C., SULLIVAN K., CASTLE R., WANNINKHOF R., LÜGER H., JOHANNENSSEN T., OLSEN A., FEELY R.A., COSCA C.E., "Recommendations for autonomous underway pCO_2 measuring systems and data-reduction routines", *Deep-Sea Research II*, vol. 56, pp. 512-522, 2009.

[REI 11] REISER V., THIERRY D, GILBERT D., "Preliminary report from the Argo-Oxygen Users Workshop – May 25-26), *Argonautics* (Newsletter of the International Argo Project), 12, pp. 10, 2011.

[RIL 65] RILEY J.P., SKIRROW G., *Chemical Oceanography*, Vol. 1, Academic Press, London, New York, 1965.

[SAK 09] SAKAMOTO C.M., JOHNSON K.S., COLETTI L.J., "Improved algorithm for the computation of nitrate concentrations in seawater using an in situ ultraviolet spectrophotometer", *Limnology and Oceanography: Methods* vol. 7, pp. 132-143, 2009.

[SCH 09] SCHUSTER U., HANNIDES A., MINTOP L., KÖRTZINGER A., "Sensors and instruments for oceanic dissolved carbon measurements", *Ocean Science*, vol. 5, pp. 547-558, 2009.

[SHA 94] SHARMA A., ROGERS K.R., "Review article: biosensors", *Measurements Science and Technology*, vol. 5, pp. 461-472, 1994.

[SØR 09] SØRENSEN S.P.L., "Enzystudien. II: Mitteilung. Über die Messung und die Bedeutung der Wasserstoffionenkoncentration bei Enzymatischen Prozessen", *Biochemische Zeitschrift*, vol. 21, pp. 131-304, 1909.

[SPA 11] SPAULDING R., DEGRANDPRE M., HARRIS K., "Autonomous pH and pCO_2 measurements in marine environments. Quantifying the inorganic carbon system with in situ SAMI technology", *Sea Technology*, Feb.pp. 15-20, 2011.

[TEN 06] TENGBERG A. *et al.*, "Evaluation of a lifetime-based optode to measure oxygen in aquatic systems", *Limnology and Oceanography Methods*, vol. 4, pp. 7-17, 2006.

[UCH 08] UCHIDA H., KAWANO T., KANEKO I., FUKASAWA M., "In situ calibration of optode-based oxygen sensors", *Journal of Atmospheric and Oceanic Technology*, vol. 25, pp. 2271-2281, 2008.

[URB 09] URBAN G.A., "Micro- and nanobiosensors – state of the art and trends", *Measurements Science and Technology*, vol. 20, 18, pp. 1-18, 2009.

[VAL 04] VALEUR B., *Invitation à la Fluorescence Moléculaire*, De Boeck, 2004.

[VUI 09] VUILLEMIN R., LE ROUEX D., DORVAl P., BUCAS K., SUDREAU J.P., HAMON M., LE GALL C., SARADIN P.M., "CHEMINI : A New in situ Chemical MINIaturized Analyser", *Deep Sea Research*, I, vol. 56, pp. 1391-1399, 2009.

[WAL 64] WALFAREN G.E., "Raman spectral studies of water structure", *The Journal of Chemical Physics*, vol. 40, pp. 3249-3256, 1964.

[WAN 99] WANG W., REIMERS C.E., WAINRIGHT S.C., SHAHRIARI M., MORRIS M.J., "Applying fiber-optic sensors for monitoring dissolved oxygen", *Sea Technology*, pp. 69-74, March 1999.

[WEI 70] WEISS R.F., "The solubility of nitrogen, oxygen and argon in water and seawater", *Deep Sea Research*, vol. 17, pp. 721-735, 1970.

[WHI 06] WHITE S.N., BREWER P.G., KIRKWOOD B.J., "Raman instrumentation for deep-sea in-situ analyses", *Sea Technology*, pp. 17-24, 2006.

[WUI 09] WUILLEMIN R., LE ROUX D., DORVAL P., BUCAS K., SUDREAU J.P., HAMON M., LE GALL C., SARRADIN P.M., "CHEMINI: a new in situ chemical miniaturized analyser", *Deep-Sea Research I*, vol. 56, pp. 1391-1399, 2009.

Chapter 3

Measurements at Sea

3.1. Oceanographic vessels

Oceanographic vessels are the essential elements of the system necessary to implement *in situ* measurement studies. Their sizes can vary from 10 m for the smaller units for coastal exploration to a 100 m for those designed for offshore campaigns. Generally, these boats are distinguishable:

– by the means at their disposal to drop or lift the measurement instruments into or out of the environment to be explored;

– by their means of positioning and eventually probing or their surface measurement system;

– by the means for the storage and transmission of information acquired;

– by the laboratories and work zones provided; and

– by the carriers or means for oceanographic measurements they are equipped with.

3.1.1. *Ways of launching instruments into the water*

The weight and volume of instruments to be placed into the water can vary from several kilos and several tons of cubic centimeters to several tons and several cubic meters. For heavier and more cumbersome loads, oceanographic vessels are equipped with loading cranes, *davits* or rotating cranes and container cranes (see Figure 3.1). These cranes can be moved by hydraulic jacks into and out of the boat.

Figure 3.1. *Picture of the construction of the Hydro-Oceanographic Beautemps Beaupré with its deck machinery: (1): rear rotating gantry; (2, 3): location of the winches; (4): hydraulic swivel crane; (5, 6): lateral gantries used for launching measurement equipment into the water; (7): rotating davit; (8): floating anchor of a Lagrangian buoy; (9): corer; (10): CTD profiler on a multibottle sampling array; (11): swath of the multibeam echo-sounder; (12, 13): towed fish (magnetometer, echo sounder); (14): surface buoy (Courtesy of © SHOM)*

Thus they allow the deposit of instruments at a distance far enough from the hull to avoid the boat being struck when the sea is rough. They are located at the rear and aligned with the axis of the vessel, or are on one of the sides, which are called *lateral gantries*. They are equipped with a pulley that allows the cables connecting the instruments to be guided. These cables are generally unwound from a winch mounted onto the deck. They can simply be carriers but are often electro-carriers. Their structure is then formed by a sheath composed of strands of steel and a center of copper conductors isolated from one another (see Figure 3.2).

A sealed connection is at the side of the instrument in order to link the electric conductors to a connector. To make this connection, a "sock" made of braided metallic wires is slipped over the cable in order to allow a "recovery effort" – where the traction force protects the connection from pulling efforts. This "sock" is equipped with a fixed ring to ensure the mechanical link. Any traction on this ring will tighten the stitches of the "sock" and so increase the squeezing strength on the cable.

Electrical winches used for reeling cables are equipped with devices that allow regular winding and unwinding. The winches and cables are often dedicated to particular instruments (CTD profilers, CTD velocity meters, magnetometer, etc.) Servo-control devices have been devised (but not exploited) in order to synchronize cable unwinding speed with platform movement. It is important to know that when making temperature–conductivity profiles with an instrument, such as the CTD profiler, that the acceleration and deceleration the instrument is subjected to by the boat's rolling or pitching movements can cause measurement errors, particularly when crossing zones with strong vertical temperature gradients. These errors are attributable to sensor response times, the thermal inertia of the instrument and the stirring it creates when it makes pounding movements. They are difficult to estimate, but it has been shown that they can reach 0.1°C in temperature in certain oceanic zones if no precautions are taken.

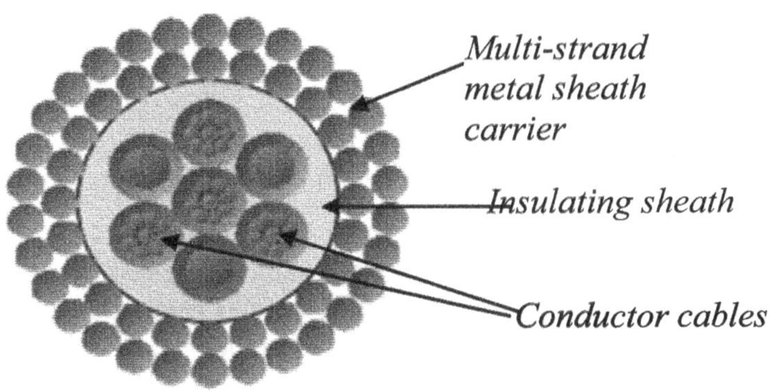

Figure 3.2. *Structure of an electro-carrier cable (Courtesy of: Macartney A/S, Denmark)*

3.1.2. *Ways of positioning and probing*

When the oceanographic instruments are launched into the water, it is important to know the depth of the sea bed and the position of the carrier. To determine the depth, it is not always sufficient to rely on marine charts. Boats are equipped with shallow echo sounders that allow them to move close to shores or to establish sea bed maps if their mission includes a hydrographic part, and possibly deep sea echo sounders enabling the production of offshore maps. These *echo sounders* can be *mono-beam*, where the transducer placed beneath the hull measures the distance between the intersection of the emission cone of an acoustic wave and the sea bed. This wave is formed using a matrix of piezoelectric transducers (see section 2.3.2) that make up the acoustic antenna. They can also be *multi-beam* (see Figure 3.1), where the transducer produces a transverse scan of the sea bed and quickly establishes a map of zones crossed.

3.1.2.1. *Multi-beam echo sounders*

Multi-beam echo sounders, or MBESs, generally consist of an acoustic emission antenna placed on the axis of the boat and a receiving antenna placed perpendicularly to the axis of the boat, which picks up echoes coming from multiple directions. These antennae are made of piezoelectric ceramic cylinders assembled in a matrix. If d is the distance between two cylinders and θ is the angle of return of the echo, the difference Δ of the acoustic path between these two transducers will be:

$$\Delta = d \sin(\theta) \quad [3.1]$$

If sound travels at speed c, this corresponds to a delay between two transducers of:

$$\Delta t = d \sin(\theta) / c \quad [3.2]$$

For the nth transducer in the same row:

$$\Delta t_n = (n - 1)\, d \sin(\theta) / c \quad [3.3]$$

Knowing these delays, if we sum up the values issued from n transducers, it is possible to find the signal corresponding to the opening angle θ. This is what we call *beamforming*. This technique enables a cross-scan that is transversal to the axis of displacement of the boat. It requires a good knowledge of the propagation speed of sound at the moment of measurement. The distance relative to the sea bed can be calculated for angles corresponding to the angular resolution of the echo sounder either by detecting the amplitude of the echo (this being larger when the wave reflects off the sea bed) if the speed of sound in the column of water is known, or by detection of the phase variation between two transducers that are close to each other.

This variation can be distinguished from echoes coming from the sea bed. Phase detection is used when the sea bed echo is too large to allow precise position measurement.

Deep sea echo sounders (50 to 12,000 m) work at frequencies between 10 and 30 kHz, medium depth echo sounders (5 to 1,000 m) work at frequencies between 90 and 100 kHz and shallow sea echo sounders (1 to 100 m) work at frequencies between 300 and 400 kHz. For example, the echo sounder EM3002 of the Konsberg Society works at 300 kHz and has a maximum aperture of ±65°, made with 160 transducers with an aperture of 1.5° × 1.5°. The width of the swath that it realizes, can be 4.3 times the depth.

Whatever the technology used, these transducers are positioned with great accuracy relative to the axes of the vessel, and measurements are taken to determine, and then to correct, the "lever arms" and the angular errors from this positioning by using software. The tolerances of the vessel's reference coordinates can be ±5 mm and ±0.005° on the angles of the axes. The echo sounders can be positioned with heading, pitch and roll angular errors smaller than ±0.025°. The distance between the echo sounder and the surface is also determined with accuracy.

An algorithm for signal processing based on the sonar equation [2.145], allows the echoes from waves reflecting off the sea bed to be extracted from the noise. It calculates propagation times and deducts from them the distance relative to the sea bed, knowing the propagation speed of sound in the layers of water crossings. The MBES feature a clock, synchronized with global positioning system (GPS) time (section 2.7.2), which dates the *probes* or pulses returns and data from remote sensors. Synchronization of this information is hugely important for the correction of heading, pitch, roll, position, heave or velocity sensor biases. It can be checked by passing over an obstacle twice, at two different speeds v_1 and v_2, with the same heading and the same beam apertures. The time difference dt introduced by synchronization error causes a difference in position dx, if it is not zero. This can be calculated as follows:

$$dt = dx / (v_2 - v_1) \qquad [3.4]$$

A hull velocity meter is associated with the multi-beam echo sounders. These velocity meters, whose operation is described in section 2.4, allows the direct measurement of the propagation speed of sound at the transducer level and hence enables the beam forming previously discussed. These meters are sometimes replaced, or even complemented, by thermosalinometers (section 2.1.1) and temperature sensors fixed to the hull at the point of water entry for the thermosalinometers. These allow the calculation of surface or shallow water salinity. These data, which are associated with temperature measurement, provide knowledge of the local speed of sound. Use of this device has the advantage of providing oceanographers with temperature and surface salinity data that will complete measurements made at depth. Bearing in mind the multiple refractions, due to the temperature differences and water salinity layers crossed that the acoustic wave is subjected to during its propagation, in order to evaluate the height of the water with accuracy it is necessary to establish velocity profiles. These profiles are obtained with the help of velocity profilers (section 2.4.2) or from XBT probes or XCTD (section 2.1.1.3). They allow the acoustic path to be encountered.

The water heights measured are then referenced to hydrographic zero (see section 2.8). The International Hydrographic Organization (IHO) published a standard called *Special Publication No. 44*, which specifies the measurement uncertainties tolerated for water height measurements, according to hydrographed depths. These specifications are classified according to orders (exclusive, special, 1, 2, 3 or 4), the exclusive order being the most demanding and the fourth order corresponding to the greatest measurable depths.

3.1.2.2. *Ways to position and biases*

Hydrographic probing must be associated with positioning data in order to geographically identify mapped areas. Different means were used to achieve this until the advent of the GPS global positioning system. The freedom of access to this system, the disappearance of "s.a." (section 2.7.2) and the accuracy it provides has led the older means to fall out of use, except perhaps for the Loran C system (section 2.7.2) which continues to be used as a backup system in case GPS fails. The vessels are hence equipped with GPS antennae and receivers that allow them to position themselves in real time at geographical landmarks. These receivers are usually associated with an *inertial navigation or positioning system* whose function is to measure the boat heading, pitch and roll movements with high accuracy (see Table 3.1) relative to a 'still point' on the boat that corresponds to its center of gravity.

There are two inertial navigation system technologies. The oldest technology is composed of a platform on gimbals that are controlled by using gyroscopes to find the horizontal. The more recent *strap-down systems* are secured to the structure of the carrier. The horizontal is found using algorithms supplied with data from laser gyrometers. Some of these pieces of equipment perform the hybridization or fusion of data and are called *fused inertial navigation systems*. They are equipped with three laser gyrometers that allow the calculation of carrier rotation relative to the rotation of the Earth and three accelerometers (see section 2.9.2.1) that, by derivative calculations, allow us to find the speed and displacement within the three axes. The GPS provides position data with low accuracy and distance data with high accuracy. The combination of these two instruments and filtering adapted to the data

(Kalman filters) allows the acquisition of hybridized data at any time where the GPS errors are detected. Positioning is then permanently and continuously available. The PHINS system (fused inertial navigation system) produced by iXBlue Society is an example of a hybridization system (see Table 3.1).

Heading data accuracy can be improved if we have two GPS antennae to carry out differential positioning (see sections 2.7.2.1 and 2.7.2.3). The measurement of phase difference generated by these antennae – associated with heading, pitch and roll data provided by the inertial navigation system – enables an accuracy of 0.01° over this quantity, whereas it is usually 0.25°. However, this receiving system is sensitive to multiple reflections, which create multi-paths. To guard against this, the antennae of "*shock-ring*" technology should be used.

Parameters	**Accuracy**	**Measuring range**
Dynamic positioning	0.01°	Up to 260°/s
Pitch	0.01°	± 180°
Roll and yaw	10 cm or 10% of the read value	± 180°
Heading	0.01° (with GPS diff.)	± 180°
Position	5–15 m in GPS mode, 0.5–3.0 m in DGPS mode	-

Table 3.1. *Metrological characteristics of the M-phins position inertial navigation system produced by iXBlue society*

The calibration of these systems remains delicate. It can be done in the laboratory using vibration exciters (for adjusting accelerometers) or rotation measurement systems (for gyrometers), but these require the inertial navigation system to be dismantled from the hull of the boat.

There are other techniques that can be used to measure heading, pitch and roll bias and to globally correct for echo sounder data.

The bias on the roll is generally characterized by an error in the vertical positioning of deep measurements or "probes". This bias increases with water height. To detect biases, the boat must make two passes over a sea bed known to be flat, one passage being perpendicular to the other (opposite heading). The passes must be made at the same speed and with a large beam aperture. Profiles returned by the echo sounder must be flat if the roll is well measured and well compensated, and their point-to-point difference must be zero. Error increases with increasing echo sounder opening angle θ (see Figure 3.3), so the difference between the two profiles generally results in non-zero values at the extremities of the range being scanned.

Similarly, biases from pitch can be detected and corrected for by making two passes in opposite directions over an obstacle or a slope at constant speed using a small beam aperture. They are essentially characterized by errors in positioning according to the x axis (see Figure 3.3), which increase with rising water level. Thus, if the pitch biases are badly corrected, the obstacle will be positioned with an error δx in one direction and an error δx in the other, to make $2\delta x$.

The heading biases can be detected by passing over the same obstacle twice on each side, but with the same heading, constant speed and high beam aperture. These biases result in a greater error on the x axis. If the "misalignment" in the heading is $\delta\alpha$, the error will be $\delta x = z\ \delta\alpha\ tan(\theta)$, and will result in an obstacle positioning error of $2\delta x$.

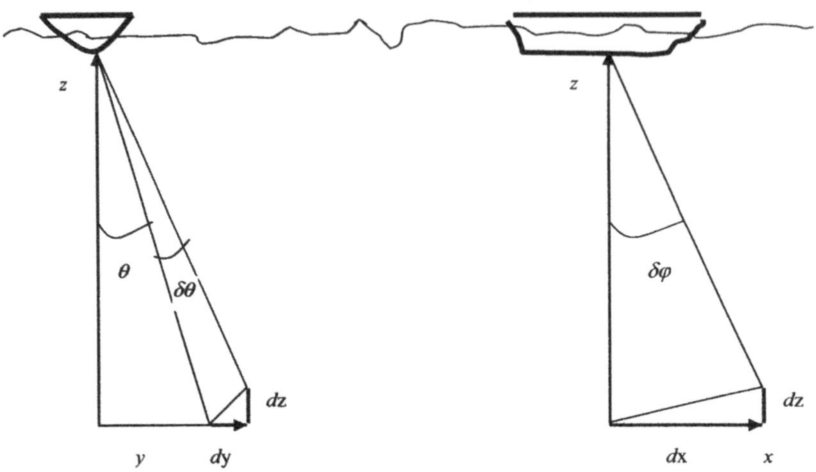

Figure 3.3. *a) Illustration of the bias in roll. For a beam aperture of θ, if the boat rolls with an angle of $\delta\theta$, we will get an error dz on vertical positioning and dy on the horizontal positioning. dz is proportional to water height z: $dz = z\ \delta\theta\ tan(\theta_i)$. b) Similarly, if the boat rocks at an angle $\delta\varphi$, positioning will essentially be erroneous on the x axis. We will have: $dx \approx z\ \delta\varphi$ in this instance*

As heading, roll and pitch are not independent quantities, an error in one will affect the others, even if the others have been corrected for. In order to verify the corrections for bias, certain agencies immerse targets of known dimensions at known positions, and pass over these targets to qualify their echo sounders in order to match the requirements of the IHO *Special Publication No 44*.

3.1.2.3. *Lateral sonars*

When work is necessary in shallow waters (depth <100 m) in order to detect objects or recognize submerged shapes, boats tow instruments called *lateral sonars*. These instruments, which use transducers working at frequencies from 450 to 900 kHz, carry out the imaging. They are light in air (15–30 kg) and are small (with a length close to 1.30 m and a height close to 20 cm), so they can be easily submerged from a small boat and towed by a similar boat. Software provides real-time sea bed imaging. Lateral sonars are not strictly speaking measurement instruments, as calibration of the images obtained is delicate, but they are part of the panoply of instruments regularly deployed to analyze the sea bed.

3.1.3. *Ways to transmit data*

Measured data is transmitted from a submerged instrument or from a fixed post on the boat to one or several computers for pre-treatment or for treatment and storage. The data are then optionally sent to the ground via satellite links. On more modern boats, these computers communicate by networks such as the Ethernet.

The data acquired *in situ* must be dated and positioned. For this it is necessary to be able to combine information from different devices. Transmission of this information is often done using another type of network called *NMEA 0183* and produced by the *National Marine Electronics Association*. In the 1980s Robert Bosh GmbH and Intel Society sought to develop a network communication series that offered a reliable and inexpensive solution for automobile manufacturers. From this concept, the standard NMEA 0183 was developed to interconnect equipment on board boats and to exchange data. It is an asynchronous network that works at a speed of 4,800 bauds by carrier detection. It does not have a central controller. On one bus, only one piece of equipment can be a transmitter. The other equipment can only be receivers. The network has multiple access points and contains a mode to arbitrate collisions. Its architecture, based on a *controller area network (CAN)* type communication protocol, was normalized by the *International Standards Organization/Open Systems Interconnect (ISO/OSI)*. The CAN protocol can be implanted on most micro-controllers on the market, such as Intel's AN8051 or AN87C196 and Motorola's MC68HC05 or MC68HC705. It is described in the ISO-11898 manual, but this description is also free to access on the Internet. In principle, when a message needs to be transmitted by an instrument, it allows us to listen to the bus (by carrier detection) and to transmit the message if the line is not busy (see the message format in Figure 3.4).

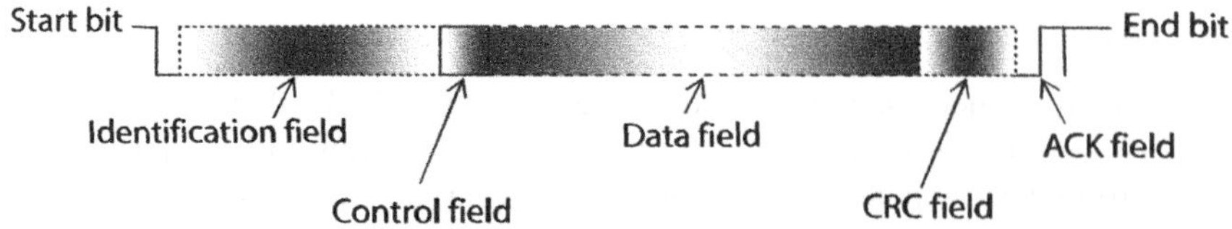

Figure 3.4. *Format of an NMEA data message. This message consists of a start bit, an identification field that is 29 bits long in the standard NMEA 2000, a control field showing which CAN format is used, a data field of varying length, a cyclic redundancy check (CRC) field, a reception control field (ACK) and an end bit*

The instruments with GPS type receivers or inertial navigation systems are equipped with interfaces and outputs with a NMEA 0183 format. Oceanographic instruments such as CTD profilers, thermosalinometers, etc., are equipped with interfaces that allow the reception of NMEA 0183 messages. The pieces of software incorporated in these instruments are designed to read, interpret and associate position data with physical measured data. For example, in the case of an interface with a geographic position system (GLL), the data field is symbolized in the following format:

$--GLL,lll.l, a, yyyyy.yy,b, hhmmss.ss,A*hh<CR><LF>

In this message:

– "lll.l" is a number of the form "degree | minute.fraction of a minute in decimals" that codes for latitude;

– "a" is a simple character indicating north (a = N) or south (a = S);

– "yyyyy.yy" is a number of the form "degree | minute.fraction of a minute in decimals" that codes for longitude;

– "b" is a simple character indicating east (b = E) or west (b = W);

– "hhmmss.ss" is a number of the form "hour | minute | second.fraction of a second in decimals" that codes for time;

– "A" is a simple field character indicating whether the data are valid (A = Y) or invalid (A = N);

– "*" is an optional checksum delimiter;

– “hh” is an optional checksum field; and

– “<CR><LF>” are the return and return new line characters, which are useful when printing or storing messages.

Once dated and located, some data must be rapidly transmitted to land to provide for bases or for meteo-oceanographic prediction models. There are, at present, numerous data transmission systems (see section 2.7.2.7), but historically the most common system on boats is the *INternational MARitime SATellite organization (INMARSAT)*. As suggested by its name, the INMARSAT implements a satellite transmission system. This system, which has been running since 1982, was initially developed to provide a powerful way to convey distress calls, safety messages and to manage some boats. In the case of oceanographic vessels, these messages can be used to transmit data.

Each INMARSAT signatory country (84 to date) contributes to its operation by providing the interface with a terrestrial telecommunication network. It is accessible at sea but also on the ground and in the air, thanks to four active geostationary satellites (and seven emergency ones), each of which covers an oceanographic region of the globe. In each country, numerous INMARSAT partner groups operate between satellite receiver stations and terrestrial networks.

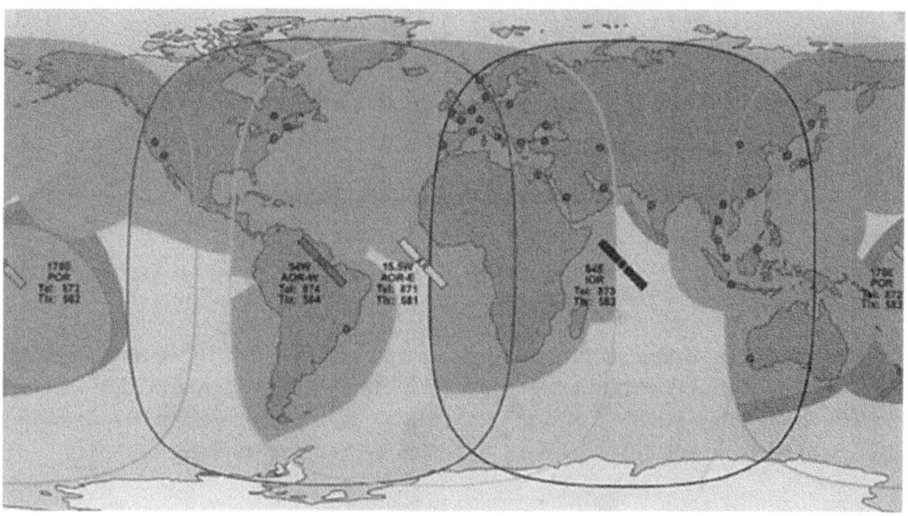

Figure 3.5. *Zones covered by INMARSAT-C network satellites (Courtesy of France Telecom Mobile Satellite Communication SA)*

Figure 3.6. *INMARSAT radome antenna (Courtesy of © SHOM)*

To be able to transmit, vessels must be equipped with:

– a main control unit (MCU) to which telephone handsets can be linked;

– a router for the transfer of digital data;

– an NMEA interface box providing heading and position data;

– optionally, a printer; and

– an INMARSAT antenna.

This can be a simple conic omni-directional antenna but if we want optimal transmission reliability regardless of the state of the sea, this antenna will in fact be a radome containing a parabola and a system allowing automatic orientation towards a satellite to perform automatic tracking after linking by "self-tracking".

3.1.4. *Ways to take oceanographic measurements by boat*

In terms or oceanographic measurement, boats are equipped with the instruments described in section 2.1.1: CTD profilers, multi-bottle sampling array, expendable bathythermographs or hull thermosalinometers. Thermosalinometers are fixed instruments because they require work to be planned for their installation; whereas as the placing of other instruments depends on the study requirements.

CTD profilers and expendable thermographs permit temperature–conductivity profiles to be created at fixed points, over an entire column of water. The production of a CTD profile can require several hours (depending on the depth to be explored). Exploration of a region requires "*radials*" to be made, with consecutive fixed points in a same alignment so that field sections of temperature or salinity can be acquired. To more rapidly carry out this sectioning, with eventually degraded uncertainty measurements, boats can sometimes be equipped with *tow fish* on which scientific instruments are installed.

These fish are in the shape of towed submarine "gliders" and are controlled from the boat using an electro-carrier cable. This cable contains command lines made to operate the fish so that it ascends and descends as the boat moves, and it contains lines that allow the acquisition of data measured. The profile obtained is a wave starting from a few meters below the surface to a depth that can reach 500 m, depending on the settings made.

The wave is regulated by a control unit installed on the boat. This control unit contains algorithms and actuators. The algorithm calculates the distance between fixed set pressures that are dependent on the desired profile and values provided by the pressure sensor located in the fish. When the fish approaches the minimal and maximal pressures to be achieved, the algorithm commands the actuator to modify the inclination of the rudder to allow the fish to begin an ascent or descent. To control the towing effort which the cable is subjected to, particularly during inclination changes, it is equipped with a *force sensor* (a load cell equipped with strain gauges). In order for the rate of change depth to be maximal, it is necessary to streamline the cable with flexible fins which modify its buoyancy.

The SeaSoar fish produced by the Chelsea Technologies Group, can reach speeds of 3 m/s with a faired cable and peaks at 1 m/s without fairing. Similarly, the maximum depth of 500 m peaks at 100 m without fairing can be obtained. It is possible to install different scientific instruments on this fish, including a CTD profiler. The body of the CTD profiler is fixed in the fish; whereas its sensors are installed on the flanks. Adaptations are required to reduce the effects of speed and its variations on the sensor's temperature and conductivity response time and to remove the artifact introduced in salinity profiles (sections 2.1.4 and 2.2.4). Other sensors are often interfaced with the CTD profiler and fixed on the fish. These sensors are instruments allowing the study of optical properties in the first layers of water, which are rich in plankton and particles of all sorts. Fluorimeters, transmissometers, bioluminescence or irradiance sensors are then installed (section 2.10).

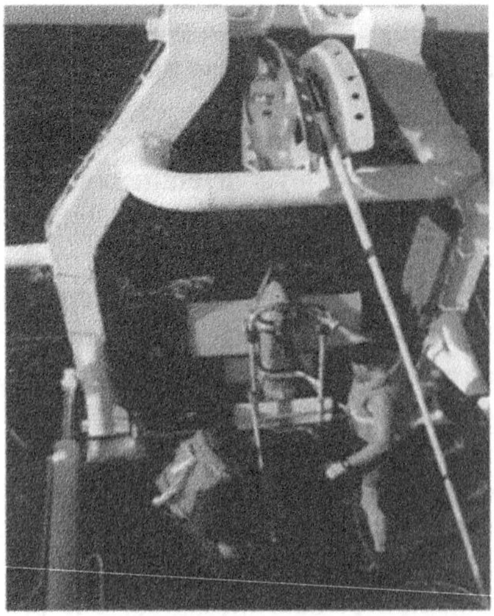

Figure 3.7. *Launching of a SeaSoar-type tow fish via the stern gantry of an oceanographic boat (Courtesy of © SHOM)*

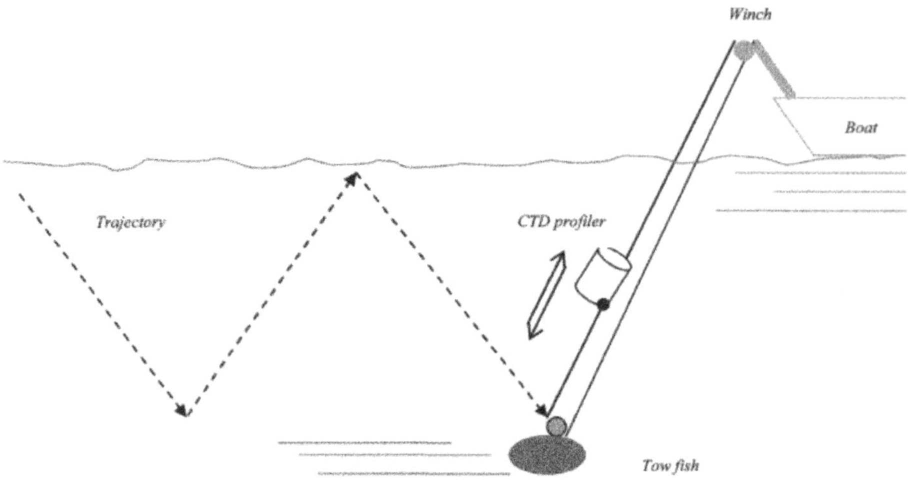

Figure 3.8. *Illustration of the operation of a Hi-Re-TOS tow fish system. The CTD profiler is secured to a cable that is wound and unwound using a winch. The trajectory obtained forms a regular oscillation that allows the measurement of the bathythermic profile in the first layers of the ocean*

There are other processes that can be used to create profiles during the progress of the boats. These processes attempt to measure the fine structures in the superficial layers of the ocean. We can cite, for example, the Hi-Re-TOS or *high-resolution towed oscillating system* put in place by NURC (previously Saclantcen) from Spezzia in Italy in the 1980s. The fish is then ballasted and provided with an assembly of pulleys allowing it to be towed and for the measuring instrument to be raised and lowered using a cable wound or unwound by a winch fixed to the stern gantry (see Figure 3.7). This system attains profiles of the same type as those made by SeaSoar. Its maximum depth is between 20 m and 300 m for a boat moving forward at a speed of 3 knots.

3.2. Moorings

When seeking to understand the evolution of a parameter over long time periods, for matters of cost but also due to meteorological conditions or the state of the sea, it is not possible to take measurements from an oceanographic vessel. We must resort to leaving the instruments submerged or having them at the surface during periods of measurement. This is only possible with the manufacture of *moorings*. A mooring is an instrumented line designed to maintain a fixed position in a marine environment. There are *Eulerian* moorings, referenced to Euler reference points that are fixed relative to land, where the connected instruments take measurements from fixed points in a moving environment; and *Langrangian* moorings, referenced to Lagrange mobile points where the observation instrument is "anchored" in the moving water mass and follows its movement. Eulerian and Lagrangian moorings can also be distinguished by the fact that they have a surface floating system, called surface moorings, or by the fact that they are entirely underwater. They are then called *subsurface* moorings.

3.2.1. *Constraints of mooring implementation*

The deployment of a mooring requires preliminary study of the environment at the site chosen, the local legislation and the type of instrument that will be used. The study of the environment involves the physical parameters of the deployment area. Knowing these elements will allow us to size certain components and prevent a number of problems, such as those linked to the swinging radius, verticality or bad anchoring. To do this it is necessary to know or to be able to estimate the density of the medium, the characteristics of the current, the tide and the nature of the sea bed (bathymetry) or the meteorology predicted for the time of launch into the water. Although marine charts are an essential tool when determining the location for a deployment, they are not themselves sufficient for this. It is often useful to conduct a finer bathymetric survey in order to find out whether there are any natural or artificial obstacles (wrecks, underwater cables, etc.) that the sea bed may contain, the exact nature of the sea bed (rocks, mud, etc.) or the slope of the terrain.

The preliminary study also involves the biological and human environment. Coastal areas are often rich in nutrients that promote phytoplankton development and biological chains. These allow other organisms to reproduce and settle on the moorings, thereby affecting their reliability. Diverse techniques that are part of the "art" of the design of moorings must therefore be implemented to limit the deposits of living organisms and hinder the development of this biological chain, which has the additional drawback of accelerating corrosion and hence the risk of rupture. These zones are also favorable for the development of human activities linked to fishing, which implicate the use of trawlers or dredgers and the risk of losing a mooring. Similarly, channels that have heavy maritime traffic must be avoided, as it is difficult and dangerous to carry out mooring operations or lifting and the moorings themselves may be hazardous to navigation.

When mooring zones are close to the coast, it is necessary to get agreement from riparian states or the maritime authority responsible for the actions of the state or country at sea. A request to carry out work must be made and exercise zones called *ZONEX* must be reserved. Depending on the type of mooring, the inclusion of a beacon must also be planned to avoid collisions. This security can consist of *beacon transmitters* or *flash lamps*, but it is also necessary to disseminate information to navigators regarding the dimensional characteristics of the installation and its exact position through official notices, urgent notices to mariners (*AVURNAV*) or warnings by zone (*NAVAREA* messages). Security can be ensured through the establishment of Argos transmitters or GPS receivers that allow slightly deferred time tracking from the mooring position. The Argos transmitters can also be used to control slipping or accidental rise to the surface subsequent to rupture or dredging.

Finally, instruments installed on the mooring line must satisfy different constraints. If they cannot be clamped onto the cable, they must be designed to be resistant to traction. The submersion time may be several months or years, so they must have autonomous energy and sufficient memory capacity. They must above all be protected against corrosion to limit the risks of rupture or irretrievable deterioration.

3.2.2. *Generalities on the implementation of moorings*

The design of oceanographic *moorings* essentially rests on empirical laws that call upon a great deal of scientific and technological knowledge. The development of offshore techniques has led to the creation of simulation software that is widely used today. Software are based on the implementation of equations described next.

3.2.2.1. *Principles of buoyancy calculations*

All bodies immersed in water are subjected to the famous Archimedes principle: when an object is immersed in a fluid, it pushes an amount of fluid equal to its volume out of the way. This pushing force $\vec{P}_A$ is applied at a point called the center of pushing and is directed upwards:

$$\vec{P}_A = -\rho V \vec{g} \quad [3.5]$$

Here ρ is the density of the fluid, V is the volume of the body and $\vec{g}$ is gravity. In the case of sea water, the value of ρ is determined by relationships [1.64] to [1.74] or by calculation modules from the TEOS-10 (see section 1.3).

The weight $\vec{P}$ of the body of mass M opposes this force. It is applied to the center of gravity and is directed downwards depending on the well-known relationship:

$$\vec{P} = M \vec{g} \quad [3.6]$$

Supposing that the centers of gravity are identical, the resulting amplitude of the two forces gives the weight $\vec{P}_{water}$ of the body in water, which is equivalent to its carrying capacity:

$$\vec{P}_{water} = (\rho V - M)\, \vec{g} \quad [3.7]$$

In a static marine environment, the density of sea water has a general tendency to increase from top to bottom (although exceptional cases do exist), so objects will position themselves in such a manner that:

– if $P > P_A$, they will settle on the sea bed as they have negative buoyancy;

– if $P_A > P$, they will stay at the surface as they have positive buoyancy; and

– if $P = P_A$, they will place themselves in hydrostatic equilibrium at an estimable depth where buoyancy is zero.

On this basis, to make mooring lines it is therefore possible to design:

– weights that will be placed on the sea bed to allow a set of objects to remain at a fixed point; and

– buoys that exert an upward traction force to keep the assembly in a vertical position.

Shape	***Orientation***	***Reynolds number R_e***	***Drag coefficient C_t***
Spherical	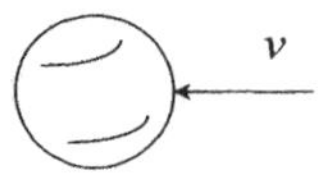	10^3 to $1x10^5$	0.50
		$3\ x10^5$	0.20
Convex hemisphere	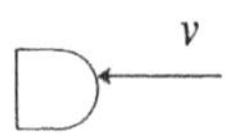	$> 10^3$	0.34
Cylinder perpendicular to current	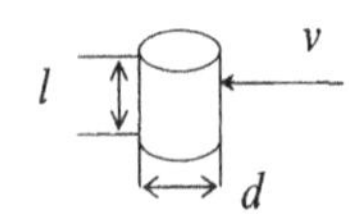	10^5	0.63 if $l/d = 1$ 0.74 if $l/d = 5$ 0.82 if $l/d = 10$ 0.90 if $l/d = 20$
Rectangle perpendicular to current	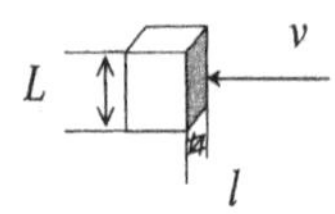	$> 10^3$	1.16 if $L/l = 1$ 1.20 if $L/l = 5$ 1.50 if $L/l = 10$ 1.90 if $L/l = \infty$
Hydrodynamic profile		> 200,000	0.06–0.1

Table 3.2. *Examples of drag coefficient values. For the definition of Reynolds number, see Chapter 2 (values extracted from M. H. Berteaux, "Coastal and oceanic buoy engineering")*

The ocean is not a static environment; other elements influence the maintenance of the system. The presence of marine currents causes displacement forces $\vec{F}_d$ for which direction is arbitrary. In fact, $\vec{F}_d$ can be decomposed into a vertical lifting force and two perpendicular horizontal forces that constitute the drifting force and the drag force $\vec{F}_t$ for which the direction is the same as that of the current. In order for the mooring to remain immobile, an equal and opposite force must be applied to $\vec{F}_d$. In the case of oceanographic moorings, however, it is the force $\vec{F}_t$ that is dominant. Its study shows that it can be split into two other forces:

– that of pressure drag $\vec{F}_{tp}$, which is due to pressure exerted by the fluid onto the mooring front and a slight depression created at its rear; and

– that of rubbing or friction $\vec{F}_{tf}$, which follows from the friction of fluid on the surface of the mooring elements.

The amplitude of $\vec{F}_{tf}$ depends on the submerged or moored surface S_m of these elements. This can be estimated from relationship [3.8] where $\frac{\partial z}{\partial t}$ is the amplitude of the speed of the fluid and C_f is the coefficient of friction of the objects.

$$\vec{F}_{tf} = \frac{1}{2}\rho C_f S_m \left|\frac{\partial z}{\partial t}\right| \overrightarrow{\frac{\partial z}{\partial t}} \qquad [3.8]$$

The amplitude of $\vec{F}_{tp}$ involves the coupling of the surface S_c between mooring elements and the direction of current, and the pressure *drag coefficient* C_d of these elements.

$$\vec{F}_{tp} = \frac{1}{2}\rho C_d S_c \left|\frac{\partial z}{\partial t}\right| \overrightarrow{\frac{\partial z}{\partial t}} \qquad [3.9]$$

There are several books that allow us to determine values for C_f and C_d. These values are dependent on shape, the surface dimensions of the objects perpendicular to the current (see Table 3.2) and on the Reynolds number (see relationship [2.41] for the definition of this number). It should be noted, however, that generally in the case of oceanographic moorings, the force $\vec{F}_{tp}$ is large compared with $\vec{F}_{tf}$, to the point that only $\vec{F}_{tp}$ is taken into account in calculations. Uncertainties exist on the values that can be attributed to C_d, causing large errors in the dynamic behavior model of submerged objects; Chan and Kang developed a simple experimental system to measure C_d [CHA 11]. The object is submerged in a pool and connected by a cable assembly including pulleys with suspended mass. The equality between F_{tp} and the force of gravity on the mass allows the measurement of F_{tp} and from it the deduction of the value of C_d.

Knowing the submersion depth of the measurement instruments, their geometric characteristics and their orientation relative to the dominant current, the calculation of the different forces for each submerged object will allow components whose role will be to anchor or ensure the buoyancy of the assembly to be sized. This sizing can be determined in static with a manual calculation, starting from a point where efforts are known and then propagating calculation by iteration. To address problems in dynamics, however, we must pass through a model based on the principle of finite elements. There are software programs that carry out this type of modeling: Moordyn, CABLE (created by WHOI), Deepline and ORCAflex have been developed especially for offshore applications.

For more expensive or large volume moorings, it is also possible to create models and test their behavior in *trial pools*. These pools are generally equipped with wave generators that create periodic waves with a specific amplitude. If the model is equipped with strain gauges, it is possible to measure the effort induced by the movements of the medium.

3.2.2.2. *The different types of moorings and anchored buoys*

Depending on the nature of the observations to be made, it is possible to design different types of moorings. The type will depend on the measurement principle (Eulerian or Langrangian) and the necessary anchoring method. The simplest to design is probably "*sea bed*" *mooring* (see Figures 2.64 and 3.9). It consists of a cage in or onto which measurement instruments (current meters, tide meters, etc.) are fixed, which must be installed close to the bottom. It is the only type of mooring that guarantees total immobility of the instruments. The cage is kept at depth by using a weight that is fastened or can be released. In this second case, the release system offers the possibility of activation through the use of acoustic pulses emitted from the surface to dissociate the 'dead body' part of the part of the instrumented cage.

The rise of this cage can also be facilitated by the addition of buoys at the side or suspended above. These systems, having to stay on the sea bed for long periods of time, can be subjected to dredging, depending on where they are deposited. To decrease this risk, certain "anti-trawling" cage shapes have been designed. To limit its height and allow its recovery, the weights and buoys are then replaced by automatic inflation systems that modify the cage's buoyancy and allow it to return to the surface through the use of simple acoustic pulses.

Acoustic releases are instruments that couple electronic and mechanical action. One end is integrally fixed to the part of the mooring to be recovered, while the other end consists of a hook with an opening that is controlled by a motor activated by the receipt of a train of ultrasonic pulses, emitted from the surface by a remote control. The gradual opening of the hook releases a submerged part of the assembly that can then rise to the surface and be recovered. These are key elements of mooring because their non-function causes permanent loss of the equipment and measurements.

Figure 3.9. *Example of a cage used for moorings placed on the sea bed. In the foreground, we can see a Nortek AquaPro Doppler current meter; in the middle, an acoustic release; and at the bottom buoys to allow the cage to rise when the release system is activated and a concrete weight held by the release system (Courtesy of © SHOM)*

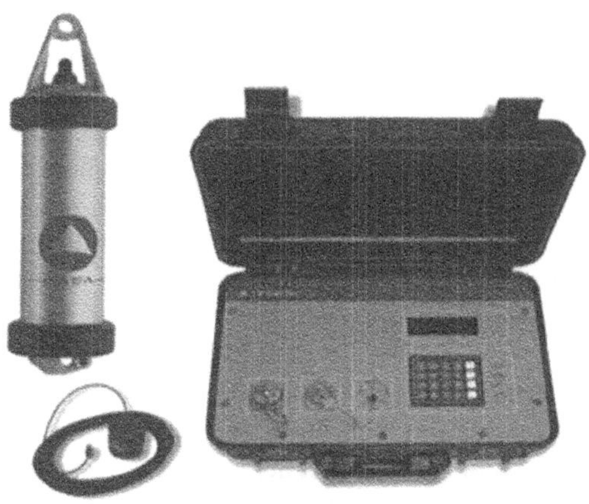

Figure 3.10. *Equipment used to recover submerged materials. Top left: The acoustic release. Bottom left: an acoustic transmitter. Right: the remote control case to which the acoustic transmitter is connected (Courtesy of IXblue)*

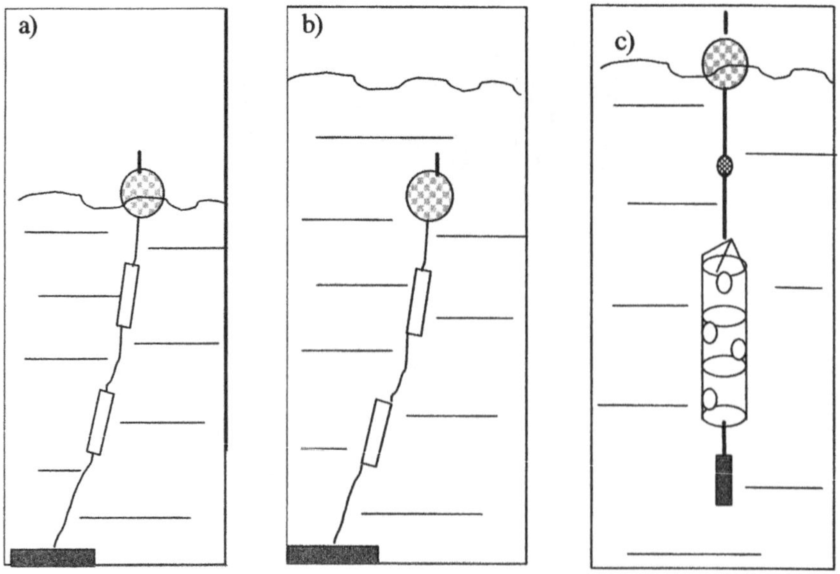

Figure 3.11. *The different types of moorings: a) surface mooring; b) subsurface mooring; and c) pendulum mooring*

When it is necessary to deposit instruments in the layers of water close to the surface, we use *surface moorings* (see Figure 3.11a). These consist of a weight or "dead body" deposited on the sea bed, a line on which measurement instruments are fixed, and a buoy located at the surface that shows where the line is placed and keeps the instruments in the top layers of water. This is one of its advantages. The depth and submersion position of devices vary as a function of the tide, current amplitude and action of the wind on the sea and the buoy, which is a disadvantage of this type of mooring.

The shape of these moorings can have different variants depending on the position of the measurements to be made. For measurements that are only to be made at the surface, a semi-rigid line called a "fender" can be fixed to the buoy. A satellite buoy is linked to this line, from which the measuring instrument and its weight are suspended (see Figure 3.13).

For measurements distributed along the water column, the line usually has a steel cable on its outer part in order to resist fish bites. Its inner part is made of synthetic fiber cables that are insensitive to corrosion. The length of these cables is calculated as accurately as possible, as a function of knowledge of the depth of the sea bed, in order to keep the whole assembly in permanent tension. A heavy chain is placed between the weight and the last elements of the line to reduce the incline that could be adopted by the mooring and to absorb vertical movements due to waves at the surface.

Finally, for measurements close to the sea bed, instruments are placed in a cage, like in the case of sea bed moorings. This cage is connected by a rope to a surface buoy whose role is simply to mark the position of the moorings. The length of this rope must be around 1.5 times that of the height of the water. It can be kept under tension by an intermediate buoy midway between the sea bed and the surface so that the cable does not tangle around the cage. Depending on the type of instrument used, precautions must be taken to prevent this cage from disturbing the sensors or leaking, thereby introducing errors into the measurements. Depending on the type of sea bed, there are also siltation risks of the assembly and a rope rupture risk when rising.

Figure 3.12. *Surface buoy equipped with a flash lamp (Courtesy of © SHOM)*

This type of mooring is sensitive to human activities (fishing, shipping, etc.). To limit this sensitivity, the surface buoy is equipped with a radar reflector or radio beacon that will allow boats with the appropriate instruments on board to locate it. It will also have a *flashing light* that makes it visible at night and allows visual identification during recovery operations.

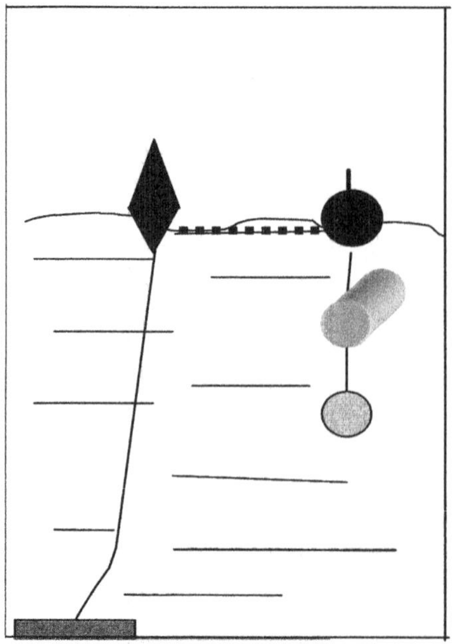

Figure 3.13. *Surface mooring variant: case where measurements are made close to the surface*

Surface mooring is also sensitive to the effects of waves that cause the buoy to rock and lead to additional traction forces and risk of rupture. It has the advantage, however, of being fully recoverable and does not require the systematic use of acoustic releases (depending on the weight of the ballast required).

If it is not possible to produce a surface mooring, due to the depth of the sea bed for example, we use "*subsurface*" *moorings* (see Figure 3.11b). Being permanently underwater, the buoy has the advantage of being able to overcome vertical and horizontal movements: rocking due to waves and movement due to the wind. Moreover, the buoy is not an obstacle to shipping. This system also allows the reduction of the length of line to be deployed. The drag is hence reduced as is the swinging radii. Cable technologies are used to meet the same constraints as those encountered with surface moorings: a metallic cable over 1,000 m, and a synthetic cable below this depth.

With this type of mooring, instruments' submersion depth no longer depends on surface height variations: it remains approximately constant. This is one of its advantages. They do, however, have the drawback that their lifting requires the insertion of acoustic releases that are fixed to the end of the line in order to separate the assembly part from the weight part. This part will stay on the sea bed while the other elements rise to the surface under the effect of the Archimedes' force exerted on the buoy.

Depending on the number of instruments we want to moor, the submersion depths, current, nature and depth of the sea bed, water launch and possibilities of raising, it is possible to create more complex moorings than we have discussed so far. For example, the *U-shaped mooring*, which is the combination of a surface mooring and a subsurface mooring, the two being connected by cables at the level of weights. The assembly can then be raised starting from the surface buoy. The subsurface weight, being secured to the line, does not need to be equipped with an acoustic release. This is one of its advantages. Moreover, the surface buoy signals the presence of the assembly. It does, however, have a mount that is difficult to launch into the water and to raise again. It also presents non-negligible risks of material loss and can only be used at small depths.

Finally, there are pendulum or *surdrift* (*surface drifter*) moorings (Figure 3.16). They work using the Lagrangian principle, as they were designed to move with the water mass being studied. They consist of a surface buoy. Under this buoy, we can find:

– a mechanical system for damping the vertical movements from waves;

– another intermediate buoy of smaller dimensions, to ensure tension of the line;

– a floating anchoring system; and,

– below the anchoring system a weight often consisting of chain links whose role is to keep the anchor vertical.

The surface buoy is supposed to offer little control to the winds, as the mooring assembly must move with the marine currents and not the meteorological conditions. It is generally equipped with an Argos transmitter and a GPS receiver, which can be used to follow its movements and its speed. It can also be equipped with a *submergence sensor* consisting of two electrodes placed on the upper sphere. When it is submerged, electrical contact is established between the two electrodes and the duration of the submersion can be recorded. This sensor can also be used to detect the loss of the floating anchor. Detection is achieved by observing a change in immersion information. The buoy can also be equipped with a temperature sensor on its underside so that we can determine the evolution of the value of this quantity at a near-surface immersion or, more recently with a conductivity sensor to retrieve temperature and surface salinity values that are useful for the calibration of SMOS satellites' radiometers, which are used to measure surface salinities (at 1 cm depth in band L).

The function of the floating anchor is to really "anchor" the mooring in the water mass whose movement is to be followed. It therefore has a lot of drag. Studies have been carried out to optimize the anchor's shape and drag ratio to the surface buoy. The most common shape is called *holey socks* (see Figure 3.14). Another type of anchor called *Tristars* is also used. Tristars have a radar reflecting shape but their large dimensions (wingspan of around 4.70 m) make their deployment at sea difficult. For this reason, *holey socks* are often preferred.

Figure 3.14. *A holey socks floating anchor connected to a thermistor chain. In the background, we can see an surface velocity profiler-type surface buoy (Courtesy of Marlin-Yug, Sebastopol, Ukraine)*

The mooring drag ratio has to be above 30 in order to minimize error due to the surface buoy and the extension line. This constraint leads to a slightly negative anchor buoyancy calculation. To compensate for this, it is necessary to fix an intermediate buoy that is sometimes integrated with the anchor. It is difficult to evaluate the efficiency of this device with the movement of the water mass to be followed, especially since – depending on the drop zones – the elements nearest the surface (buoy, rope, etc.) are progressively loaded with algae and barnacles over time.

Depending on sea conditions, the surface buoy can be temporarily submerged. The movements it is subjected to from the waves cause great tension on its fastening to the mooring line and ruptures due to this are frequent. To detect these ruptures and treat the position measurements as a consequence, an *anchor presence sensor* consisting of a load cell equipped with strain gauges is arranged under the buoy and the information it provides is transmitted by an Argos link with the other sensors. Recently a new system was developed to detect the loss of drogues. It is based on the analysis of speeds measured by the drifters. Geostatic currents and Ekman currents, which are known for the area, are subtracted and the remainder is correlated with the local wind speed (see [RIO 12]).

There are specific formats of Argos transmissions recommended by the *Data Buoy Cooperation Panel* (*DBCP*) for each type of drifting buoy. The DBCP depends on the *Global Drifter Center* based in Miami in Florida, which itself depends on the National Oceanic and Atmospheric Administration. This center is in charge of managing all of the drifting buoy deployments across the globe. Two international programs, the *Surface Velocity Program* and the *Global Drifter Program*, were put into place in the 1980s with the aim of collecting current measurements made at a nominal depth of 15 m.

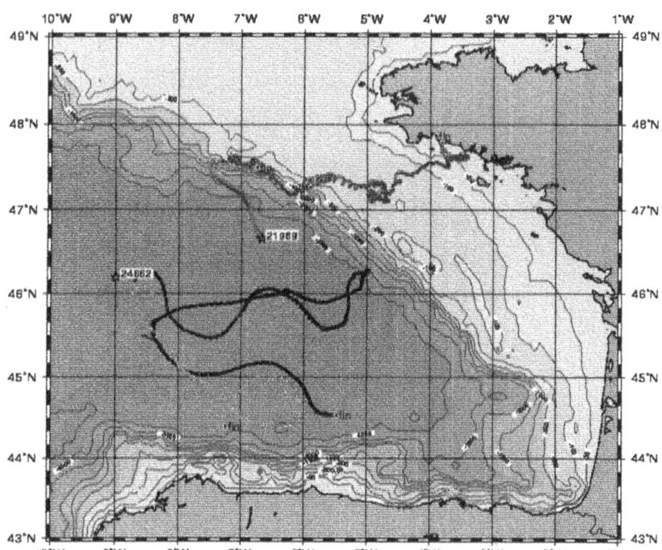

Figure 3.15. *Example of a trajectory chart made from GPS positioning data of "Surdrift" drifters, transmitted by ARGOS. The numbers written at the beginning of each trajectory are the Argos beacon numbers (Courtesy of © SHOM)*

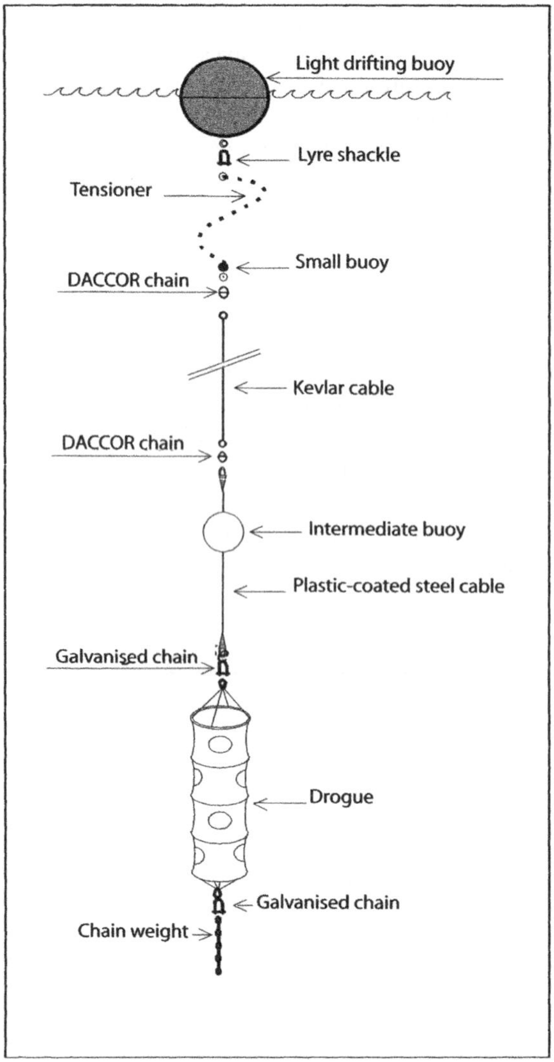

Figure 3.16. *Detailed composition of a surdrift mooring with a holey sock (Courtesy of © SHOM)*

3.2.2.3. *Measurement errors caused by surface and underwater mooring lines*

Mooring lines are designed so that the measurement instruments secured to them remain vertical; however internal tide current and surface currents produce forces whose influence is rarely accounted for in mooring calculation models. These forces can cause curvature of the line, particularly in the case of subsurface mooring lines, and even displacement of the line centered on its anchor point. If the submerged instruments are current meters, the curvature of the lines leads to errors in recorded speed amplitudes and directions. These errors, which can be negligible relative to the precision of instruments close to the anchoring point, are especially important if the instrument is located a long way from the anchoring point, as shown in [LAN 90].

Under the effects of waves, vertical movements on the surface buoy can introduce measurement errors, particularly in the case of current meters. Vibrations that are propagated along the line superimpose a measured noise on the data recorded. In the case of mechanical current meters used in areas with a current of small amplitude, these vibrations can modify the threshold of the instruments and considerably bias measurements. Studies are being carried out to study the effects of these vibrations on acoustic current meters.

Vertical movements of the mooring lines and curvature also cause errors in the measurements obtained from temperature sensors placed along these lines. These sensors are supposed to measure the temperature of the layer of water at their immersion depth. If the line curves at "upper" of the immersion depth, the sensors are lifted and measure the temperature of a higher layer than they are supposed to. In 2008, Meinen suggested a technique based on the use of bathymetric data to correct these measurement errors [MEI 08], but it is often preferable to associate a pressure sensor with a temperature sensor in order to *a posteriori* correct the temperature data.

3.2.2.4. *Corrosion: the enemy of moorings*

As moorings consist of an assembly of elements (cables, loops, shackles, measurement instruments, etc.), which are mostly metal and can have different molecular compositions, electric couples occur when immersed in sea water. Certain materials play the role of an anode, others a cathode and the sea water plays the role of electrolyte. The phenomenon of *electrolysis* can be produced, which causes electrons and other matter to migrate from anodes towards cathodes. This phenomenon can be represented by the following simplified reactions, where M is a metal, O an electrolyte species and R the product of reduction:

– anodic reaction: $M \rightarrow M^{n+} + n.e^{-}$;

– cathodic reaction: $O + n.e^{-} \rightarrow R$.

This migration results in the progressive destruction of anodic metals and can result in breakage or loss of the mooring.

Depending on the nature of the metals present, the loss of material is of varying importance. It can be evaluated from the difference in potential created, which is called the *corrosion potential* or abandonment. It can be numbered, separately for metals and alloys, relative to the standard hydrogen reference potential. In this case, the more negative it is, the greater the reaction will be. For example, for platinum the potential is around +0.2 V, for titanium and alloys such as Ni-Cr-Mo it is 0 V on average, for Ni-Al-bronze alloys it is around –0.15 V, for copper it is –0.35 V and for zinc it is –0.95 V. For certain alloys or stainless steels, it moves towards more negative values after oxidation in the environment.

Stainless steels are defined as iron–chromium (Fe-Cr) and iron–nickel–chromium (Fe-Ni-Cr) type alloys. They can be enriched with carbon, nitrogen or even molybdenum or titanium, but they only qualify as stainless when they have a minimal chromium concentration of 11–12%. They are classified according to their composition and molecular structures into martensitic, ferritic, austenitic or austenitic–ferritic types of steel. Austenitic–ferritic types that are enriched with molybdenum generally have good corrosion resistance characteristics in the marine environment.

The factors that influence corrosion are of a chemical nature (dissolved oxygen, salinity, pH, carbon dioxide) but the physical environment is also important relative to water speed, temperature and pressure. It is also known that biological factors are responsible, as all materials immersed in the ocean are covered in a layer called a *biofilm*. This biofilm consists of microbial cells that adhere to surfaces, reproduce, and develop a series of actions and reactions causing the evolution of the film's composition. The biofilm is very heterogeneous. It modifies oxides formed on the surface of stainless steels, favoring the development of corrosion. The bacteria, which are often involved in these phenomena, can produce sulfates (sulfate-reducing bacteria) and acids that modify the pH of the environment. It has also been demonstrated that the enzymatic activity of certain bacteria accelerates the cathodic reaction speeds that participate in the initiation of localized corrosion.

Different forms of corrosion can therefore appear and present themselves as generalized, localized or as the result of mechanical effects. Localized corrosions can be galvanic, cavernous, by holes (or pricks) or by bacteria. Biocorrosion is characterized by deep, localized holes, (crevices) and very rapid attacks.

There are few solutions to guard against it. Obviously, huge attention must be paid to the metals involved. Some of them develop an oxide film on their surface giving them a good resistance to corrosion once submerged. The state of instrument surfaces is also of importance, the adhesion of microorganisms being promoted by irregularities. Microorganism development is also reduced in the absence of light, so at greater depths this is less of a problem. At such depths, however, pressure can have a greater negative influence. It is also important that water is able to flow easily around the instrument, which usually slows down the phenomenon of corrosion. It is often necessary to have sacrificial *anodes* in certain places. Consisting of a metal with a high corrosion potential (usually zinc), their oxidation provides efficient cathodic protection for the other metals present, for which the corrosion potential is much lower. Finally, biofouling and corrosion – being phenomena of similar origins – can be tackled or minimized by the use of biocidal products and coatings called "active protection" (see section 2.2.5).

3.2.3. *Deployment and recovery of moorings*

Deployment is an important phase in the implementation of a mooring. The longevity and reliability of measurements to be carried out can strongly depend on it because it is during the water launching operations and raising that the strains exerted on the line are strongest. The deployment phase is usually preceded by a bathymetric survey in order to precisely determine the depth and nature of the sea bed at the location it will be deposited. This survey allows us to modify the mooring point and better adjust the length of the line. A bathymetric survey can also be made using expendable probes or CTD profilers, in order to determine the distribution of different water layers and to optimize sensor immersion depths.

Before their immersion, the different components of the mooring are checked and placed on the boat deck. The chronology of this launch is decided based on meteorological conditions and the maneuverability of the vessel. There are two possibilities for launching:

– *The buoy is dropped first*. In this case, the line must be fully rigged and laid out on the deck, ready to be lowered. The boat must be positioned relative to the mooring point, at a distance slightly greater (about 1.5 times) than the length of the line to be moored. The speed at which it is lowered must be able to be modified while the elements are being launched in order to keep the line stretched and not create knots so the ballast can be dropped at the desired location. This operation is carried out when the line is completely stretched out and the boat has passed the mooring point. This method has the advantage of not introducing excessive strain on the elements on the line. These strains are maximal when the "dead body" is lowered into the water, but are still below its weight in water. It is safer for the crew but the method is difficult to implement.

– *The ballast can be dropped first*. In this case, the boat is positioned near the mooring point and the elements are assembled as they are launched into the water. The ballast is dropped into the water and the line is unwound until the first instrument to be inserted (e.g. the release system). It is equipped with loops that allow tension recovery for each cumbersome element so that the elements can be introduced on the mooring without having to pass through the launching gantry pulley. Once inserted, the line continues to be unwound until the next instrument, and so on. The advantage of this type of launch essentially rests with the fact that it requires very little space for deployment. Its drawbacks are that the instruments are permanently under tension and the maneuvers are more dangerous for the crew.

All such launches are usually carried out using the gantry at the rear of the boat (see Figure 3.1 and Figure 3.7), which allows the elements to be deposited sufficiently far from the hull and the propellers.

The technique for lifting depends on the type of mooring. In the simplest case, the surface buoy is recovered first and the line is gradually wound up by hand or by using a deck crane, the instruments being unhooked as they are raised. If the ballast is too heavy or if none of the mooring elements appear at the surface, however, an acoustic release is called for. If it is activated, the releaser frees the end of the line, which then rises. Recovery occurs only when everything has risen. Recovery can also be started with the surface buoy or by the end of the mooring. In both cases, good coordination is necessary between the boat's commander and the crew so that contact with the hull does not damage the mooring elements close to the surface.

3.3. Drifters

If it is necessary to study the evolution of marine currents over large time and space scales, drifters are used. Pendulum moorings described in section 3.2.2.2 include surface drifters. Equipped with GPS receivers and Argos transmitters, it is easy to follow their movement due to currents close to the surface and to winds. For deep current studies, however, another type of float, called a *subsurface float*, is required. Subsurface drifters are currently the main Langrangian tools used for studying the movement of deep water mass.

3.3.1. *History and operating principles*

The first subsurface drifters were designed in a rudimentary manner in 1955 by John Swallow of the *Institute of Ocean Sciences* (United Kingdom). From aluminum tubes, originally intended for use in scaffolding, he built the first floats capable of stabilizing themselves at a given immersion. He also managed to acoustically monitor them from a vessel that was present in the immersion area. These early instruments revealed that, at the chosen depths of 2,000 and 4,000 m, current speeds were close to 10 cm/s, whereas scientists were expecting values in the order of 1 cm/s. This forced a modification of their observation strategies and acoustic monitoring.

To do this, at the end of the 1970s Rossby and Webb developed drifting acoustic emission sources whose position was detected by hydrophones in fixed positions. This system was already operating *SOFAR* channel sound characteristics (see section 3.3.2.1), which is in fact a natural wave guide found at certain depths in the ocean. Hence the name SOFAR given to these floats; it was possible to detect at over 1,000 km with an accuracy in the order of 3–5 km, through acoustic slopes that were emitted and correlated at listening stations with the expected signal.

Requirements and technologies have evolved, particularly with the appearance of the Argos system; in the 1980s, listening stations were replaced with acoustic sources and drifting sources with floats. The positioning system was reversed to track floats from acoustic waves emitted from the sources in fixed positions. As the positioning system was reversed, the floats were named *RAFOS* (a simple inversion of the term SOFAR). The first RAFOSs, which were created by Rossby at the University of Rhode Island, were built in glass tubes 1.60 m long and 10 cm in diameter. This is still the case. They are programmed to be able to drift up to 4,000 m and rise to the surface after 12–18 months by releasing ballast. They are equipped with temperature and pressure sensors. At the surface, the position data and annexed sensor data are transmitted by an Argos antenna included in the glass tube.

Other developments followed, using the same principle, especially in France with the SIVOR range. Unlike the American RAFOS, however, the SIVOR envelope was metallic and the Argos antenna was external. As these floats were lost after their descent, drift and rise cycle, the *multi-cycle RAFOS* was devised. The first of this range, developed by Ifremer in industrial partnership with Tekelec, were called *Marvor* (the Breton word for "sea horse"). Equipped with an inflatable bladder, these floats could descend to a programmed depth, drift for two or three months, rise to the surface to transmit their data and begin the cycle again between 10 and 50 times. Some of this equipment has functioned for over five years.

The establishment and maintenance of acoustic source networks is expensive, it requires substantial logistics and does not allow full coverage of oceans. Another concept was therefore developed in the 1990s where a multi-cycle float is brought up to the surface more often, so that it can be positioned by the Argos system. The movement observed between two localizations is considered to correspond to the immersion drift. The trajectory is therefore less precise, but the hydrophone and its electronics can be replaced more usefully with other sensors. Similarly, more temperature data and data from other quantities can be stored and transmitted in the Argos messages. This concept named *Autonomous LAgrangian Circulation Explorer* (*ALACE*) appeared in the 1980s. It was devised by Davies and Webb [DAV 92]. In 1999 it allowed the establishment of a network of global operational floats called *Argo*, an analogy with Greek mythology. Argo was the boat belonging to the hero Jason, the name given to the observation satellite dedicated to oceanography.

3.3.2. *The concept and evolution of the Argo program*

The Argo program aimed for the deployment of a network of 3,000 buoys between 2001 and 2006 (see Figure 3.17), placed at distances of 3° on all of the ice-free surfaces of the deep ocean. Each country was to make a contribution to the network by dropping the buoys, recovering and analyzing their data in order to make them available within 24 hours for any user who may need them. It was part of the development of *operational oceanography*, whose aim is to create a forecasting system for ocean currents and climatic variations based on a daily scale up to decades, from a pool-scale to that of a global scale.

Currently, several profilers of this type are available on the market. The first of these is the *Autonomous Profiling EXplorer* or *APEX* (see Figure 3.19a) developed by the Webb Research Corporation and the University of Washington. It has another American competitor, *Sounding Oceanographic Lagrangian Observer* (*SOLO)* developed by the *Scripps Institution of Oceanography* and a French competitor, the *PROVOR* developed by Ifremer and NKE Electronics based on the MARVOR float. The PROVOR has a successor called the ARVOR-C, which is smaller (h = 2.1 m, ϕ = 11 cm) and lighter (20 kg), and is capable of staying in a standby position on the sea bed. The Japanese also have their profiler, the *NINJA* or *New Profiling Float of Japan*, developed by Tsurumi Seiki Corporation and the Japan Marine Science and Technology Center.

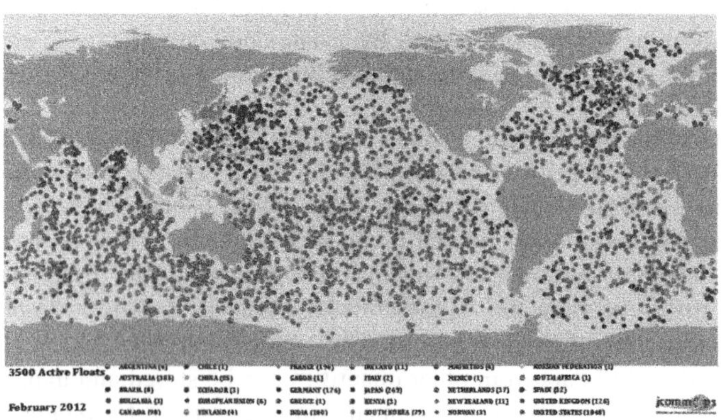

Figure 3.17. *The Argo network status in February 2012, with the contribution of each participating country expressed in number of moored buoys ([ARG 12])*

Operating cycles programmed into the ALACE type profilers have a typical duration of 10 days in high seas and some tens of hours in coastal areas. They usually consist of a descent to the specified depth, often between 400 and 2,000 m (Argo recommends 1,000 m), drift for several days (nine days for Argo buoys) or several hours in coastal areas, then descend to maximum immersion (1,500 m or 2,000 m for the Argo network), followed by an ascent to transmit data (see Figure 3.18). The descent and the ascent occur at speeds between 10 and 20 cm/s (see [SCH 07]). This cycle can be repeated 100 to 250 times. Depending on their programming, their theoretical lifespan, which depends primarily on the energy they carry, varies from three to four years.

During the ascent, they take measurements with the sensors they are equipped with. Initially temperature and pressure were the only quantities accessible to measurement with this type of carrier; they now integrate conductivity sensors, and even dissolved oxygen sensors (see Figure 3.19b), fluorimeters or wind or rain sensors. Developments are being oriented towards the ability for profilers to carry optic underwater sensors (see section 2.10), and physicochemical sensors (to measure nutrients, nitrate, pH, etc., see sections 2.11.3 and 2.11.4). Only weight and energy capacity limit this type of integration.

ALACE, which were mainly used to make temperature–conductivity–pressure profiles, have been renamed *PALACE* (*Profiling ALACE*). As it is impossible to re-calibrate the sensors during their operating life, the accuracy requirements for measurements are lower than those initially recommended by the WOCE program. They are, however, still very high: ±0.005°C for temperature, ±5 dbar for pressure and ±0.01 for salinity.

Given the drift of conductivity sensors, this latter tolerance is difficult to maintain and verify. Salinity data are the object of special treatments carried out by the organizations that make them available, delaying the time at which they are generally available. They are compared to older measurements or measurements made from boats when campaign areas coincide with the positions of floats.

Several publications outline techniques that can be used to make readjustments, intercomparisons or *in situ* calibrations. Brechner Owens and Wong [BRE 09] use salinity data issued from climatology and an objective statistical analysis method. They are also able to assess the uncertainty of this adjustment by using a Monte Carlo simulation method. Another technique developed by Durand and Reverdin [DUR 05] also uses an objective analysis method, but includes a least squares technique to make the adjustment.

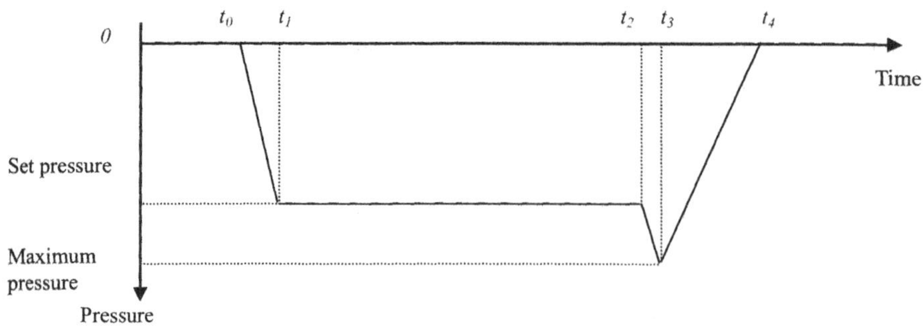

Figure 3.18. *Operating cycle of a typical autonomous multicycle profiler. The t_0t_1 phase corresponds to the descent until the set pressure programmed into the instrument. The t_1t_2 phase corresponds to the subsurface drift at the set depth. It ends with a descent to the maximum depth of the float (t_2t_3 phase) and by an ascent to the surface (t_3t_4 phase) for the positioning and transmission of data recorded during the ascent profile*

The evolution of the concept of drifting floats does not only occur in the matter of integration of sensors. It is also done in matters of positioning and data transmission. Hence, the Argos system, which is still used by 95% of floats (see section 2.7.1.3), is rivaled by the bi-directional Iridium system, which allows data transmission modes with different flow rates (see section 2.7.2.7). ARGOS-3 is capable of the same performance, but the transmission systems on the floats must be adapted to use it: installation of PMT (platform messaging transceiver) cards and new bi-directional antennae that are resistant to pressure, and the development of transmission protocols and data compression techniques.

In addition to the increasing number of loaded sensors and the speed of data transmission, the new generations of buoys are oriented towards a more compact size with further miniaturization of onboard electronics. Similarly, certain oceanic areas below 2,000 m need to be explored, and new developments are ongoing to permit certain buoys to descend to 3,500 m.

Ice-covered areas cannot yet be surveyed with these types of floats; adaptations have been devised or are still under development to better understand the variations in temperature and salinity that can be observed in these regions of the globe. In 2007, Kikuchi *et al.* [KIK 07] devised a system consisting of a platform stuck in the ice underneath which a cable is stretched to a depth of 1,000 m. A PROVOR type float moves along this cable by the intermediary of two collars fixed on the float and slicing on the cable, one of which has an inductive transfer system of data from the float in the cable. The platform surface is equipped with an inductive modem for retrieving and transmitting via an Iridium antenna. It is also equipped with temperature and atmospheric pressure sensors. The buoy makes descent–ascent profiles that are nearly identically to those in Figure 3.18 (without drift), but with an additional level at 300 m where it is in standby.

The French *Ice, Atmosphere, Ocean Observing System* (IAOSS) project involves the installation of 15 platforms similar to Kikuchi's system in the Arctic Ocean [PRO 12]. The difference is in the equipment placed on the surface, which is an instrumented buoy (see section 3.4.1) capable of floating when ice is absent. It is also equipped with a GPS/iridium antenna, a LIDAR (light detection and ranging – an optical remote sensing technology) temperature and air pressure sensors, and a 6 m thermistor chain (see section 2.1.1.5) equipped with sensors every 2 cm to measure the thickness of the ice.

The melting of Arctic ice and the low depth measurement coverage of this oceanic area has encouraged the development of drifting floats that are capable of taking measurements in these cold waters. To do this, they must be able to detect floating ice before reaching the surface to transmit their data. This detection is under study using hydrophones (see section 2.4.1) as depth gauges or optic sensors to measure sunlight. It implies the ability to auto-adapt the mission programmed in the buoy depending on the ice conditions encountered. The melting of ice allows light to penetrate into the environment. Its ecological equilibrium is modified with the development and production of plankton. These buoys are or will therefore be equipped with biogeochemical sensors, such as the fluorimeters described in section 2.10.5, in order to understand the evolution of this environment.

a) *b)*

Figure 3.19. *a) An APEX float with a conductivity sensor on its upper side. At the base, we can see the mechanical system used to adjust its buoyancy (courtesy of © SHOM). b) Head of an APEX buoy equipped with Sea Bird temperature–conductivity sensors and a Clarck polarimetric dissolved oxygen sensor (Courtesy of S.C. Riser, Univ. of Washington)*

3.3.3. *Principles for positioning by acoustic sources*

The principle of detection and positioning of drifters was initially based on mooring in a fixed position around the oceanic area for the study of autonomous listening stations. These stations, constituting of hydrophones and batteries to power them, had the role of detecting and recording the arrival time of sonic signals transmitted by floats entering the area. These floats, which are acoustic sources, were named SOFAR which is short for *SOund Fixing And Ranging*.

3.3.3.1. *The SOFAR channel*

The SOFAR was originally the name given to an acoustic channel located at the bottom of the main thermocline, where sound propagates at its maximum speed. When emitted at this level, the channel behaves as a wave guide, where signals can propagate themselves over long distances (up to 2,000 km or more) by refraction and reflection on the "walls".

It was established that the average intensity, I_r, of a continuous signal detected by a hydrophone placed in the channel takes the form:

$$I_r = I_s \frac{r}{r_0} 10^{-\left(\frac{\alpha}{10}\right)r} \qquad [3.10]$$

where:

– I_s is the intensity at the source;

– r the propagation distance;

– α the absorption coefficient of the environment, which depends on the frequency of the signal; and

– r_0 a factor called transition length of a spherical dispersion to a cylindrical dispersion.

r_0 depends on the depth of the source and the receiver. It can be estimated if we know the variation of the speed of sound propagation as a function of depth. Its characteristic value is 3 km. It is also possible to estimate the value of α from relationship [2.146]: $\alpha \approx 1\text{x}10^{-5}$ dB/m at a frequency of 260 Hz. In fact, it is useful to calculate and determine the level of the signal-to-noise ratio *S/N*:

$$S/N = IA - PP - NB + GD \qquad [3.11]$$

where:

– IA represents the acoustic intensity emitted (in dB);

– PP the losses by propagation;

– NB the level of ambient noise expressed in dB in the signal bandwidth; and

– GD the detector gain which is often a correlator.

The value of PP is calculated from a relationship issued from:

$$PP = 10\log_{10}(r_0) + 10\log_{10}(r) + \alpha r \qquad [3.12]$$

The value of NB is highly variable depending on the region, the meteorology and especially maritime shipping. A value of 65 dB, referenced to 1 μPa in the signal bandwidth of 1 Hz, corresponds to intense traffic combined with a wind speed of 10 knots.

By transmitting acoustic signals at regular intervals, the position of the SOFAR can be determined at distances of 2,000 km by a series of listening stations located several hundred kilometers from one another (Bermuda, Bahamas, Grand Turk Island and Puerto Rico for the North-West Atlantic area, for example). These stations open listening windows with a periodicity corresponding to that of buoys, and detect the pulses emitted.

Currently, buoys are still in the SOFAR channel, but their roles are reversed as they are destined to permanently listen during their drift, and then to detect and identify different events such as underwater seisms in order to predict their occurrence and the tsunami risks they engender, or to monitor whale calls (the Mermaid float project). This application was made possible with the advent of Iridium type broadband satellite transmission systems (see section 2.7.2.7).

3.3.3.2. *The positioning of RAFOS floats*

In the 1980s, there was a need to develop lighter and cheaper floats that were able to be scattered in large numbers. The SOFARs, requiring the transport of a fairly powerful acoustic transmission system, consumed a huge amount of energy and had to be powered by heavy battery packs. Rossby *et al.* of the University of Rhode Island then had the idea of designing a reverse SOFAR system, which they named RAFOS [ROS 86]. The RAFOS buoys became simple receivers of acoustic pulses emitted from sources immersed in the SOFAR channel. This localization principle persisted until the rise of the Argo network.

The acoustic sources are located in the middle of the ocean, in places where it is possible to deploy and raise subsurface moorings (section 3.2.2). They come in the form of two large parallel cylinders (see Figure 3.19). One of the cylinders contains enough battery packs for the system to be autonomous for two to three years. The other contains the hydrophone emitter. The acoustic intensity of signals emitted is around 175 dB, with reference to 1 µPa. Table 3.3 gives some values of *S/N* that we can obtain with this acoustic power and a receiver that has a gain *GD* of 20 dB. It appears that detection is possible at distances of over 2,000 km. However, the quality of the receiver depends partly on the difference in depth that can exist between the float and the source. In order to favor good reception, the detection radius is usually reduced to distances of less than 1,500 km.

This distance determines how long the floats should be listening for. Sound, propagating itself at an average of 1,500 m/s, it takes 1,000 s or at most 17 minutes to reach the float. This also determines the time delay required between each emitter source so that the buoy does not simultaneously receive two signals from different places.

Distance (km)	500	1000	1500	2000	2500
PP (dB)	96.8	104.8	111.5	117.8	123.8
S/N (db)	30.2	22.2	15.5	9.2	3.2

Table 3.3. *Propagation loss values (PP) obtained from relationship [3.8] and signal-to-noise ratio (S/N) calculated from relationship [3.7] for an acoustic intensity of 175 dB emission referenced to 1 µPa*

The positioning of floats is calculated by detecting the *Times Of Arrival* (*TOA*) of signals emitted by sources. These signals are sinusoidal pulses 80 s long, for which the frequency is centered around 260 Hz and linearly increased by 1.523 Hz. The floats listen periodically, at set times (one to three times a day), for sounds emitted by three different sources offset by about 20 minutes. The TOA are determined by the float's clock, which is set and verified to the second at the time of launch. This mode of operation involves the use of clocks of high stability and is compensated for temperature effects inside the floats (acceptable relative stability: 7×10^{-8}). Readers can refer to section 2.6 for more details on oscillators and clock referencing.

From their conception, the electronics of buoys have evolved towards increasing digitalization. However, the principle of detection of acoustic positioning signals is usually based on amplification then initial filtering of the signal received with a bandwidth centered around 260 Hz. This is followed by a second low-pass filtering with a width of 1.5 Hz. The narrowness of this bandwidth, which should be enlarged, limits the accuracy and repeatability of the RAFOS positioning system. An analog–digital converter samples the signal issued from the first or second filter. The digitalized signal undergoes lopping before the processor makes an autocorrelation calculation on 800 points. From this calculation, phase information is extracted that allows precise determination of the TOA and duration of transmission.

During the listening phase, the buoy receives a main signal that is made noisy by numerous echoes caused by multiple reflections it is subjected to during its journey across the mass of water. The correlator allows the extraction of a signal that is the result of the combination of paths that have the greatest phase stability. This combination of paths is one of the biggest sources of uncertainty in the positioning of buoys. A second source of uncertainty comes from the approximation made on the estimation of the propagation speed of sound, which is done close to 2–3%, which represents between 3.0 and 4.5 m/s of error.

Finally, the float being in motion during its listening phases needs correction for the Doppler effect (section 2.5.2), particularly in areas with strong currents. The problem with this correction is that speed v of the float is not known at the time of positioning. It can only be estimated from knowledge of the variation frequency of the TOA from one transmission to another. The frequency shift engendered by this effect is given by:

$$\Delta f = \frac{v}{c - v} f_0 \tag{3.13}$$

where f_0 is the frequency transmitted and c the speed of sound.

As the signal being sent is a ramp of frequencies, the TOA offset Δt is obtained by:

$$\Delta t = f_0 \frac{v}{c - v} \frac{80}{1.523} \tag{3.14}$$

For a current of 0.5 m/s, the correction Δt is around 4.4 s.

The calculation of position is, in fact, an iterative procedure initiated by the knowledge of the position of launch and the calculation of distances between this position and those of emitter sources. This calculation is repeated until the sum of absolute values of differences is below 1 km. This estimated position is then used to initiate determination of the next position and so on throughout the lifespan of the float. Deviations or errors currently observed between the launch position and the first acoustic position, and then between the last acoustic position and the first Argos location are in the order of 1–3 nautical miles or 1.8–5.5 km.

Figure 3.20. *Launching an acoustic source for RAFOS floats (Courtesy of © SHOM)*

3.3.4. *Design and ballasting of drifters*

To balance a float in open water, it is possible to alter its weight or its volume. In the case of mono-cycle floats, it is mainly the weight that is altered. In the case of multi-cycle floats, it is volume that is varied.

3.3.4.1. *Mono-cycle floats*

The description of mono-cycle floats in this section corresponds to those that were used in conjunction with positioning by acoustic sources, described in section 3.3.3. The concepts that govern the buoyancy conditions of mono-cycle instruments are discussed in section 3.2.2.1.

These buoys are equipped with ballasts whose mass is calculated so that the assembly meets the zero buoyancy conditions at a set depth, of which the value is usually between 400 and 2,000 m. These ballasts are connected to their lower screw-cap by an intermediate Inconel wire. This material is chosen for its excellent resistance to marine corrosion.

After their descent to a set depth, during their underwater drifting phase the mono-cycle buoys record the TOA of signals emitted by acoustic sources moored in their drop zone. They can also record temperature and pressure data. Pressure data will subsequently be used to estimate their actual immersion depth.

After a pre-programmed time, which can be several days, several months or even several years, a current is sent down the Inconel wire so that it dissolves by electrolysis. The ballast is then freed and the float rises to the surface to transmit Argos messages containing the data it has been recording throughout its immersion phase. These messages are emitted until the battery is exhausted. These floats are not usually recovered, although it is possible if a boat is in the area at the time when it transmits its messages. In the case of recovery, they can be repackaged for a new mission.

Ballasting or weight adjustment to obtain zero buoyancy is an important operation in the fabrication cycle and implementation of these buoys. The zero buoyancy conditions can be calculated for a certain depth or for a certain surface density. These are known as *isobaric floats* or *isopycnal floats*.

The isobaric floats must have a compressibility that is less than that of sea water (30–50%) so that when moving to a depth that is less than that of their equilibrium surface, they take up less volume than the water surrounding them, and vice versa. They are said to have a force of return to equilibrium greater than that of water. In this way, the risks of them rising to the surface or running aground are decreased. The amplitude of this force can be calculated from the variation in volume ΔV caused by a vertical displacement Δz engendered by a variation in pressure Δp and temperature ΔT:

$$\Delta V = \frac{\partial V}{\partial p}\Delta p + \frac{\partial V}{\partial T}\Delta T \qquad [3.15]$$

If we denote the compressibility coefficient of the shell of the float by γ and its coefficient of thermal expansion by α, they can be defined by:

$$\gamma = \left(\frac{1}{V}\frac{\partial V}{\partial p}\right)_T \quad \text{and} \quad \alpha = \left(\frac{1}{V}\frac{\partial V}{\partial T}\right)_p \qquad [3.16]$$

We can then write equation [3.15] in the form:

$$\Delta V = V\gamma\frac{\partial p}{\partial z}\Delta z + V\alpha\frac{\partial T}{\partial z}\Delta z \qquad [3.17]$$

The terms $\frac{\partial p}{\partial z}$ and $\frac{\partial T}{\partial z}$ are the hydrostatic and temperature gradients of the environment. Equation [3.17] can also be rewritten to calculate the change in isobaric volume ΔV_B per unit of displacement and the float volume V_f:

$$\frac{\Delta V_B}{V_f \Delta z} = \gamma\frac{\partial p}{\partial z} + \alpha\frac{\partial T}{\partial z} \qquad [3.18]$$

A similar equation can be established for a change in the volume ΔV_e of water with constant salinity:

$$\frac{\Delta V_e}{V_e \Delta z} = \gamma_e \frac{\partial p}{\partial z} + \alpha_e \frac{\partial T}{\partial z} \qquad [3.19]$$

The isobaric force F_B of return to equilibrium, expressed per unit of volume and displacement, can be calculated from the hydrostatic relationship and relationships [3.18] and [3.19]:

$$F_B = \rho g\left[(\gamma_e - \gamma)\frac{\partial p}{\partial z} + (\alpha_e - \alpha)\frac{\partial T}{\partial z}\right] \qquad [3.20]$$

Here ρ designates the density of the water and g the acceleration due to gravity.

This principle of buoyancy was adopted by the first generations of buoys. However, in the 1980s, isopycnal technology was preferred. In this case, the compressibility of the instrument has to be adjusted to that of sea water in such a manner as to cancel forces induced by pressure variations. If we allow the equality $\gamma_e = \gamma$, relationship [3.20] then becomes:

$$F_P = \rho g\left[(\alpha_e - \alpha)\frac{\partial T}{\partial z}\right] \qquad [3.21]$$

where F_p designates the isopycnal force of return to equilibrium.

It is possible to carry out this compressibility correction operation by adding a volume-correcting element to the buoy consisting of an assembly of piston–cylinders and a spring by counter-reaction, whose role is to compensate for loss of volume as a function of pressure. However, this system cannot compensate for the variation in density induced by variations in temperature. It was for this reason that the first RAFOS buoys developed by Rossby *et al.* [ROS 86] consisted of a glass casing. Borosilicate type glasses have a coefficient α that is about 20 times less than α_e, in other words their expansions under the effect of heat are negligible in the range of oceanic temperatures.

Isopycnal technology requires rigorous ballasting if sufficient accuracy on the position of the buoy at depth is desired. For example, for a buoy with a constant volume V of 0.01m^3, to get an error of less than $\pm 0.1 \times \rho_m$, ρ_m being the density of the environment, we must know and be able to adjust its weight to within almost 1 g. $0.1\times\rho_m$ corresponds to an uncertainty of 30–40 m in the vertical positioning or to a temperature change of 1°C in the main thermocline of the Gulf Stream, according to Rossby.

To carry out the ballasting operation, the float is placed in a bath of distilled water whose temperature T_b and pressure p_b have been measured so that density ρ_b can be calculated. Distilled water is preferred to tap water because the latter can introduce errors in the order of 0.01% in ρ_b, due to the dissolved matter it contains. The excess mass ΔM to be added for the ballasting is calculated according to:

$$\Delta M = V_m \rho_m - V_b \rho_b \qquad [3.22]$$

This is based on the acquired knowledge of the temperature T_m, pressure p_m and density ρ_m of the environment where the buoy will be launched. V_m is the volume of the buoy in an oceanic environment and V_b its volume in the bath of distilled water. V_m is determined by using γ and α:

$$V_m = V_b[1 + \gamma(p_m - p_b) - \alpha(T_m - T_b)] \qquad [3.23]$$

If equation [3.23] is integrated into equation [3.22], we get:

$$\Delta M = V_b[(\rho_m - \rho_b) + \rho_m\gamma(p_m - p_b) - \rho_m\alpha(T_m - T_b)] \qquad [3.24]$$

In the case of glass buoys, this ballasting operation is carried out in a hyperbaric chamber because of the nonlinear behavior of the volume corrector element at low pressure and the error it can engender upon the calculation of ΔM, and also because of inhomogeneities in the fabrication of the glass tubes that introduce uncertainties on the knowledge of γ. As tolerances on the value of γ are reduced in the case of metal tubes, it is no longer necessary to test them in a pressure tank.

3.3.4.2. *Multi-cycle floats*

The buoyancy of these instruments is ensured by varying their volume. To do this, they are equipped with an internal reservoir, an electric motor and a hydraulic valve pump assembly (in the case of PALACE and PROVOR) or rob–piston mobile pumps (in the case of APEX, SOLO or NINJA). Activation of the pump leads to the transfer of fluid into an external bladder that swells and allows the volume of the float to be adjusted to obtain the desired density.

The variation in volume ΔV_f necessary to go from surface water with density ρ_0 to deep water with density ρ_m can be calculated from a relationship analogous to expression [3.22]:

$$\Delta V_f = \frac{M}{\rho_0} - \frac{M}{\rho_m} \quad [3.25]$$

The variation in float volume ΔV_{pT} engendered by variation in pressure ($p_m - p_0$) and temperature ($T_m - T_0$), however, should be taken into account in order to calculate the variation in volume ΔV to be generated in order to pass from pressure p_0 to pressure p_m. We then have:

$$\Delta V_{pT} = \frac{M}{\rho_0}\left[\gamma(p_m - p_0) - \alpha(T_m - T_0)\right] \quad [3.26]$$

Knowing that $\Delta V = \Delta V_f - \Delta V_{pT}$, *by* combining relationships [3.25] and [3.26] we get:

$$\Delta V = M\left[\frac{1}{\rho_0}\left[1 - \gamma(p_m - p_0) + \alpha(T_m - T_0)\right] - \frac{1}{\rho_m}\right] \quad [3.27]$$

If we consider a float with a mass of 33 kg, whose tube is made of Alu 6061 T6 alloy (as is the case for PALACE and PROVOR floats) to be dropped to 2,000 m, it is necessary to generate a change in volume between 120 and 320 cm^3, depending on differences in density that exist between the surface and deeper layers of water.

As the variation in volume is controlled automatically by a pressure sensor located in the float, it is not constant during the descent. In fact, it changes as a function of local density changes encountered by the float on its vertical path. This is illustrated in Figure 3.21. Positioning accuracy at depth depends not only on the ability to generate sufficient changes in volume, but also on the accuracy of the pressure sensor and its drift characteristics over time and depending on temperature.

This measurement of pressure is also vital for the assessment of the amount of energy stored in the oceans and their changes in volume, which generate some of the changes in sea level. An error in vertical positioning generates errors in the evaluation of the relative warming or cooling of oceanic areas.

It should be noted that certain profilers require ballasting during their preparation phase, depending on the environment in which they will evolve, in order for them to reach their programmed set depth (usually 2,000 m) and so that they can ascend to the surface in any season, regardless of variations in the density of the environment.

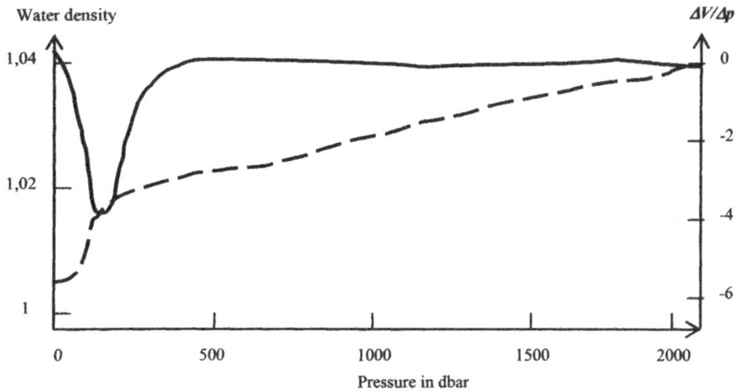

Figure 3.21. *Representation of an oceanic density profile and variations in volume (in cm^3/dbar) that must be generated as a function of pressure for a buoy descend from the surface to a depth of 2,000 m*

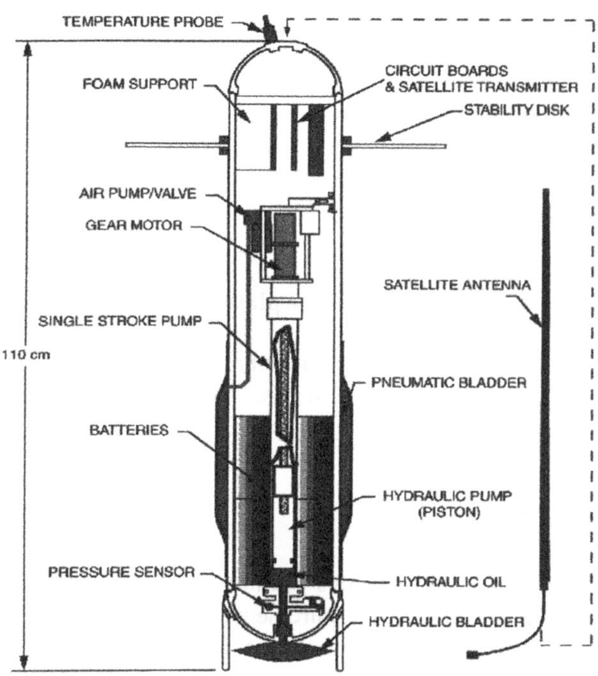

Figure 3.22. *Sectional view of a SOLO multi-cycle buoy with its principal constitutive elements including the hydraulic pump, and external bladders that allow variation of its volume (Courtesy of S. Riser, University of Washington, USA)*

3.3.4.3. *The equation for the movement of subsurface drifters*

The fundamental principle of dynamics allows us to model the hydrodynamic behavior of drifters and therefore better predict their behavior. If we suppose the drifting float is in an environment that is only animated by horizontal movements and is perfectly Lagrangian, the result of forces exerted on it is interpreted by equation [3.28], in which $\overrightarrow{P_a}$ is the Archimedes force, $\overrightarrow{F_{tp}}$ the drag force of pressure and $\overrightarrow{F_{tf}}$ the force of friction.

$$M\overrightarrow{\frac{d^2z}{dt}} = M\vec{g} + \overrightarrow{P_a} + \overrightarrow{F_{tp}} + \overrightarrow{F_{tf}} \qquad [3.28]$$

The amplitude of $\overrightarrow{P_a}$ is defined by equation [3.5], that of $\overrightarrow{F_{tf}}$ by equation [3.8] and that of $\overrightarrow{F_{tp}}$ by equation [3.9], all of which are provided in section 3.2.2.1.

Relationship [3.28] is a second-order nonlinear differential equation for which the coefficients are not constant. They involve the profiler volume, which is dependent on the temperature and pressure of the environment, the density of water, which is also dependent on temperature and pressure, and salinity. They involve friction and drag coefficients that depend on the Reynolds number. The resulting model is therefore complicated to set up.

3.4. Instrumented buoys and underwater platforms

In addition to moorings and drifters, heavier means have been developed to take measurements on the surface and at depth. These are instrumented buoys and underwater platforms.

3.4.1. *Instrumented buoys*

These are floating and autonomous platforms that are equipped with diverse instruments, destined to perform measurements of air and water parameters at a fixed position. Their role is to ensure constant sampling of these parameters and to transmit these data to a ground station. Their energetic autonomy requires that they are:

– connected to land by an electric cable if they are near the coast and consist of equipment that consume a lot of energy;

– equipped with batteries if they are a long way from any coast (this requires a means by which the batteries can regularly be changed); or

– equipped with solar panels or energy recovery systems to recharge the batteries, when autonomy is necessary.

They are placed in a fixed position by using the mooring technologies described in section 3.2, which allow anchoring on the sea bed. They transmit their data using the systems described in sections 2.7.1 or 2.7.2.7, or more simply by using *Wi-Fi* or *Global System for Mobile communications* (*GSM*) connections if they are near the coast.

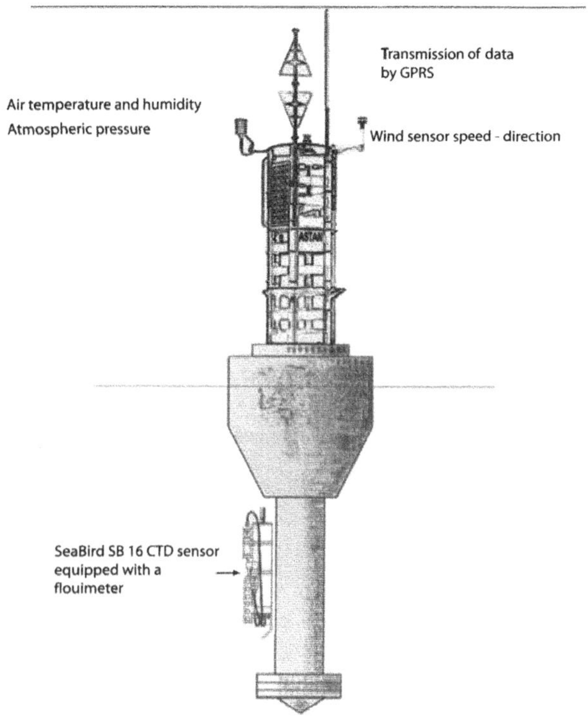

Figure 3.23. *Representation of an ASTAN buoy installed in the English Channel. Sensors are fixed onto the submerged part, allowing continuous measurement of temperature, conductivity, pressure and chlorophyll fluorescence. The section above includes solar panels that recharge the batteries that are located inside the mast, wind, temperature, humidity, and atmospheric pressure sensors. There is an antenna for the transmission of data by General Packet Radio Service (GPRS) protocol (Courtesy of CNRS Roscoff Biological Station, CETMEF, DT-INSU)*

Wi-Fi is a communication protocol currently used on most micro-computers. It is governed by the standards produced by the IEEE 802.11 group or ISO/IEC 2201-11. It allows data rates of 8 Mbit/s over ranges up to 300 m in an obstacle-free area. It is used to interconnect computers, routers, personal digital assistants (PDAs) or internet decoders.

GSM is a communications standard established for cell phones. The scope of a GSM antenna can be 50 km, which therefore allows its use at points that are distant from the coast. GSM networks transmit spoken communications but also data in *Wireless Application Protocol* or *WAP* mode. WAP allows access to the internet from a mobile phone or PDA. Thus, *File Transfer Protocol* type files can be directly transmitted to the WEB server from a laboratory. GSM has an extension called *General Packet Radio Service*, which sends data in packets. This technique economizes on transmission time, with transmission only occurring when a determined assembly of data is available.

These buoys can be used to perform directional spectrum measurements of waves, atmospheric flux of the surface of the water, solar radiation in radiance or irradiance (section 2.10.2) coupled with underwater optical measurements (sections 2.10.3 to 2.10.5) or physicochemical parameters (section 2.11). In the case of solar radiation measurements, the buoys must be designed to be "transparent" to ocean swell, that is to say they must remain vertical regardless of surface current and waves. When it is too deep and the buoy cannot be fixed to the sea bed, it is necessary to design fixtures that minimize the drag coefficient of the structure and maximize Archimedes push (section 3.2.2.1). This is what was done with the "Boussole" buoy, installed in the Mediterranean to calibrate data on water color measured by satellites (see [ANT 08]).

3.4.2. *Underwater platforms*

These underwater platforms are also called *underwater observatories*. This is, officially, an infrastructure that provides a number of services to instruments used underwater for long-term measurement operations. The list of services includes the providing, powering, bi-directional transmission of data from the platform to the coast or to a boat, the distribution of reference times and the synchronization of measurements.

They can be used to observe one-time events, such as volcanic eruptions or underwater earthquakes, hydrothermal vents, etc., but they are primarily used for observations over periods of several months to several years. *Cabled observatories* are included in this infrastructure category. These observatories are used in astrophysics to detect neutrinos, such as ANTARES or NEMO, which are deployed in the Mediterranean, and also MARS in Monterey Bay in the USA and NEPTUNE in Canada. They meet the same constraints as observatories dedicated to oceanography and they are also instrumented for oceanographic observation.

3.4.2.1. *Materials used*

To meet long-term objectives, research is still under way to improve the characteristics of materials required to make underwater platforms. These materials are common to all instruments used in oceanography, and in general they must meet corrosion constraints linked to biofouling and the pressure or hydrothermal environment. They must also meet constraints connected to prolonged immersion at great depths, i.e. they must perform under pressures between 2,000 and 6,000 dbar.

Few metals can be used for deployments for more than 10 years. Steel with cathodic protection is a solution that is often used as it is inexpensive (see section 3.2.2.4). It has a disadvantage, however, in that the zinc anode needs to be changed regularly, which leads to additional operating costs. Stainless steels of type 316L are rarely used because they are susceptible to crevice corrosion and high-quality steels are expensive. Nickel alloys, such as Inconel 625 or Hastelloy C22, are safer and more interesting solutions, but titanium alloys are preferred. Once again, it is cost that limits their use. Their other drawback is that their electrochemical potential can also cause corrosion of other metals present. Copper alloys have good resistance to corrosion and have the advantage of freeing copper ions, which protect against biofouling (see section 2.2.5). Finally, aluminum alloys from the 6000 and 7000 series are frequently used. These alloys can be protected by anodization or with aluminum–indium anodes.

The use of thermoplastic materials has also been studied. They have the advantage of being insensitive to corrosion, but their long-term immersion in water depends on the ratio between their crystalline and amorphic structures. They are used to mold cables, electrically isolate or make joints, but their use is limited by water infiltration into the material, which is always possible and from their mechanical fragility. Thermoplastics that have the best survival underwater are *PEEK* (polyetheretherketone) and *PCTFE* (polychlorotrifluoroethylene). PEEK is a semi-crystalline polymer that retains its properties regardless of temperature. PCTFE is a polymer formed from fluorocarbon molecules. They are generally used due to their resistance to solvents, acids and bases.

Composite materials also have characteristics that are useful for underwater applications: no corrosion and good mechanical resistance. They have been used for a number of years in the telecommunications industry to protect cables. Repeaters are manufactured out of epoxy glass. They can also be used in the form of resins (epoxy, vinylester), fiberglass for reinforcing structures, carbon fiber to lighten the structures or syntactic foams to make buoys.

Brittle glass or ceramics-based materials can also be used to electrically isolate connectors, build portholes or spherical buoys like those of the ANTARES project, where neutrino detectors are placed in glass spheres. Glass and ceramic materials are very resistant to compression, but they have poor resistance to tensile or shear forces, which result in ductile ruptures of the materials.

3.4.2.2. *The different types of underwater observatories*

The platforms or underwater observatories can have different forms depending on their distance from the coast, the means of data transmission or their size and the number of instruments to be submerged. For these reasons and due to the possibility of the observatories being electrically powered, these platforms can be simple mooring cages identical to those used to place current meters (or tide gauges) on the sea bed (see Figure 3.9), or more imposing tubular structures.

The simplest form of observatory consists of an underwater platform placed on the sea bed that is connected to a ground station by a cable via a junction box that forms the link between the submerged part and the land part of this connection. This is convenient for observatories that are near to the coast. The power supply of the instruments is carried by a cable, which involves energy losses depending on the distance from the platform to the coast. Data can also pass through this cable but it is common to use fiber optics, especially when there are a large number of instruments to consider. Research is under way to try to channel power through a fiber optic cable using laser power installed at the ground station.

The complexity of the shape of an observatory can be increased by the addition of several other platforms also resting on the sea bed linked by the submerged part of the cable. These platforms are nodes that can lead to other platforms, cages or instrumented mooring lines. The overall assembly then forms a network of instruments that can stretch over large geographical areas. These instruments are equipped with communication functions allowing them to exchange or fuse data. New protocols and communication languages are being developed to improve the functions of these networks.

The number of instruments needed, their spatial dispersion and the use of hydrophones or seismometers introduces a new problem: that of time distribution for the synchronization of measurements. The IEEE 1588 standard can be used to resolve this problem, as demonstrated in [DEL 12]. This standard is best known as PTP (Precision Time Protocol). It defines a real-time synchronization method for clocks located in the nodes of a system spread in space, and communicates via an Ethernet type protocol. Ethernet is a network protocol used for local networks that sends data in packets. To work, the PTP requires a primary clock that is placed in the ground station. This is synchronized to UTC (see section 2.6.1). The time signals can be transmitted to the instruments by the Ethernet network. The PTP allows synchronization with accuracies of well below a millisecond.

If these observatories are located a long way from the coast, cable connections are no longer possible. The platforms must then be equipped with acoustic tele-measurement systems. These observatories are called *Stand Alone Acoustic Observatories*. These tele-measurements allow distant platforms to communicate between each other, and to send their data to surface drifters equipped with acoustic receivers and Iridium-type satellite communication systems (see section 2.7.2.7). The surface drifter must be anchored with a ballast on the sea bed to avoid drifting (see section 3.2.2.2 and Figure 3.11a).

Finally, if it is necessary to take measurements under the ice, a particular type of observatory called a *Stand Alone Winch Observatory* can be put into place. The platform on the sea bed is connected by a cable to the mooring line drawn between the sea bed and the surface, where a winch buoy keeps the line vertical. Another procedure involves equipping the platform with an acoustic modem. This modem transmits data when a boat remotely searches for it from an area that is not covered by ice.

3.5. Bibliography

3.5.1. *Oceanographic vessels*

[DES 82] De Stobel F., Cairns J.L., "A small inexpensive towed system for automatically measuring vertical CTD cross-sections", *International STD Conference and Workshop Proceedings*, Marine Technology Society, La Jolla, California, February 1982.

[HAR 95] Hare R., "Budget d'erreurs de profondeur et de position pour les sondages par échosondeur multifaisceaux", *Revue Hydrographique Internationale*, Monaco, vol. LXXII, no. 2, pp. 39-75, 1995.

[IHO 08] IHO, Standards for Hydrographic Survey, 5th edition, Special Publication No. 44 (S-44), IHO, February 2008.

[KAP 96] Kaplan E.D., (Ed.), *Understanding GPS Principles and Applications*, Artech House Publishers, Boston, London, 1996.

[NAT 07] National Oceanic and Atmospheric Administration, Office of Coast Survey, Hydrographic Survey Division, Field Procedure Manual, NOAA, March 2007.

[REY 88] Reynolds M., Hendershot R., Jungck M., Reid, B. "The Zeno Alliance network: a dual-loop fiber optic instrumentation network for ships", *Proceedings of the OCEANS' 88 Conference*, Oct. 31-Nov. 2, pp. 1560-1568, 1988.

[TRU 83] Trump C.L., "Effects of ship's roll in the quality of precision CTD data", *Deep-Sea Research*, vol. 30, 11A, pp. 1173-1183, 1983.

3.5.2. *Moorings and anchored floats*

[BER 76] BERTEAUX M.H., *Coastal and Oceanic Buoy Engineering*, Woods Hole Oceanographic Institution, Wiley Interscience, New York, 1976.

[BRU 91] BRUGG B., DENGG J., "Differences in drift behavior between drogued and undrogued satellite-tracked drifters", *Journal of Geophysical Research*, vol. 96, no. C4, pp. 7249-7263, 1991.

[CAR 05] CARTER OHLMAN J., WHITE P.F., SYBRANDY A.L., NIILER P.P., "GPS-cellular drifter technology for coastal ocean observing system", *Journal of Atmospheric and Oceanic Technology*, vol. 22, pp. 1381-1388, 2005.

[CHA 90] CHARACKLIS W. G., MARSHALL K.C., *Physical and Chemical Properties of Biofilms*, John Wiley and Sons, New York, 1990.

[CHA 11] CHAN W.L., KANG T., "Simultaneous determination of drag coefficient and added mass", *IEEE Journal of Oceanic Engineering*, vol. 36, no. 3, pp. 422-429, 2011.

[CHE 89] CHERESKIN T., NIILER P.P., Poulain P., "A numerical study of the effects of upper ocean shear on flexible drogued drifters", *Journal of Atmospheric and Oceanic Technology*, vol. 6, no. 2, pp. 243-253, 1989.

[CHH 77] CHHABRA M.K., "Correction of vector averaging current meter records from the MODE-1 central mooring for the effects of low frequency mooring line motion", *Deep-Sea Research*, vol. 24, pp. 279-287, 1977.

[GOU 74] GOULD W.J., SAMBUKO E., "The effect of mooring type on measured values of ocean currents", *Deep-Sea Research*, vol. 22, pp. 55-62, 1974.

[HOW 88] HOWE B.M., MUNK W. H., "Deep-sea moorings in a tidal current", *Deep-Sea Research*, vol. 35, pp. 111-119, 1988.

[HUR 07] HURLEY J., DE YOUNG B., WILLIAMS C.D., "Reducing drag and oscillation of spheres used for buoyancy in oceanographic moorings", *Journal of Atmospheric and Oceanic Technology*, vol. 25, pp. 1823-1833, 2007.

[LAC 90] LACOMBE P., BAROUX B., BÉRANGER G. (eds), *Les aciers inoxydables*, Les éditions de physique, France, 1990.

[LAN 90] LANGLOIS G., MAZÉ R., "Influence of mooring line motion on current measurements", *Deep Sea Research*, vol. 37, no. 8, pp. 1363-1374, 1990.

[MAR 79] MARICHAL D., Contribution à l'étude statique et dynamique des câbles sous-marins, doctoral thesis, Université de Nantes, 1979.

[MAR 95] MARCUS P., OUDAR J., *Corrosion Mechanism in Theory and Practice*, Marcel Decker Inc., New York, 1995.

[MEI 08] MEINEN C.S., "Accuracy in mooring motion temperature corrections", *Journal of Atmospheric and Oceanic Technology*, vol. 25, pp. 2293-2303, 2008.

[MOR 12] MORISSET S., REVERDIN G., BOUTIN J., MARTIN N., YIN X., F. GAILLARD F., BLOUCH P., ROLLAND J., FONT J., SALVADOR J., "Surface salinity drifters for SMOS validation", *Mercator Ocean-Coriolis Quarterly Newsletter*, Special Issue, vol. 45, pp. 33-37, 2012. [NII 95] NIILER P.P., SYBRANDY A.L., BI K., POULAIN P., BITTERMAN D., "Measurements of the water following capability of holey-sock and TRISTAR drifters", *Deep-Sea Research*, vol. 42, 1951-1964, 1995.

[REV 07] REVERDIN G., BLOUCH P., J. BOUTIN J., NIILER P.P., ROLLAND J., SCUBA W., LOURENÇO A, RIOS A.F., "Surface salinity measurements – COSMOS 2005 experiment in the bay of Biscay", *Journal of Atmospheric and Oceanic Technology*, vol. 24, pp. 1643-1654, 2007.

3.5.3. *Drifting floats*

[AND 10] ANDRE X., LE RESTE S., ROLIN J. F., "Arvor-C: a coastal autonomous profiling float", *Sea Technology*, Feb. pp. 10-13, 2010.

[BAC 01] BACON S., CENTURIONI L.R., GOULD W.J., "The evaluation of salinity measurements from PALACE floats", *Journal of Atmospheric and Oceanic Technology*, vol. 18, pp. 1258-1266, 2001.

[BAL 09] BALLABRERA-POY J., MOURRE B., GARCIA-LADONA E., TURIEL A., FONT J., "Linear and non-linear T-S models for the eastern North Atlantic from Argo data: role of surface salinity observation", *Deep Sea Research I*, vol. 56, pp. 1605-1614, 2009.

[BAR 12] BARKER P.M., DUNN J.R., DOMINGUES C.M., WIJFFELS S.E., "Pressure sensor drifts in Argo and their impacts", *Journal of Atmospheric and Oceanic Technology*, vol. 28, pp. 1036-1049, 2012.

[BIT 90] BITTERMAN D.S., NIILER P.P., AOUSTIN Y., A. DU CHAFFAUT A., Drift Buoy Intercomparison Test Results. NOAA Data Report ERL AOML-17, NOAA, Miami, USA, 1990.

[BÖH 05] BÖHME L., SEND U., "Objective analyses of hydrographic data for referencing profiling float salinities in highly variable environments", *Deep-Sea Research II*, vol. 52, pp. 651-664, 2005.

[BRE 09] BRECHNER OWENS W., WONG A.P.S., "An improved calibration method for the drift of the conductivity sensor on autonomous CTD profiling floats by θ-S climatology", *Deep Sea Research I*, vol. 56, pp. 450-457, 2009.

[DOR 12] D'ORTENZIO F. *et al.*, "Autonomously profiling the nitrate concentrations in the ocean: the PRONUTS project", *Mercator Ocean-Coriolis Quarterly Newsletter*, Special Issue, vol. 45, pp. 8-12, 2012.

[DUR 05] DURAND F., REVERDIN G., "A statistical method for correcting salinity observations from autonomous profiling floats: an ARGO perspective", *Journal of Atmospheric and Oceanic Technology*, vol. 22, pp. 292-300, 2005.

[GEY 89] GEYER W.R., "Field calibration of mixed layer drifters", *Journal of Atmospheric and Oceanic Technology*, vol. 6, pp. 333-347, 1989.

[GOU 05] GOULD W.J., "From Swallow floats to Argo-the development of neutrally buoyant floats", *Deep Sea Research II*, vol. 52, pp. 529-543, 2005.

[JAM 03] JAMSTEC, NOAA, *First ARGO Science Workshop*, Tokyo, Japan, November 12-14, 2003.

[KIK 07] KIKUCHI T., INOUE J., LANGEVIN D., "Argo-type profiling float observations under the Arctic multilayer ice", *Deep-Sea Research I*, vol. 54, pp. 1675-1686, 2007.

[KÖR 05] KÖRTZINGER A., SCHIMANSKI J., SEND U., "High quality oxygen measurements from profiling floats: a promising new technique", *Journal of Atmospheric and Oceanic Technology*, vol. 22, pp. 302 – 308, 2005.

[MAC 89] MACKA D.L., W.R. CRAWFORD W.R. *et al.*, "A performance comparison for 2 Lagrangian drifter designs", *Atmosphere-Ocean*, vol. 27, no. 2, pp. 443-456, 1989.

[OHL 05] OHLMANN, J.C., WHITE P.F., SYBRANDY A.L., NIILER P.P., "GPS-cellular drifter technology for coastal ocean observing systems", *Journal of Atmospheric and Oceanic Technology*, vol. 22, pp. 1381-1388, 2005.

[OLL 94] OLLITRAULT M., LOAËC G., DUMORTIER C., "MARVOR: a multi-cycle RAFOS float", *Sea Technology*, February 1994.

[PRO 12] PROVOST C., PELON J., "Ice, Atmosphere, Ocean Observing System: the EQUIPEX-funded IAOOS project", *Mercator Ocean-Coriolis Quaterly Newsletter*, Special Issue, vol. 45, pp. 5-7, 2012.

[RIO 12] RIO M.-H., "Use of altimetric and wind data to detect the anomalous loss of SVP-type drifter's drogue", *Mercator Ocean-Coriolis Quarterly Newsletter*, Special Issue, vol. 45, pp. 28-32, 2012.

[ROS 83] ROSSBY T., DORSON D., "The deep drifter: a simple tool to determine average ocean currents", *Deep Sea research*, vol. 30, pp. 1279-1288, 1983.

[ROS 85] ROSSBY H.T., LEVINE E.R., D. CONNORS N., "The isopycnal swallow float – A simple device for tracking water parcels in the ocean", *Progress in Oceanography*, vol. 14, pp. 511-525, 1985.

[ROS 86] ROSSBY T., DORSON D., FONTAINE J., "The RAFOS system", *Journal of Atmospheric and Oceanic Technology*, vol. 3, pp. 672- 679, 1986.

[SCH 07] SCHMID C., MOLINARI R.L., SABINA R., DANESHZADEH Y.-H., "The real-time data management system for Argo profiling float observations", *Journal of Atmospheric and Oceanic Technology*, vol. 24, pp. 1608-1628, 2007.

[STE 05] STEIN D.W.J., "Statistical characteristics of moving acoustic sources in ocean waveguides", *Journal of the Acoustical Society of America*, vol. 98, pp. 3, 1995.

[SYB 91] SYBRAND A.L., NIILER P., WOCE/TOGA Lagrangian Drifter Construction Manual. WOCE Report No. 63; SIO Report No. 91/6, Scripps Institution of Oceanography, USA. 1991.

[THA 12] THADATHIL P., BAJISH C.C., BAHERA S., GOPALAKRISHNA V.V., "Drift in salinity data from Argo profiling floats in the sea of Japan", *Journal of Atmospheric and Oceanic Technology*, vol. 29, pp. 129 – 138, 2012.

[WEB 70] WEBB D.C., TRUCKER M.J., "Transmission characteristics of the SOFAR channel", *Journal of the Acoustical Society of America*, vol. 48, pp. 768-769, 1970.

[WON 03] WONG A.P.S., JOHNSON G.C., OWENS W.B., "Delayed-mode calibration of autonomous CTD profiling float salinity data by θ-S climatology", *Journal of Atmospheric and Oceanic Technology*, vol. 20, pp. 308-318, 2003.

[XIN 11] XING X., MOREL A., CLAUSTRE H., "Combined processing and mutual interpretation of radiometry and fluorimetry from autonomous profiling bio-Argo floats: chlorophyll a retrieval", *Journal of Geophysical Research*, vol. 116, C06020, 2011.

3.5.4. *Buoys and instrumented platforms*

[ANT 08] ANTOINE D., GUEVEl P., DESTÉ J.F., BÉCU G., LOUIS F., SCOTT A.J., BARDEY P., "The 'BOUSSOLE' buoy – A New transparent-to-swell taut mooring dedicated to marine optics: design, tests, and performance at sea", *Journal of Atmospheric and Oceanic Technology*, vol. 25, pp.968- 989, 2008.

[CHA 12] CHAFFEY M, BIRD L., ERICKSON J., GRAYBEAL J., HAMILTON A, HEADLEY K., KELLEY M., MCBRIDE L., MELLINGER A., MEESE T., O'REILLY T., PAUL W., RISI M., RADOCHONSKI W., MBARI's buoy seafloor observatory design, MBARI, www.mbari.org/rd, 2012.

[DEL 12] DEL RIO J., TOMA D., SHARIA-PANAHI S., MANUEL A, GEIRINHAS RAMOS H., "Precision timing in ocean sensor systems", *Measurements Science and Technology*, vol. 23, pp. 1-7, 2012.

[GRA 00] GRABER H.C., TERRA E.A., DONELAN M.A., PETERS D.B., "ASIS-A new air-sea interaction spar buoy: design and performance at sea", *Journal of Atmospheric and Oceanic Technology*, vol. 17, pp. 708- 720, 2000.

[MAN 10] MANUEL-LAZARO A., NOGUERAS M., DEL RIO J., "OBSEA: an expandable seafloor observatory", *Sea Technology*, vol. 51, pp. 37-39, 2010.

[VUI 10] VUILLEMIN R., SANFILIPPO L., "A compact, low-power in situ flow analyser for marine applications. A new tool for in situ analysis implemented on an automated buoy in the Mediterranean sea", *Sea Technology*, March, pp. 29-32, 2010.

Chapter 4

Evolutions and other Measurement Concepts

4.1. Other processes for measuring salinity and density

Until 1978, salinity was evaluated based on water chlorinity measurements. Technological advances in the 1960s to 1970s allowed the development towards more accurate techniques based on electrical conductivity measurements, which led to the Practical Salinity Scale (PSS-78) being defined, which is still in use (section 1.2.3). *In situ* conductivity measurements are, however, strongly correlated with temperature and relatively little with salinity. Moreover, this measurement introduces errors in the calculation of density (section 1.3). Thus, Millard and Seaver showed that a density calculation made from a *refraction index* measurement introduces an error four to five times smaller than if the calculation were carried out based on a conductivity measurement [MIL 90]. Hence the concept of using index measurements to directly determine sea water density.

The refractive index is defined as being the ratio between the propagation speed of light in a vacuum and that in the environment it passes through. This difference in speed is reflected in geometric optics by a deviation in the propagation direction of light and hence an angle of refraction r, which is proportional to the ratio between indexes n_1 and n_2 of the environments it passes through (see Figure 4.1). One of the Descartes-Snell laws connects the angle of incidence i of light to r. We have:

$$n_1\sin(i) = n_2\sin(r) \qquad [4.1]$$

The idea of using the refractive index to determine the properties of sea water is not new, as it was first implemented in 1877. At that time, Hilgard who worked for the U.S. Coastal Survey Office converted his sextant into a goniometer to make density measurements with a prism at minimum deviation. Since, the absence of sensors that can be used *in situ* and the lack of accuracy of the algorithms linking the refractive index to salinity, temperature and density have hampered the development of this method of measurement. The publication of new algorithms or polynomial relationships in the 1990s, in particular those created by Millard and Seaver, the evolution of technology and, more importantly, the adoption of the *TEOS-10* relationships (see section 1.2.3.1), which use the notion of absolute salinity, have revived interest in refractive index measurements.

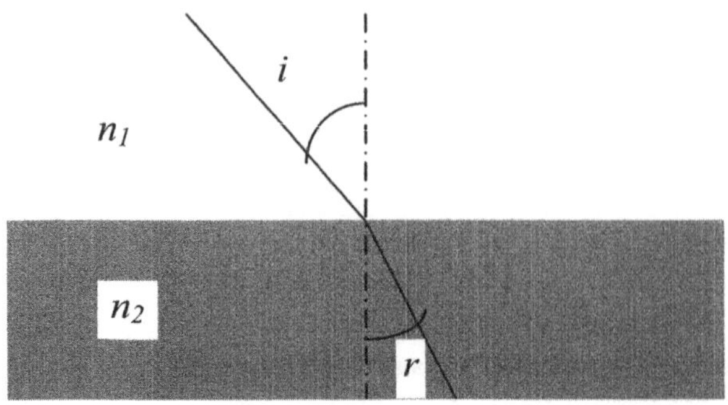

Figure 4.1. *Representation of the angles of incidence i and refraction r of a beam of light passing into two media with different indexes n_1 and n_2*

4.1.1. *Relationship between density and refractive index*

In atomic theory, matter is perceived as a set of particles (atoms or molecules) that interact in a vacuum. These particles produce an internal field that can be altered by all other external fields applied to them. Maxwell's theory, as applied to them, describes the effects of the electromagnetic field propagation of light, regardless of the nature of the environment traversed. The link between these two theories was found, almost simultaneously and independently in 1880–1881 by two physicists, Lorentz and Lorenz. They defined the concept of average polarizability α of an electric field in contact with matter, from the number N of molecules per unit of volume and dielectric constant of matter ε:

$$\alpha = \frac{3}{4\pi N}\frac{\varepsilon - 1}{\varepsilon + 2} \qquad [4.2]$$

ε is connected to the refractive index, n, by Maxwell's relationship: $\varepsilon = n^2$. A connection was hence established between the physical characteristics of matter and light waves.

We know that N, being equivalent to density ρ, varies with temperature, pressure and the different species of molecules present in matter and, in our case, dissolved into sea water. The measurement of n, associated with the *Lorentz–Lorenz relationship*, allows characterization of the environment traversed by light. However, this relationship does not account for oscillatory effects engendered by light wave frequency on particles of matter. These effects result in a variation of the value of n as a function of the wavelength λ of light. This phenomenon is called dispersion or fluctuation. Relationship [4.2] can be rewritten to reflect this, and it leads to approximate relationships in the following form [4.3], which we owe to Sellmeier:

$$n^2 - 1 = \sum_k \frac{\rho_k \lambda^2}{\lambda^2 - \lambda_k^2} \text{ with } \lambda << \lambda_k \qquad [4.3]$$

Here, λ_k represents the wavelength of the oscillation of particle k. These relationships can be broken down and presented in different forms. Thus, in the case of water, Briot showed that the relationships could be written as the empirical formula:

$$n^2 = -A'\lambda^2 + A + \frac{B}{\lambda^2} + \frac{C}{\lambda^4} \qquad [4.4]$$

where A', A, B and C are experimentally determined constants.

More recently, many works have been devoted to the study of the refractive index of distilled water and that of sea water. From statistical analyses of experimental measurements of n as a function of T, p, S and λ, made by different laboratories, these studies have helped to develop polynomial relationships inspired by that of Briot, or algorithms that directly connect n to T, p, S and λ. For fresh water, the most comprehensive study was published in 1990 by Schiebener *et al.* [SCH 90]. It presents a new formula valid for: $0.2 < \lambda < 2.5$ µm, $-10 < T < +500$°C, $0 < p < 100$ MPa and $0 < \rho < 1{,}045$ kg.m^{-3}. The uncertainties obtained vary depending on the different sub-domains from 1×10^{-2} to 5×10^{-6} on the absolute determination of n. This formula was revised in 1998 by Harvey *et al.* [HAR 98].

For sea water, the publication by Millard and Seaver is currently an essential reference [MIL 90]. The algorithm they developed consists of 27 terms that were determined from 428 data points and cover the fields of temperature, salinity and pressures usually encountered in oceanography: $0 < T < 30°C$, $0 < S < 40$ and $0 < p < 11{,}000$ dbar. It can be used for $0.5 < \lambda < 0.7$ µm and its accuracy is also variable, depending on the sub-domains being studied, the absolute standard deviation of n varies from 4×10^{-7} to 8×10^{-5}. It comes in the following form:

$$n(T, p, S, \lambda) = n_1(T, \lambda) + n_2(T, \lambda, S) + n_3(p, T, \lambda) + n_4(S, p, T) \quad [4.5]$$

$$n_1(T, \lambda) = A_0 + L_2\lambda^2 + LM_2/\lambda^2 + LM_4/\lambda^4 + LM_6/\lambda^6 + T_1T + T_2T^2 + T_3T^3 + T_4T^4 + TL\,T\lambda + T_2LT^2\lambda + T_3LT^3\lambda \quad [4.6]$$

with:

$A_0 = 1.3280657$ $\quad T_2 = -0.0000030738272$

$L_2 = -0.0045536802$ $\quad T_3 = 0.000000030124687$

$LM_2 = 0.0025471707$ $\quad T_4 = -2.08831178 \times 10^{-10}$

$LM_4 = 0.000007501966$ $\quad TL = 0.000010508621$

$LM_6 = 0.000002802632$ $\quad T_2L = 0.00000021282248$

$T_1 = -0.0000052883907$ $\quad T_3L = -0.000000001705881$

$$n_2(T, \lambda, S) = S_0S + S_1LM_2S/\lambda^2 + S_1TST + S_1T_2ST^2 + S_1T_3ST^3 + STLST\lambda \quad [4.7]$$

with:

$S_0 = 0.00019029121$ $\quad S_1T_2 = 0.0000000089818478$

$S_1LM_2 = 0.0000024239607$ $\quad S_1T_3 = 1.2078804 \times 10^{-10}$

$S_1T = -0.00000073960297$ $\quad STL = -0.0000003589495$

$$n_3(p, T, \lambda) = P_1p + P_2p^2 + P\lambda M_2\,p/\lambda^2 + PT\,pT + PT_2\,pT^2 + P_2T_2\,p^2T^2 \quad [4.8]$$

with:

$P_1 = 0.0000015868383$ $\quad PT = -0.0000000094834486$

$P_2 = -1.574074 \times 10^{-11}$ $\quad PT_2 = 1.0100326 \times 10^{-10}$

$P\lambda M_2 = 0.000000010712063$ $\quad P_2T_2 = 5.8085198 \times 10^{-15}$

$$n_4(S, p, T) = P_1S\,pS + PTS\,pTS + PT_2S\,pT^2S \quad [4.9]$$

with:

$P_1S = -0.0000000011177517$ $PT_2S = -1.5460458 \times 10^{-12}$

$PTS = 5.7311268 \times 10^{-11}$

A numerical application of these gives:

– for $\lambda = 6.328 \times 10^{-7}$ m, $T = 22°C$, $S = 0$ and $p = 0$: $n_e = 1.3315614 \pm 0.0000004$; and

– for $\lambda = 6.328 \times 10^{-7}$ m, $T = 22°C$, $S = 35$ and $p = 0$: $n_e = 1.33788 \pm 0.00001$.

By carrying out measurements of the refractive index n, it is possible to extract salinity values with these relationships for a given temperature, wavelength and pressure. According to the authors, the standard deviation on this salinity value will be 0.024.

Moreover, from this algorithm and the EOS-80 relationships, they have developed another relationship that determines the density anomaly γ of sea water for D lines of sodium ($\lambda_{average} = 0.58926$ μm). Lorentz and Lorenz had independently determined that the refractive index was correlated to density at 99.7%. The density anomaly that interests oceanographers is within the remaining 0.3%. The relationship devised by Millard and Seaver, which allows this calculation, consists of only six terms:

$$\gamma = 12.2571 + 3089.4201(n - 1.33) + 0.062370\,T + 0.2891 \times 10^{-8}\,p^2 - 0.3624 \times 10^{-5} p\,S + 0.19347\,S \quad [4.10]$$

It is valid in the ranges: $0 < T < 30°C$, $0 < p < 11{,}000$ dbar, $0 < S < 40$. It allows the determination of γ with a standard deviation of 0.049 kg m^{-3}. For information, an uncertainty of $dn = 1 \times 10^{-6}$ in the measurement of n only leads to an uncertainty of $d\gamma = 0.003$ kg m^{-3} in the calculation of γ, according to relationship [4.10].

4.1.2. *Measurement instruments of the refractive index*

Various optical techniques allow us to perform refractive index measurements. Direct measurement of the angle of deflection of a beam passing between two environments of different index or refractometry is the oldest technique. It can be broken down into different methods:

– that of minimum deviation, which requires the use of a goniometer;

– that of the critical angle, for which the principle has been the subject of commercial development following works by Mahrt and Waldmann; and

– that of a bi-prism, where the angle of incidence is optimized.

In the second method, a beam of light penetrates water of index n_e then passes through a prism of index n_v by refraction. It emerges at an angle s, and can then be measured (see Figure 4.2).

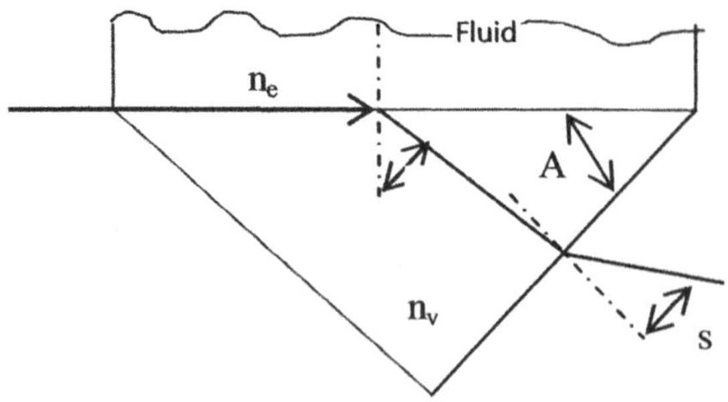

Figure 4.2. *The critical angle method. A beam of light is guided to illuminate the prism at grazing incidence. It is deflected according to the relationship: $n_e = n_v \sin(\theta)$*

The value of n_e, expressed as a function of the output angle s, is then obtained from:

$$n_e = \sin(A)\sqrt{n_v^2 - \sin^2(s)} + \cos(A)\sin(s) \quad [4.11]$$

The device developed by Mahrt and Waldmann uses this relationship. It is equipped with a charge-coupled device (CCD) detector that continuously measures the variations in angle s. Its compactness and performance is remarkable. The optical–electronic assembly is contained in a cylinder that is 5 cm in diameter and 53.6 cm in length. The resolution limit that can be achieved is 5×10^{-7} over n and its compensation in temperature and pressure, according to the inventors, permits a relative uncertainty of measurement of 1×10^{-6} for a sampling frequency of 10,000 samples/s.

The third method is the most recent and it is still under development today. It stems from the production of a prototype consisting of a diode laser, an assembly of two prisms whose concept has been patented, and an analog detector of the type *position sensitive detector* (PSD) (see Figure 4.3). The laser beam passes through the first prism and then the sea water contained in a "V" shaped by the junction with the second prism on which the PSD is set. The refraction undergone by the beam leads to a displacement in the position of the spot of light detected by the PSD. The relation obtained between the variations of the index and position is complex, but proper calibration of the instrument shows that it is linear, with minor corrections for systematic errors linked to variations in temperature, pressure and wavelength of the laser diode (see [LEM 11]).

This principle of measurement was validated by trials at sea carried out with a prototype where the laser beam was reflected by mirrors fixed onto prisms, such that the laser diode and the PSD were in the same plane and gained compacity. This prototype allowed salinity profiles to be obtained by using the Millard and Seaver relationships described in section 4.1.1.

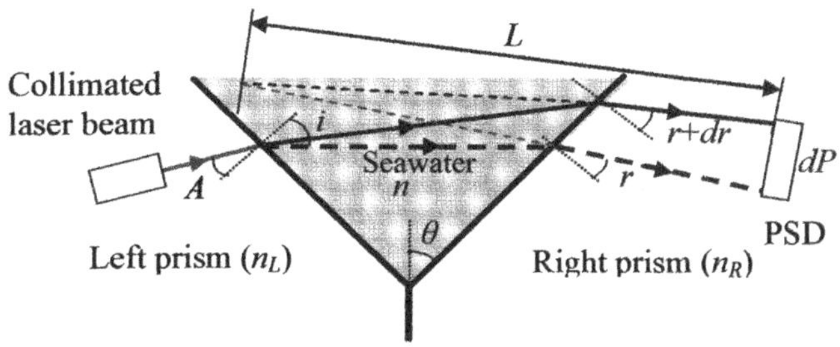

Figure 4.3. *Operating principle of the bi-prism refractometer (Courtesy of © Télécom Bretagne)*

However, refractometry is not the only technique used to trace the refractive index. Interferometry has also been under development in oceanography. There are two-wave interferometers and multiple-wave interferometers. In the case of two-wave interferometers, the fluid to be analyzed is placed on the optical path of one of the waves. Two general structures can be used:

– one is known as Jamin's interferometer and the other is the so-called Mach–Zehnder.

In the Jamin structure, the separation of beams is achieved by total reflection on the rear face of a blade whose entrance is semi-reflective. This type of interferometer was used by Rusby in 1966 to carry out measurements of index variation relative to salinity at constant temperature [RUS 66]. Vlasov and Kostianoy [SEA 97] also used it to build an instrument that was calibrated in a laboratory and used at sea to make some index profiles up to a depth of 4,000 m. In the instrument they made, a reference cell containing distilled water was placed in the gap between two oblique mirrors. It was traversed by a reference beam, while the probe beam passed directly through the medium whose index was to be determined (see Figure 4.4). In this instrument, phase measurement is carried out using a heterodyne modulation. The device is temperature compensated and its sensitivity to vibrations is reduced. If α designates the angular difference between the two beams and l the difference in length between their respective optical paths (20 mm), index variations δn can be deduced from the following relationship based on the equality of optical paths traveled:

$$l\,\delta n = \lambda\,\delta\alpha \qquad [4.12]$$

The optical and electronic elements of this device have been integrated in a cylinder that is about 20 cm in diameter and 1 m in length. Its calibration, according to the authors, has allowed temperature measurements to be made with an uncertainty of just ±0.04°C and salinity measurements with an uncertainty of ±0.005°.

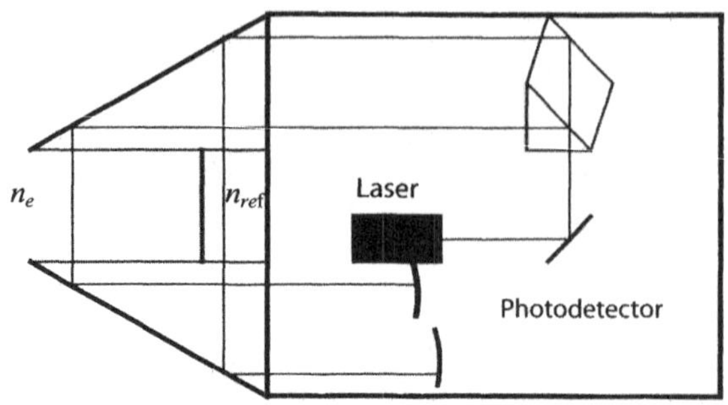

Figure 4.4. *Operating principle of the Vlasov and Kostianoy interferometer*

However, the most generally used structure is that devised by Mach-Zehnder (see Figure 4.5), as it allows greater freedom in choice for the dimensions of the fluid-containing cell. If l is the length of this cell, the phase difference φ between the emerging rays is given by:

$$\varphi = \frac{2\pi}{\lambda} n_e l \qquad [4.13]$$

The measurement of φ can therefore be brought back to the index value of water, n_e. Moreover, index variations n are directly related to the number k of fringes that pass:

$$n = \frac{k\lambda}{l} \qquad [4.14]$$

Relationship [4.14] allows the use of measurement techniques of fractional excess for more precise measurements of n. When combined with the use of numerous wavelengths, these techniques allow us to make highly accurate measurements of absolute index.

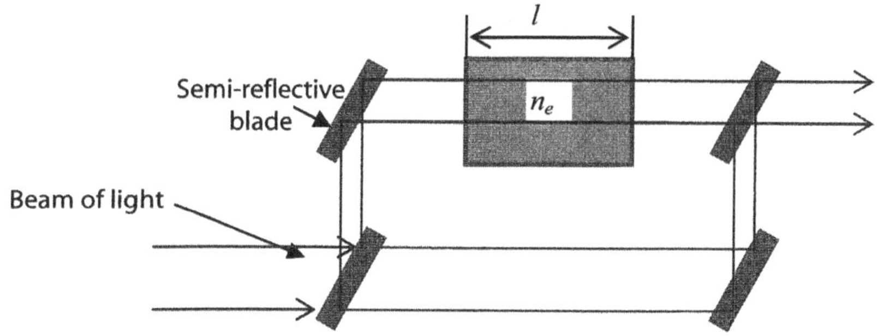

Figure 4.5. *Operating principle of a Mach-Zehnder interferometer*

Mahrt and Kroebel [MAR 84] used this structure in 1984 to build a laboratory salinometer. This device, consisting of two thermostated cells 40 mm in length, compares the salinity of a solution to that of a reference water. Their fringe-counting technique, offering a resolution of 1/100th, the authors estimated that they could determine salinities with an absolute value better than 4×10^{-4}. Lu and Worek also used it to develop a piece of laboratory apparatus capable of carrying out precise index measurements of salt solutions at different temperatures.

The Fabry-Perot type multiple-wave interferometer has, in the same way, been the object of improving the refractive index measurements of fluids. The mirrors, obtaining multiple reflections, are fixed on either side of a cylindrical cavity that can be pressurized. Thus in 1970, Stanley [STA 70] carried out refractive index measurements of sea water at various temperatures (from 0 to 30°C) and under various pressures (from 0 to 1,378 bar), with a standard deviation of $\pm 2 \times 10^{-5}$ and an experimental error estimated to be $\pm 6 \times 10^{-5}$ over n. The diameter of the cavity of this interferometer was only 5 mm and its length was about 3 mm.

Other methods have been and are still being explored to make refractive index measurements of sea water. These include the use of capillary tubes or cubes to create interference fringes and to make measurements of index variation. Measurements made by different experimenters suggest the possibility of determining these variations with an uncertainty level of 1×10^{-6} or 0.05°C in temperature and 0.01 in salinity. We can also include the optode technique (see section 2.10.2.3). Achievement of this relies on the use of polished fiber optics covered with a very thin metal layer (8 nm) that, in turn, is covered with a dielectric film. The width of the metal layer determines the measurement range of the sensor.

A single-fiber polarization mode can excite the surface deposit, leading to the creation of a transverse magnetic mode. Polarization of the beam must be controlled. The parameter measured is the power transmitted. Power is attenuated according to the refractive index of the external environment in contact with the metal deposit. Measurements carried out by Esteban *et al.* [EST 99] showed that the sensor can determine the salinity of solutions to ±0.1 salinity units in the oceanographic range ($0 < S < 40$).

4.2. Acoustic tomography of oceans and acoustic measurements

4.2.1. *General principles*

In some civil and military applications, it is important to be able to quickly acquire information about the physical and dynamic characteristics of oceans and their evolution in space over long distances. In 1978, Munk and Wunsch proposed using a technique called acoustic tomography to determine these characteristics [MUN 78]. As shown in section 1.1.1, the speed of sound in water depends on its density, which itself varies with pressure, temperature and salinity. This variation is at most 5% around 1,500 m/s. When acoustic waves are propagated in the ocean, on their travels they will meet water masses of different temperatures and salinities. In addition to this, wave planes pass through layers whose pressure increases with depth. All of these variations create differences in the acoustic refractive index n. In a similar way to light waves (section 4.1), n is defined by the relationship:

$$n = \frac{c_0}{c} \qquad [4.15]$$

Here, c_0 denotes a reference velocity. The value of n is close to 1 in weakly heterogeneous environments. According to *Snell-Descartes' law* of refraction, the path of these waves, which could be horizontal at the start, is partly modified. From this, we can define the notion of *acoustic rays* (analogous to light rays). *An acoustic ray can be represented by a curve that connects a transmitter E to a receiver R, for which the shape depends on variations in velocity encountered and corresponds to the shortest travel time between E and R.* Hence, if θ_0 is the transmission angle of a ray and c_0 is its initial velocity, in a homogeneous ocean its velocity and direction will depend on the depth z. If we apply Snell-Descartes law, we get:

$$\frac{c(z)}{\cos[\theta(z)]} = \frac{c_0}{\cos[\theta_0]} = constant \qquad [4.16]$$

At each ray R we can associate a travel time τ_i corresponding to the curvilinear integral along the ray in question:

$$\tau_i = \int_R c^{-1} ds \qquad [4.17]$$

Thus, when an acoustic pulse is sent, a series of pulses having followed separate paths that have different amplitudes, are received. The time taken for each ray, that is to say for each pulse received, to travel the specified distance reflects the sum of variations in the velocity of the environments passed, while their amplitudes allow quantification of dissipation of the environment as a function of the energy initially transmitted.

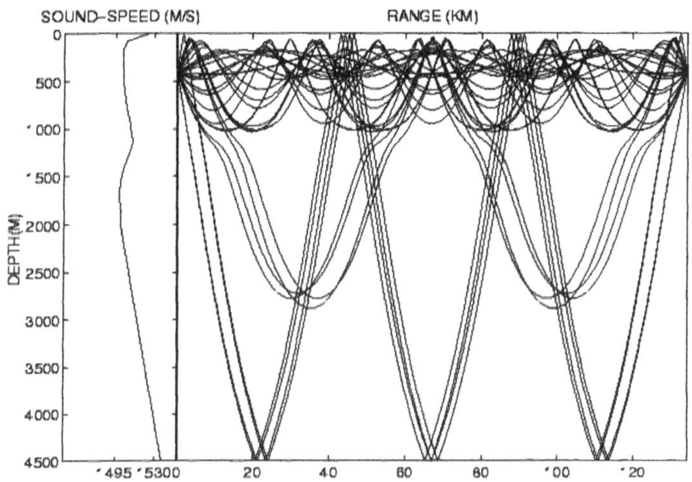

Figure 4.6. *Illustration of the concept of acoustic rays. The left panel shows a velocity profile as a function of depth. The right panel is a visualization of the corresponding acoustic paths. The environments crossed act as wave guides that allow sound propagation over long distances. The presence of minimum velocity results in the formation of natural acoustic channels that tend to guide most of the rays horizontally*

However, as the marine environment is usually in motion due to currents, particles met by waves will sometimes decrease and sometimes increase their apparent velocity, depending on the propagation direction relative to currents. It is therefore necessary to distinguish between a static component of velocity $c(S,z,T)$ and a dynamic component $u(S,z,T)$ linked to current. The static component can be broken down into a reference component $c(z)$ and an unknown perturbation $\delta c(S,z,T)$. We get:

$$C_{measured} = c(z) + \delta c(S,z,T) + u(S,z,T)\, t \qquad [4.18]$$

Here, $c(z)$ can be determined from climatological reference data for example, and the main problem then consists of estimating the value of the $\delta c(S,z,T)$ and $u(S,z, T)$ fields, from a set of τ_i. Munk and Wunsch suggested that measuring the travel time of acoustic pulses transmitted from several sources, and received by several receivers, could allow us to improve knowledge of these fields.

Hydrophones are used to carry out these measurements. As explained in section 2.4.1, a hydrophone is a reciprocal or reversible device. It can be a transmitter or receiver, depending on the way it is controlled. This property is used in tomographic instruments in order to access the current field and perturbations of the field of velocity separately. By combining equations [4.17] and [4.18], we can show that τ_i is:

$$\tau_i = \overline{\tau_i} + \delta\tau_i^c + \delta\tau_i^u \qquad [4.19]$$

where $\overline{\tau_i}$ corresponds to the integration of the component $c(z)$. $\delta\tau_i^c$ and $\delta\tau_i^u$ correspond to the integration of $\delta c(S,z,T)$ and $u(S\ z,T)$. By using the outward and return waves, we get two values for τ_i: τ_i^+ and τ_i^-. If we add these two values together, the components $\delta\tau_i^u$ being in the opposite direction, they disappear and we are left with:

$$\delta\tau_i^+ + \delta\tau_i^- = 2\,\delta\tau_i^c \qquad [4.20]$$

If we subtract them, we get:

$$\delta\tau_i^+ - \delta\tau_i^- = 2\,\delta\tau_i^u \qquad [4.21]$$

The challenge is then to invert relationships [4.20] and [4.21] in order to extract the fields of pressure, temperature, salinity and current. From these fields we can deduce and follow the temporal evolution of the hydrological properties of water masses and we can also detect vortexes, the phenomena of internal waves, etc., as well as the acoustic properties of the sea bed or sediments (natural, porosity, density, etc.). It is impossible to carry out these inversions using analytical techniques. We must use nonlinear digital numerical techniques, such as those of neuronal networks, genetic algorithms and simulated annealing. Currently, it is neuronal networks that seem to give the best results.

Acoustic tomography techniques are, to date, sometimes re-oriented towards studies based on passive acoustics. A hydrophone or an array of hydrophones are used to attempt to locate and quantify sources of noise from fish, crustaceans, volcanoes, earthquakes, waves or melting ice.

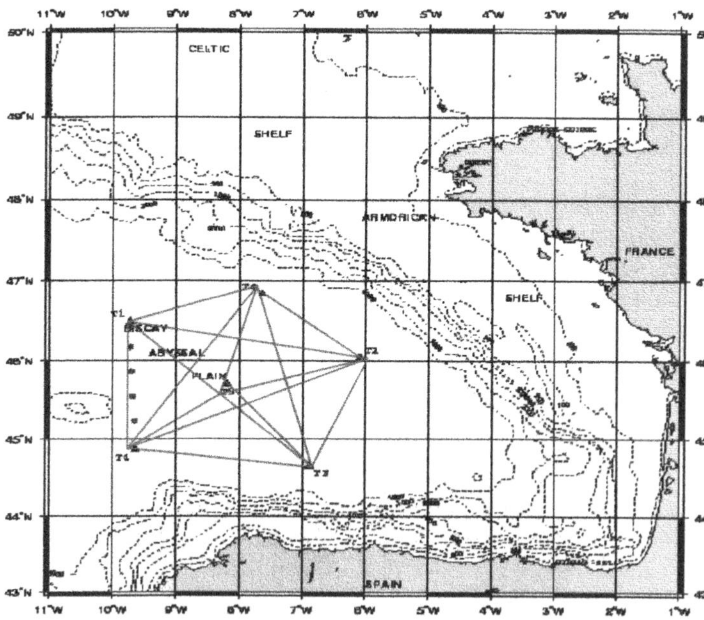

Figure 4.7. *Example of the distribution of acoustic transducers (points numbered T1 to T6) installed in the Bay of Biscay during an acoustic experiment (called GASTOM90) by the Hydrographic and Oceanographic Service of the French Navy. The dots represent the locations of CTD profiles made to exploit tomography results. The grey lines represent the possible combinations of acoustic paths (Courtesy of © SHOM)*

4.2.2. *The instrumentation used*

To ensure wave propagation over long distances, acoustic tomography experiments are traditionally carried out on the one part with low-frequency sources (between 300 and 1,000 Hz) and on the other part by arranging these sources at depths varying between 100 and 2,000 m in order to get good propagation conditions. The transducers have a bandwidth of 50–100 Hz. Their transmission power is in the order of 190 dBre (or relative to 1 μPa) to 1 m, which allows spans of about 600 km. This power is acquired using amplifiers and power supplies of 500 V A. The signal emitted is a carrier frequency whose phase is modulated by a pseudo-random code. Its transmission is not continuous but is within time windows (as in the case of acoustic sources, section 3.3.2.2).

Each transducer receives signals from other sources five to six times per day. These transducers can be placed on buoys or distributed over a cable pulled by a boat to form a hydrophone antenna.

Tomographaphic instruments deployed on moorings can remain underwater for long periods of time (from several weeks to over a year). Given the transmission power, it is necessary to have significant energy reserves. This energy is delivered by packs of alkaline batteries.

We want to increase propagation distances by working at lower frequencies (around 250 Hz) without changing the transmission power. This requires a resizing of transducers so that they retain a good sensitivity to reception and good performance on transmission. It is, however, necessary to increase the power supply (to 1 kVA for 250 Hz).

Figure 4.8. *Launch of a low-frequency transducer used for tomographic experiments where the acoustic source is towed (Courtesy of © SHOM).*

The instruments placed on moorings are subjected to movement due to marine currents. It is therefore necessary to instantly correct their position so as to not introduce additional errors in calculations. These corrections are made by acoustic triangulation using buoys fixed to the sea bed.

More recent projects have shown the feasibility of applying tomography using hydrophones placed on ALACE type drifting floats (section 3.3.2). In 2006, Simons and Nolet showed that it was possible to use SOLO floats to detect seismic waves with the help of a multichannel hydrophone [SIM 06]. Shifts in the speed of acoustic waves emitted during seisms has permitted us to detect and study them. These floats are used instead of or in addition to sea bed observatories known as OBS – for *ocean bottom seismometer*, which are usually used for geo-acoustic studies.

4.3. The unmanned underwater vehicle: a new means for ocean exploration

New concepts have been developed in recent years in terms of oceanic exploration. Classically, sensors necessary for measuring the physical or chemical characteristics of the seas of the planet are carried by the equipment described in Chapter 3. When measurements need to be carried out in a column of water at different fixed points, boats with handling tools are used; moorings are used when measurements need to be taken at fixed points in the column of water over long time spans; and drifting floats are used when we want to follow the displacement of water masses. Measurements made from boats require significant logistics and the costs involved are proportional to this. The floats, meanwhile, cannot explore all parts of the globe on demand. The idea of using autonomous and remote controlled carriers to carry out oceanographic measurements then came about. These carriers are known as *unmanned underwater vehicles* (*UUVs*).

There are three categories of UUV:

– remotely operated vehicles (*ROVs*);

– autonomous underwater vehicles (*AUVs*); and

– gliders.

ROVs are the oldest category. They are remotely operated, i.e. they are controlled and powered to carry out certain actions from the surface by a cable called an umbilicus. They are principally used for inspection operations (pipelines, boat hulls) or for mechanical work at great depths. Some can be inhabited for observation operations or the collection of objects. To perform these actions, they are equipped with cameras and lighting systems. They are also equipped with systems that inform us of their coordinates, their depth and the profile of the sea bed. Data from these sensors are sent to a control unit that displays, records and processes them in real time to allow the operator to control the engine depending on events encountered or tasks to be completed.

AUVs are more recent engines that are entirely autonomous, and were devised for navigation over long distances without any physical connection to the surface. Their use in oceanography can be developed further, as they can be equipped with Doppler effect current profilers (section 2.5.4) or a CTD (conductivity–temperature–depth) assembly (sections 2.1.1 and 2.2.1), even if their design and implementation are subject to problems linked to their energetic autonomy and to their positioning over long distances. They are often used to carry out inspections (pipelines, boat hulls) or are used in hydrography with the help of sounders (section 3.1.2). They can be used under ice or in areas close to the coast that boats cannot access. They can vary in size and weight (from 50 kg to over 1 ton) and vary in shape from small autonomous submarines through to torpedoes (see Figure 4.9).

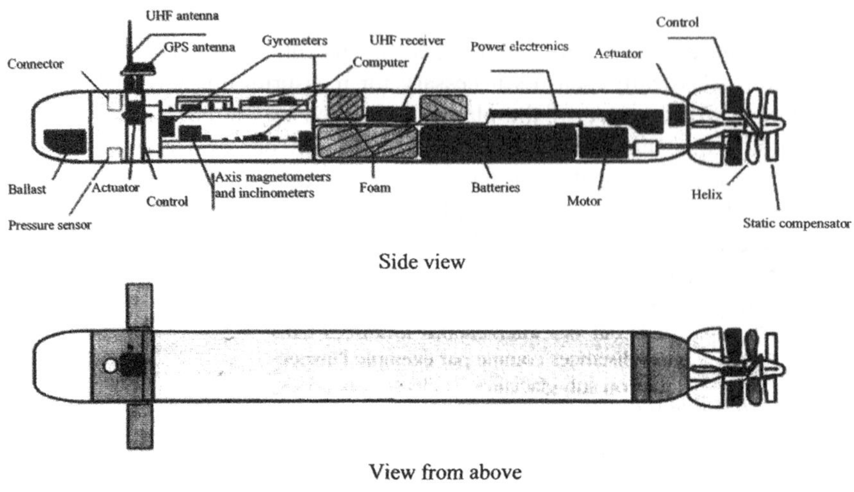

Figure 4.9. *Side and top views of the AUV Taipan developed by the Laboratory of Informatics, Robotics and Micro-electronics at the University of Montpellier. Being of low cost and low weight, this vehicle was designed to carry out physicochemical measurements and take photographs of the sea bed or bathymetric bottom. It has an autonomous range of 25 km. (Courtesy of LIRMM, CNRS - University of Montpellier 2, www.univ-mont2.fr/)*

Gliders, or underwater planers (see Figure 4.10), are a new concept and are a real change to the two earlier categories of UUV. They meet the afore-mentioned definition of an AUV, with the difference that they have no system of propulsion. They are carried by currents, but may be remote controlled, which distinguishes them from drifting floats. They are mainly used in the field of oceanography.

With autonomous surface vehicles, otherwise known as unmanned surface vehicles (USV), the AUVs and gliders are classified into the more general category of autonomous marine vehicles, whose lack of legal definition can cause problems with the authorization of navigation and maritime security.

Figure 4.10. *Launching of a prototype glider (Courtesy of © SHOM – ENSIETA)*

4.3.1. *Energetic autonomy*

The problem with energetic autonomy is common to all autonomous instruments used in oceanography (instrumented buoys, drifting floats, etc.), but is particularly important for AUVs. Battery technology is an evolving field, given the development of electric land vehicles. Oceanographic instrumentation benefits from these advances, as well as those connected to renewable energies.

Table 4.1 compares the performance in terms of the power delivered per hour and per kilo by the different technologies available. The numbers given are, however, subject to change based on research being conducted in this area. Batteries are usually reserved for small AUVs, gliders and expendable engines. Rechargeable batteries are preferred for longer missions. Generally, as the needs of the technology are considerable in terms of power sources, progress in research relating to electrodes and electrolytes is constant, with increasing energy density and specific energy, and power density.

The specific energy of a technology is not the only element to take into account when choosing an electric power source. The temperature range of operation, pressure resistance, balance of charging and discharging cells, self-heating, the possibility of transport by plane or boat, etc., are other elements to consider.

The level to which a battery is charged can be estimated by calculating their open circuit voltage (*OCV*):

$$OCV = V_{term} + I\,R \qquad [4.22]$$

where V_{term} is the final voltage of the accumulator, I the actual output current and R its internal resistance. OCV can also be directly estimated by measuring the voltage across the battery, but this value depends strongly on the discharge rate, which makes this measurement unreliable. Management of the level of charge is important for battery life. Recharging or overcharging too frequently can lead to premature wear of its elements.

In addition to these energy sources, alternative ways to power AUVs are being studied or have been developed. These include renewable sources of energy whose method of action is called "*energy harvesting*". For those destined to navigate at the surface, solar rays or the movement of waves can be used. For others, the electrolysis of sea water could be way of sourcing power.

Certain AUVs, such as the SAUV II made by Falmouth Scientific, Inc., are already equipped with solar panels. These panels recharge lithium-ion batteries when the AUV navigates the surface of the water, which in theory gives it unlimited autonomy. Such AUVs can collect between 25 and 200 W, which allows for recharge rates between 400 and 700 W/h/day at central latitudes in full sunlight (according to the manufacturers), knowing that its battery power is 2 kWh. Solar panels are also widely used on unmanned surface vehicles, such as Charlie, created by the Italian National Research Council, which uses a continuous current motor of 300 W to move forward. On this vehicle, there are 32 W flexible solar panels, which recharge four 40 Ah batteries.

In 2011, Joshi *et al.* [JOS 11] studied the applicability of the use of solar panels on AUVs in terms of power collected and the optimization of this collection based on the solar spectrum, but also its absorption according to the turbidity of the environment (see sections 2.10.1 and 2.10.4) and the depth of navigation [JOS 11]. They concluded, among other things, that the reduction and offset of the solar spectrum between 400 and 600 nm in sea water was beneficial to the performance of amorphous silicon cells. This reduction increases the conversion efficiency from 30 to 40% up to 20 cm in depth, and then efficiency decreases as a function of depth because of the absorption and reflection of the environment. Similarly, the refractive index of the water has an advantage in terms of the apparent zenith angle of the sun in water. This is lower than in air, which reduces its influence on cell performance over a full day. This advantage is, however, largely dimmed by a decrease in the intensity of solar rays in water relative to air.

Type	Maximum autonomy in Wh/kg
Lead accumulators	30
NiCd accumulators	60
NiMH accumulators	90
Lithium-ion accumulators $LiNiMnCoO_2$	115 150
Lithium-polymer	200 in air 80–120 *in situ*
AgZn accumulators	120
Lithium batteries	400
Hydrogen–oxygen fuel cell	250

Table 4.1. *The energy performances of different technologies of accumulators currently available on the market. The choice of battery technology may also depend on the nominal operating voltage, rate of self-discharge or lifespan in terms of charge–discharge cycles*

4.3.2. *ROV and AUV displacement and positioning*

The problem with positioning has been solved for ROVs and certain AUVs through the use of external references to carry out acoustic triangulation and the use of inertial navigation instruments. These systems are also completed by less traditional means of locating and nonlinear guiding to increase their autonomy. These vehicles must have *comprehension autonomy* in order to recognize their status and that of their environment. This recognition is enabled with the help of the data fusion from different sensors:

– inertial navigation systems equipped with gyroscopic or magnetic compass and inclinometers;

– pressure sensors to indicate depth and activate descent or ascent in response to surface commands;

– imaging sonars, such as the Reson 8125, or optical systems for positioning relative to the sea bed and to obstacles;

– GPS systems for repositioning upon returns to the surface;

– Doppler current meters, such as the *Workhouse Navigator* made by R.D. Instruments, to measure the displacement speed and ocean currents at the start of deviations in vehicle positioning estimations; and

– multi-beam echo sounders (section 3.1.2.1), such as the EM2000 made by Kongsberg Simrad.

According to the tasks allocated to the AUV, other systems can be used to guide its movement, such as chemical sensors when the dispersal of pollutants needs to be followed or cameras when underwater sea beds need to be explored.

For the ROV and certain AUVs, there are, usually, four types of positioning technologies.

The oldest is called *long base line* (*LBL*). It requires the use of a network of acoustic transducers or *transponders* placed on the sea bed or moored at the surface. The vehicle must then emit a signal that can be detected and translated into a position with the help of transducers. This system requires a setup phase at the site to be explored and knowledge of the propagation speed of sound in this volume to ensure positioning accuracies below 1 m.

Ultra short base line systems succeeded the LBL without replacing them. Hydrophones are fixed onto a unique antenna about 1 m in length. This antenna must be calibrated relative to data issued from the inertial navigation system of the carrier boat. This referencing is difficult to achieve and the signal processing that is used to obtain position data is complex. Moreover, the accuracy of positioning that we get is proportional to the measuring range desired.

Previously reserved for military applications, inertial navigation systems (INS) are now used on AUVs. An INS contains an *inertial measurement unit*, which is composed of three laser gyrometers and three accelerometers. The INS, independently of the marine environment, thus provides position but also heading and incline values. However, its sensors can also be subjected to drifts and it is necessary to periodically verify their calibration.

These problems of drift can be reduced with a *fused INS* approach. These systems fuse data from other instruments (GPS, acoustic positioning systems, Doppler current meter, etc.) with INS data. Errors from each means of measurement can be tracked and compensated for. The GAPS system made by IXblue, for example, does this. GAPS has a range of 4,000 m and a positioning accuracy of

0.2% of the full scale. It can perceive acoustic signals coming from transducers with an aperture of 200°. It can be launched into the water from a boat using a simple pole or placed on a buoy. This tool, which uses software using a method of trajectory tracking based on Kalman filtering, can follow AUV or ROV movement.

4.3.3. *Autonomy in decision-making and communication*

In order for systems used for positioning to work with each other and for the vehicle to adapt to the mission it was programmed for, and eventually to reprogram this mission depending on observations made, AUVs must be equipped with *autonomy of decision*. The use of nonlinear guiding systems based on neuron networks are currently being analyzed with this in mind. These tools allow structures and methods of fusing data from the environment and means of mobile status recognition (see [NAP 05]). Artificial intelligence techniques have also been adopted to adapt missions depending on the probabilities of detecting events. Hence, the T-REX (*Teleo-Reactive Executive*) software has been implemented in the MBARI center's Dorado AUV. Depending on the data provided by sensors installed on the vehicle, the software generates a schedule of actions that the AUV will carry out by choosing the best solution to accomplish the objectives it was programmed to do. For example, when searching for thermal fronts, an AUV carries out radials and reduces the sampling frequency of its measurements if no front is detected so as to cover the maximum amount of space. It automatically increases the frequency as soon as it detects the presence of a significant temperature gradient.

During their journeys, AUVs and gliders must also be able to record and transmit the data collected. They must therefore be equipped with *communication autonomy*. This autonomy can be ensured through the inclusion of sufficient memory that can be emptied at the end of the mission. In this case, processing of data is delayed and it is impossible to interact and redirect the AUV according to the discoveries made. On certain AUVs, communication can be ensured by a fiber optic deployed bit-by-bit throughout its progress. This connection, which remains a source of constraint, then limits the displacement autonomy by several tens of kilometers. AUVs and gliders are therefore obliged to regularly ascend to the surface and transmit their data via radio or satellite channels using Iridium or Argos systems. Acoustic communication systems are under development, however, but they are energy intensive and are not very suitable for the small dimensions of AUVs. The marine environment is not adapted to the propagation of communication signals, as its stratification in temperature and salinity provokes refractions and reflections, and hence multiple echoes. Transmission can only be carried out in bandwidths of a few tens of kHz, which is ridiculously low compared to classic aerial communication systems.

Some systems have been tested based on different techniques, of which the *multiple-input multiple-output* or MIMO technique works differentially by sending the signal over several emission sources and listening with several receivers (see [SON 11]). The transmitter is the AUV, equipped with three transducers and the receiver is an antenna equipped with eight transducers, arranged vertically in the water and connected to a buoy or boat. A hybrid acoustic–optic system has also been created at the Woods Hole Oceanographic Institution. The AUV is equipped with acoustic transmitters, as before, but the signals emitted are sensed by spheres floating a small distance from the surface (<200 m in clear water). They are transmitted to a surface carrier equipped with a photomultiplier, by modulation of light emitted by powerful LEDs located in the spheres.

4.3.4. *Gliders*

We owe the concept of gliders to the Americans Stommel and Webb, who designed the first glider prototype in the 1980s. These underwater gliders behave like floats whose trajectory can be controlled remotely. They are able to dive down to a depth of 1,500 m and unlike propelled AUVs are designed to have considerable autonomy. The first and most well-known is the *Slocum*, which is named after an American skipper, Joshua Slocum, who was the first to circumnavigate the world on his own in a small sailboat. It has a wingspan of about 1.5 m with a weight of 50 kg. It can move at a horizontal speed of 30 cm/s and contains three to ten months' energetic autonomy.

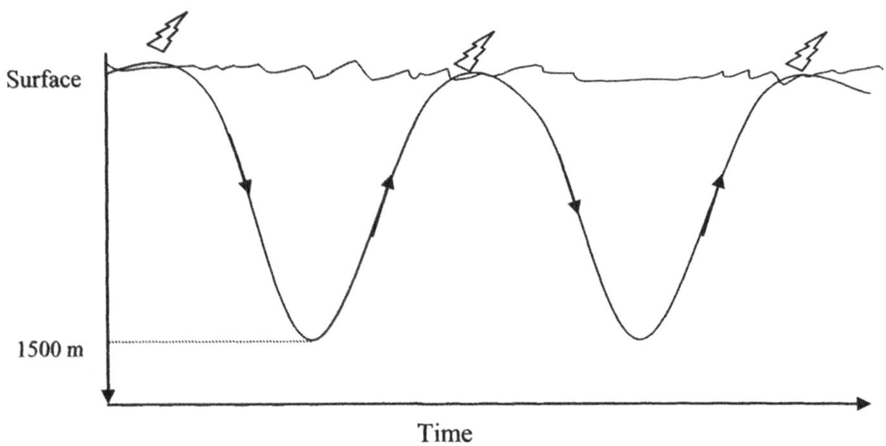

Figure 4.11. *Typical profile movement of a glider during the exploration of a water mass. During the phase when it is near the surface, it is positioned and data recorded at depth are transmitted*

It can be vertically displaced by varying its buoyancy (see section 3.3.4.2). The automated movement of an internal mass (usually its battery block) modifies its center of gravity, which controls its inclination. A hydraulic pump or piston empties or fills an internal reservoir, which allows the glider to ascend or descend. In most oceans, the vertical displacement speeds of water masses are small (around 1 cm/s). However, vertical stratification in temperature is important and this characteristic is exploited to move the gliders. Their vertical displacement gives them horizontal speed. For the Slocum, it is estimated to be around 0.4 m/s. They can therefore travel between 3,000 and 5,000 km during the three to 10 months' for which they have power. Their position is determined by GPS when they rise to the surface. At the surface they then transmit the data recorded during their dive phase.

According to Merckelbach *et al.* [MEL 10], if we ignore their acceleration, the vertical displacement of gliders meets an equilibrium between the force of (net) buoyancy F_B the force of gravity $F_g = m_g g$, lift force F_L and drag force F_D. This equilibrium gives the equations:

$$\begin{aligned} &F_B - \cos(\gamma)F_L - \sin(\gamma)F_D - F_g = 0 \\ &\cos(\gamma)F_D + \sin(\gamma)F_L = 0 \end{aligned} \qquad [4.23]$$

where $\gamma = \alpha + \theta$, which denotes the angles of attack (high to low) and inclination (right to left) of the glider. F_B can be calculated with the relationship:

$$F_B = g\rho\left\{V_g\left[1 - \varepsilon p + \alpha_T\left(t - t_0\right)\right] + \Delta V_{bp}\right\} \qquad [4.24]$$

where g is the gravity acceleration, ρ the density of the environment, m_g and V_g the mass and volume of the glider, p the pressure, t the temperature, t_0 a reference temperature, ε the compressibility coefficient of the hull and α_t its coefficient of thermal expansion. ΔV_{bp} is the variation in volume corresponding to a variation in buoyancy. F_D and F_L can also be calculated:

$$F_D = \frac{1}{2}\rho C_D S U^2 \qquad [4.25]$$

$$F_L = \frac{1}{2}\rho C_L S U^2 \qquad [4.26]$$

where C_D and C_L are the coefficients of drag (see section 3.2.2.1) and lift of the glider, S the surface of its wings and its hull and U its speed. C_D and C_L are, respectively, functions of α and α^2, and from relationships [4.25] and [4.26] it is possible to get a second-order equation to numerically find the value of the gilder's angle of attack and thus its trajectory.

When the value of α is determined, equations [4.23] and [4.25] can be combined with each other to allow the calculation of speed U and deduce its horizontal component $u_g = U\cos(\gamma)$ and vertical component $w_g = U\sin(\gamma)$. As the glider is equipped with a pressure sensor, the variation's speed w_p of the pressure measured can be calculated and we can then estimate the vertical speed w of the environment crossed: $w = w_p - w_g$. This procedure was re-used and significantly improved upon in 2011 by Frajka-Williams *et al.* [FRA 11].

Although gliders can be used to estimate the displacement speed of water masses, they are also equipped to carry out measurements of temperature, conductivity, dissolved O_2, etc. The choice of equipment relating to CTD sensors is constrained by metrological specifications required in oceanography for these measurements and by behavior of these specifications in time and space (with problems in response time). The specifications required are identical to those of drifting floats (see section 3.3.2) if we want an instrument that is competitive with the latter. Sensors made by Sea Bird Electronics and NBOSI meet these constraints (see section 2.2.2). As these are fixed to the hull, their removal for calibration (see sections 1.2.1.2 and 1.2.3.6) can be tricky. Medeot *et al.* have developed a procedure to carry out this calibration without removal, by placing the glider at the surface of a calibration bath [MED 11].

The problem of thermal inertia and the alignment of the response time of temperature and conductivity sensors (see sections 2.2.3 and 2.2.4) remains difficult to solve. The GCTD produced by NBOSI was designed to limit the biases induced by thermal inertia (see section 2.2.2.2), but to date few studies have shown how this solution performs. The Sea Bird conductivity sensor is usually installed without a pumping system, taking into account the constraints of power consumption and energetic autonomy. The principles described in section 2.2.3 are therefore no longer applicable. To solve this problem, Garau *et al.* devised a solution involving the speed of displacement of the glider [GAR 11]. They used the approach given by Morison *et al.* [MOR 94], which had shown that parameters α and $\beta^1 = \tau$ of coefficients a and b of equation [2.73] depend on the speed at which fluid circulates in the cell. They generalized this approach, by considering that speed is variable over time. Instead of being constant, the values of α and τ vary according to relationships:

$$\alpha(n) = \alpha_0 + \alpha_S\, U(n)^{-1} \qquad [4.27]$$

$$\tau(n) = \tau_0 + \tau_S\, U(n)^{-1/2} \qquad [4.28]$$

where n corresponds to sampling, and the indexes O and S to the offset and slope of α and τ. They determined these four parameters (α_O, α_S, τ_O and τ_S) by calculating a function that minimizes the area between the diagrams of ascent and descent of two temperature–salinity curves obtained from four CTD profiles (two reference and two from the glider). This technique allowed them to correct for errors of thermal inertia, whose magnitude could reach 0.3 salinity.

4.4. Bibliography

4.4.1. *Other processes for measuring salinity and density*

[EST 99] ESTEBAN O, NAVARRETE M.C., GONZALEZ-CANO A., BERNABEU E., "Measurement of the degree of salinity of water with a fiber –optic sensor", *Applied Optics*, vol. 38, no. 25, pp. 5267-5271, 1999.

[GRO 10a] GROSSO P., MALARDÉ D., LE MENN M., WU Z.Y., DE BOUGRENET DE LA TOCNAYE J.-L., "Refractometer resolution limits for measuring seawater refractive index", *Optical Engineering*, vol. 49, no. 10, pp. 1-5, 2010.

[GRO 10b] GROSSO P., LE MENN M., DE BOUGRENET DE LA TOCNAYE J.-L., WU Z.Y., MALARDÉ D., "Practical versus absolute salinity measurements, new advances in high performance seawater salinity sensors", *Deep Sea Research Part I: Oceanographic Research Papers* ,vol. 57, no. 1, pp. 151-156, 2010.

[HAR 98] HARVEY A.H., GALLAGHER J.S., LEVELT SENGERS, J.M.H. "Revised formulation for the refraction index of water and steam as a function of wavelength, temperature and density", *Journal of Physical and Chemical Reference Data*, vol. 27, no. 4, pp. 761-774, 1998.

[LEM 01] LE MENN M, LOTRIAN J., "Refraction index measurement by a laser-cube-capillary technique", *Journal of Physics D: Applied Physics*, vol. 34, pp. 1256-1265, 2001.

[LEM 11] LE MENN M, DE BOUGRENET DE LA TOCNAYE J.-L., GROSSO P., DELAUNEY L., PODEUR C., BRAULT P., GUILLERME O., "Advances in measuring ocean salinity with an optical sensor", *Measurement Science and Technology*, vol. 22, (8pp), 2011.

[LU 93] LU W., WOREK W.M., "Two – wavelength interferometric techniques for measuring the refractive index of salt-water solutions", *Applied Optics*, vol. 32, no. 21, pp. 3992-4002, 1993.

[MAL 09] MALARDÉ D., WU Z.Y., GROSSO P., DE BOUGRENET DE LA TOCNAYE J.-L., Le MENN M., "High-resolution and compact refractometer for salinity measurements", *Measurement Science and Technology*, vol. 20, no. 1, pp. 1-8, 2009.

[MAR 84] MARHT K.H., KROEBEL K., "Optical interferometric bench salinometre of high precision with electronic read out", *Ocean'84, MTS/IEEE*, pp. 219-223, 1984.

[MAR 88] MARHT K.H., WALDMANN C., "Field proven high speed micro optical density profiler sampling 1000 times per second with 10^{-6} precision", *Ocean'88, MTS/IEEE*, pp. 497-504, 1988.

[MIL 90] MILLARD R.C., SEAVER G., "An index of refraction algorithm for seawater over temperature, pressure, salinity, density and wavelength", *Deep Sea Research*, vol. 37, no.12, pp. 1909-1926, 1990.

[QUA 95] QUAN X., FRY E.S., "Empirical equation for the index of refraction of seawater", *Applied Optics*, vol. 34, no. 18, pp. 3477-3480, 1995.

[RUS 67] RUSBY J.S.M., "Measurements of the refractive index of sea water relative to Copenhagen Standard Sea Water", *Deep Sea Research*, vol.14, pp 427-439, 1967.

[SCH 90] SCHIEBENER P., STRAUB J., LEVELT SENGERS J.M.H, GALLAGHER J.S., "Refractive index of water and steam as function of wavelength, temperature and density", *Journal of Physical and Chemical Reference Data*, vol. 19, no. 3, pp. 677-718, 1990.

[SEA 85] SEAVER G.A., "The index of refraction to specific volume relation for sea water – a linearised equation of state", *Journal of Physical Oceanography*, vol. 15, pp 1339-1343, 1985.

[SEA 97] SEAVER G.A., VLASOV V.L., KOSTIANOY A.G., "Laboratory Calibration in distilled water and seawater of an oceanographic multichannel interferometer – refractometer", *Journal of Atmospheric and Oceanic Technology*, vol. 14, pp 267-277, 1997.

[STA 70] STANLEY E., "The refractive index of seawater as a function of temperature, pressure and two wavelengths", *Deep Sea Research*, 18, pp 833- 840, 1970.

[WAL 96] WALDMANN H.C., THIELE S., "Results of the dynamical tests of a special designed optical microstructure density probe based on the measurement of the refractive index", *Ocean'96, MTS/IEEE*, pp. 1131-1134, 1996.

4.4.2. *Acoustic tomography of oceans and acoustic measurements*

[DUD 92] DUDA T.F., FLATTÉ S.M., COLOSI J.A., CORNUELLE B.D., HILDEBRAND J.A., HODGKISS W.S., WORCESTER P.F., HOWE B.M., MERCER J.A., SPINDELL R.C. "Measured wave-front fluctuations in 1,000km pulse propagation in the Pacific ocean", *JASA 92 (2,) pt.1*, August 1992.

[FLA 79] FLATTÉ S., DASHEN R., MUNK W., WATSON K., ZACHARIASEN F., *Sound Transmission Through a Fluctuating Ocean*, Cambridge University Press, 1979.

[GAI 06] GAILLARD F., TERRE T, GUILLOT A., "Monitoring moored instrument motion by optimal estimation", *Ocean Engineering*, vol. 33, pp. 1-22, 2006.

[JIA 10] JIANG Y.-M., CAPMAN N.R., GERSTOFT P., "Estimation of geoacoustic properties of marine sediment using a hybrid differential evolution inversion method", *IEEE Journal of Oceanic Engineering*, vol. 35, no. 1, pp. 59-69, 2010.

[MUN 78] MUNK W.H., WUNSCH C., "Ocean acoustic tomography: a scheme for large monitoring", *Deep Sea Research*, vol. 26A, pp. 123-160, 1978.

[RAJ 10] RAJAN S.D., BECKER K.M., "Inversion for range-dependent sediment compressional-wave-speed profiles from modal dispersion data", *IEEE Journal of Oceanic Engineering*, vol. 35, no. 1, pp. 43-58, 2010.

[SIM 06] SIMMONS F.J., NOLET G., BABCOCK J. M., DAVIS R.E., ORCUTT J.A., "A future for drifting seismic networks, *Eos Trans. Am. Geoph. Union*, 87)31), pp. 305-307, 2006

4.4.3. *The UUV: new means for ocean exploration*

[AN 01] AN E., DHANAK M.R., SHAY L.K., SMITH S., LEER J.V., "Coastal oceanography using a small AUV", *Journal of Atmospheric and Oceanic Technology*, vol. 18, pp. 215-234, 2001.

[BRU 11] BRUZZONE, G., BIBULI M., CACCIA M., "Improving coastal operations with Unmaned Surface Vehicles", *Sea Technology*, 2011.

[CUR 05] CURTIN T.B., CRIMMINS D.M., CURCIO J., BENJAMIN M., ROPER C., "Autonomous underwater vehicles: trends and transformations", *Marine Technology Society Journal*, vol. 39, pp. 3, 2005.

[ERI 01] ERIKSEN C.C., OSSE T.J., LIGHT T., WEN R.D., LEHMANN T.W., SABIN P.L., BALLARD J.W., CHIODI A.M., "Seaglider: a long range autonomous underwater vehicle for oceanographic research", *IEEE Journal of Oceanic Engineering*, vol. 26, no. 4, pp. 424-436, 2001.

[FRA 11] FRAJKA-WILLIAMS E., ERIKSEN C.C., RHINES P.B., HARCOURT R.R., "Determining vertical water velocities from seaglider", *Journal of Atmospheric and Oceanic Technology*, vol. 28, pp. 1641-1656, 2011.

[GAR 11] GARAU B., RUIZ S., ZHANG W.G., PASCUAL A., HESLOP E., KERFOOT J., TINTORÉ J., "Thermal lag correction on Slocum glider data", *Journal of Atmospheric and Oceanic Technology*, vol. 28, pp. 1065-1071, 2011.

[HEG 11] HEGRENAES Ø., HALLINGSTAD O., "Model-aided INS with sea current estimation for robust underwater navigation", *IEEE Journal of Oceanic Engineering*, vol. 36, no. 2, pp. 316-321, 2011.

[JOS 11] JOSHI K.B., COSTELLO J.H., PRIYA S., "Estimation of solar energy harvested for Autonomous Jellyfish Vehicles (AJV's)", *IEEE Journal of Oceanic Engineering*, vol. 36, no. 4, pp. 539-551, 2011.

[MED 11] MEDEOT N., NAIR R., GERIN R., "Laboratory evaluation and control of Slocum glider C-T sensors", *Journal of Atmospheric and Oceanic Technology*, vol. 28, pp. 838-846, 2011.

[MER 10] MERCKELBACH L., SMEED D., GRIFFITHS G., "Vertical velocities from underwater gliders", *Journal of Atmospheric and Oceanic Technology*, vol. 27, pp. 547-563, 2010.

[MOR 94] MORISON J., ANDERSEN R., LARSON N., D'ASARO E., BOYD T., "The correction of thermal-lag effects in Sea-Bird CTD data", *Journal of Atmospheric and Oceanic Technology*, vol. 11, pp. 1151-1164, 1994.

[NAP 05] NAPOLITANO F., "Current and future navigation and positioning systems for underwater applications", *International Ocean Systems*, 14-15, 2005.

[NIC 08] NICOLSON J.W., HEALEY A.J., "The present state of autonomous underwater vehicle (AUV) applications and technologies", *Marine Technology Society Journal*, vol. 42, no. 1, pp. 44-51, 2008.

[PIN 11] PINKEL R., GOLDIN M.A., SMITH J.A., SUN O.M., AJA A.A., BUI M.N., HUGHEN T, "The Wirewalker: a vertically profiling instrument carrier powered by ocean waves", *Journal of Atmospheric and Oceanic Technology*, vol. 28, pp. 426-435, 2011.

[POP 05] POP V., BERGVELD H.J., NOTTEN P.H.L., REGTIEN P.P.L., "State-of-the-art of battery state-of-charge determination", *Measurements Sciences and Technology*, vol. 16, pp. R93-R110, 2005.

[RAJ 10] RAJAN K., PY F., MCGANN C., "Adaptive control of AUVs using onboard planning and execution", *Sea Technology*, April 2010.

[ROM 02] ROMEO J., LESTER G., "Navigation is key to AUV missions", *Sea Technology*, 2002.

[SHE 01] SHERMAN J., DAVIS R.E., OWENS W.B., WALDES J., "The autonomous underwater glider 'Spray'", *IEEE Journal of Oceanic Engineering*, vol. 26, no. 4, pp.437-446, 2001.

[SON 11] SONG A., BADIEY M., TREMBANIS A.C., "High-speed acoustic communication for AUVs trough MIMO techniques", *Sea Technology*, pp. 31-35, July 2011.

[STA 04] STANLEY J.R., "The StationKeep™ function: Dynamic positioning for remotely operated vehicles", *Underwater Magazine*, March/April 2004.

[STO 89] STOMMEL H., "The SLOCUM mission", *Oceanography*, vol. 19, pp. 22-25, 1989.

[THO 04] THOMAS H., DE BELIE S., DESSUREAULT J.G., "MBARI deploys a feature-packed AUV from a small vessel", *Sea Technology*, 2004.

[VAS 11] VASCONCELOS J.F., SILVESTRE C., OLIVEIRA P., "INS/GPS aided by frequency contents of Vector observation with application to autonomous surface crafts", *IEEE Journal of Oceanic Engineering*, vol. 36, no. 2, pp. 347-351, 2011.

[WEB 01] WEBB D.C., SIMONETTI P.J., JONES C.P., "SLOCUM: an underwater glider propelled by environmental energy", *IEEE Journal of Oceanic Engineering*, vol. 26, pp. 447-452, 2001.

[WIL 09] WILSON R., SOMLYODY S.A., "Pressure-tolerant lithium – polymer batteries", *Sea Technology*, pp. 31-35, 2009.

[WOO 04] WOOD S., NULPH A., HOWELL B., "Application of Autonomous Underwater Vehicles", *Sea Technology*, pp. 10-14, December 2004.

Acronyms and Abbreviations

AC	Alkalinity of Carbonates
ADCP	Acoustic Doppler Current Profiler
ADP	Acoustic Doppler Profiler
ALACE	Autonomous LAgrangian Circulation Explorer
AMV	Autonomous Marine Vehicle
AOPs	Apparent Optical Properties
APEX	Autonomous Profiling Explorer
ASV	Autonomous Surface Vehicle
AUV	Autonomous Underwater Vehicle
AVURNAV	*Avis Urgents aux Navigateurs* (Urgent Advice to Navigators)
BIPM	*Bureau International des Poids et Mesures* (International Bureau of Weights and Measures or IBWM)
BPSK	Binary Phase-Shift Keyed
C/A	Coarse Acquisition
CAN	Controller Area Network
CDOM	Colored Dissolved Organic Matter
CIPM	*Comité International des Poids et Mesures* (International Committee of Weights and Measures)

CLIVAR	CLImate VARiability and predictability
CNES	*Centre National d'Études Spatiales* (National Center for Spatial Studies)
CODAR	Coastal Dynamics Applications Radar
CTD	Conductivity–Temperature–Depth
DBCP	Data Buoy Cooperation Panel
DCLS	Data Collection and Location System
DIC	Dissolved Inorganic Carbon
DLL	Delay Lock Loop
DOC	Dissolved Organic Carbon
DOM	Dissolved Organic Matter
DOP	Dilution of Precision
EGNOS	European Geostationary Navigation Overlay System
EOS.80	Equation of the state Of Seawater from 1980
FMCW	Frequency Modulated Continuous Wave
GDC	Global Drifter Center
GLONASS	GLObal NAvigation Satellites System
GMT	Greenwich Mean Time
GNSS	Global Navigation Satellite System
GODAE	Global Ocean Data Assimilation Experiment
GOOS	Global Ocean Observing System
GPRS	General Packet Radio Service
GPS	Global Positioning System
GPST	Global Positioning System Time
GSM	Global System for Mobile communications

GWR	Guided Wave Radar
HF	High Frequency
IAPSO	International Association for the Physical Sciences of the Ocean
IAT	International Atomic Time
IERS	International Earth Rotation and Reference System Service
IHO	International Hydrographic Organization
IMU	Inertial Measurement Unit
INMARSAT	INternational MARitime SATellite organization
INS	Inertial Navigation System
IOP	Inherent Optical Properties
IOS	Institute of Ocean Sciences
ISO	International Standards Organization
ITS.90	International Temperature Scale of 1990
IUPAC	International Union of Pure and Applied Chemistry
JAMSTEC	JApan Marine Science and TEchnology Center
LAAS	Local Area Augmentation System
LBL	Long Base Line
LORAN	LOng RAnge Navigation
MBARI	Monterey Bay Aquarium Research Institute
MCN	*Marégraphe Cotier Numérique* (Digital Coastal Tide Gauge)
MES	*Matière En Suspension* (Suspended solids)
MIMO	Multiple-Input Multiple-Output
MLM	Maximum Likelihood Method
MTSAT	Multi-functions Transport SATellite

MUSIC	MUltiple SIgnal Classification
NASA	National Aeronautics and Space Administration
NAVSTAR	NAVigation System by Timing And Ranging
NGF	*Nivellement Général de la France* (General Leveling of France)
NINJA	New Profiling Float of JApan
NMEA	National Marine Electronics Association
NOAA	National Oceanic and Atmospheric Administration
NTF	New Triangulation of France
NTU	Nephelometric Turbidity Unit
OBS	Ocean Bottom Seismometer
PALACE	Profiling ALACE
PAR	Photosynthetic Active Radiation
PCTFE	PolyChloroTriFluoroEthylene
PEEK	PolyEtherEtherKetone
PGN	Permanent Geodetic Network
PLL	Phase Locked Loop
ppm	parts par million (corresponds to 1.10^{-6})
PPS	Precise Positioning Service
ppt	parts per thousand (‰)
PRN	Pseudo-Random Noise
PSK	Phase Shift Keying
PSS.78	Practical Salinity Scale from 1978
PTP	Precision Time Protocol
PTT	Platform Transmitter Terminal

RGF93	*Réseau Géodésique Français* 1993 (French Geodetic Network of 1993)
ROV	Remotely Operated Vehicle
SA	Selective Availability
SAAO	Stand Alone Acoustic Observatory
SAWO	Stand Alone Winch Observatory
SeaWiFS	Sea-viewing Wide Field-of-view Sensor
SERS	Surface Enhanced Raman Scattering
SIO/UCSD	Scripps Institution of Oceanography/University of California, San Diego
SMOW	Standard Mean Ocean Water
SOFAR	SOund Fixing And Ranging
SOLO	Sounding Oceanographic Lagrangian Observer
SPR	Surface Plasmon Resonance
SPRT	Standard Platinum Resistance Thermometer
SURDRIFT	SURface DRIFTer
TA	Total Alkalinity
TEOS-10	Thermodynamic Equation Of Seawater of 2010
TOA	Time Of Arrival
UERE	User Equivalent Range Error
UNESCO	United Nations Educational, Scientific and Cultural Organization
USBL	Ultra Short Base Line
USNO	United States Naval Observatory
USV	Unmanned Surface Vehicles
UT	Universal Time

UTC	Universal Coordinated Time
UUV	Unmanned Underwater Vehicle
VACM	Vector Averaging Current Meter
WAAS	Wide Area Augmentation System
WAP	Wireless Application Protocol
WCRP	World Climate Research Program
WERA	WEllen RAdar
WGS84	World Geodetic System 1984
WHP	WOCE Hydro-graphic Program
WHPO	WOCE Hydro-graphic Program Office
WMO	World Meteorological Organization
WOCE	World Ocean Circulation Experiment
XBT	eXpendable Bathy Thermograph
XCTD	eXpendable Conductivity Temperature Depth profiling system
ZONEX	ZONe of EXercise

Index

D

E

Q, R

S

X, Z

Printed and bound by CPI Group (UK) Ltd, Croydon, CR0 4YY
13/09/2026
14961075-0006